TITANIUM DIOXIDE

Edited by **Magdalena Janus**

Titanium Dioxide
http://dx.doi.org/10.5772/66559
Edited by Magdalena Janus

Contributors

Fu-Quan Bai, Wei Li, Hong-Xing Zhang, Lifang Xu, Jing Xu, Jian-Tao Wang, Annabella Selloni, Sergio Ricardo De Lazaro, Renan Augusto Pontes Ribeiro, Luis Henrique Da Silveira Lacerda, Alberto Adriano Cavalheiro, Silvanice Aparecida Lopes Dos Santos, Lincoln Carlos Silva De Oliveira, Yeliz Gurdal, Marcella Iannuzzi, Fumiaki Amano, Barbara Bonelli, Serena Esposito, Francesca Freyria, Vesna Đorđević, Bojana Milićević, Miroslav Dramicanin, Marcela Achimovičová, Christoph Vonderstein, Bernd Friedrich, Katarzyna Siwińska-Stefańska, Teofil Jesionowski

Published by InTech
Janeza Trdine 9, 51000 Rijeka, Croatia

Publishing Process Manager Dajana Pemac
Technical Editor SPi Global
Cover InTech Design team

First published July, 2017
Printed in Croatia
Legal deposit, Croatia: National and University Library in Zagreb

Additional hard copies can be obtained from orders@intechopen.com

Titanium Dioxide, Edited by Magdalena Janus
p. cm.
Print ISBN 978-953-51-3413-8
Online ISBN 978-953-51-3414-5

For EU product safety concerns:
IN TECH d.o.o., Prolaz Marije Krucifikse Kozulić 3, 51000 Rijeka, Croatia,
info@intechopen.com or visit our website at intechopen.com.

Meet the editor

Dr. Magdalena Janus is currently an associate professor at the Department of Civil Engineering and Architecture, West Pomeranian University of Technology, Szczecin. She graduated from the Department of Chemical Technology and Engineering, Szczecin University of Technology (from 2009 in West Pomeranian University of Technology, Szczecin). Her research interest includes photocatalysis, water and wastewater treatment technologies, photoactive building materials, and nanomaterials. She has published more than 50 research papers in international journals; results of her studies were present in more than 60 national and international conferences. In 2014, she was awarded by the Ministry of Science and Higher Education of Poland for science achievements. She is one of the editors of an international journal on the latest advances in the science, engineering, and application of miniature and ultraminiature structures.

Contents

Preface

Titanium dioxide is mainly used as a pigment and photocatalyst. It is possible to find it in food, cosmetics, building materials, electric devices, and others. This book contains chapters about characteristics of anatase and rutile crystallographic structure of titanium dioxide and the use of DFT method for photoactivity calculation. The book *Titanium Dioxide* contains ten chapters written by researchers and experts of the field. The book chapters are organized in four sections. Section 1 is focused on characteristics of anatase structure of TiO_2. Section 2 contains chapters about rutile structure of TiO_2. Section 3 presents different types of TiO_2 materials. Section 4 is focused on theoretical calculation of titanium dioxide.

I hope that this book will be useful for a wide circle of people dealing with TiO_2.

Magdalena Janus
West Pomeranian University of Technology
Szczecin, Poland

Section 1

Anatase Structure of TiO2

Chapter 1

The Reactivity of Anatase TiO_2 (211) Surface and the Bond-Charge Counting Model

Jing Xu, Li-Fang Xu, Jian-Tao Wang and
Annabella Selloni

Additional information is available at the end of the chapter

http://dx.doi.org/10.5772/intechopen.69141

Abstract

In this chapter, we intend to present a generic understanding of surface reactivity and water dissociation on TiO_2 surfaces through a study of anatase TiO_2 (211) surface—an idea model surface containing both four-coordinated Ti atom (Ti_4) and five-coordinated Ti atom (Ti_5). Our first-principles calculations show that the (211) surface is a high reactivity surface and reveal that water molecule can be easily dissociated on a Ti_4 site while it hardly dissociates on Ti_5 site. Furthermore, we introduce bond-charge counting model to clarify the mechanism. More generally, after an intensive investigation of literature, we found that the bond-charge counting model is applicable to all anatase and rutile TiO_2 surfaces including step edges and vacancies where the reactivity of surfaces enable to dissociate water attribute to the existence of Ti_4 atom or equivalent Ti_4 atom.

Keywords: TiO_2 surfaces, surface reactivity, water dissociation, first-principles calculations, bond-charge counting model

1. Introduction

Titanium dioxide is a semiconductor-based heterogeneous photocatalysis material which received more and more interest. In nature, TiO_2 crystallizes in three different structures: rutile, anatase, and brookite, all formed by TiO_6 octahedra connected by shared edges and/or corners; rutile is the thermodynamically most stable bulk phase, while anatase is very common and stable in nanomaterials. As a major polymorph of TiO_2, anatase TiO_2 is the most widely studied phase and is extensively used in many industrial applications such as photovoltaic cells, photo and electrochromics, photocatalysis, photonic crystals, smart surface coatings, and sensors [1–9]. In all applications, the surface structure plays a key role, as the surface reactivity and physicochemical properties depend strongly on the exposed crystallographic facet. Therefore, the search for high reactivity TiO_2 surfaces is a topic of great interest and an area of intense activity.

The rutile (110) surface has been investigated early [10–12]. Anatase is less stable than rutile, but more efficient than rutile for applications. Many studies of the anatase TiO_2 surface focus on the (001) and (101) surfaces [13–20]. Theoretical studies [13–17] show that the (101) surface is the thermodynamically most stable surface with a small surface energy of 0.49 J/m^2, while the (001) surface is the highest reactivity surface with a high surface energy of 0.98 J/m^2. More recently, using different dopants, adsorbates, or solvated species, anatase TiO_2 nanocrystals exposing various crystalline facets have been prepared, including low-index facets such as (001) facets [18, 19], (100) [21–24], (010) [25, 26], (101) [27–32], (110) [33], and (111) [34], and high-index facets such as (103), (105), (106), (201), (301), and (401) [35–39]. Besides, many researchers have made extensive investigation on surface defects since those defects, e.g., step edges [40–42] and O vacancy [43, 44] are intrinsic on the surfaces of crystalline materials, which strongly influence the surface chemistry.

Since the discovery of photocatalytic splitting of water on a TiO_2 electrode in 1972 [1], the interaction of TiO_2 surfaces with water is of special interest. The structure of the hydrated surfaces is important not only because water is always present on TiO_2 surfaces, but also because it can help understand and control the catalytic and (photo) electrochemical properties of this material. In fact, different water adsorption states have been found; for example, water favors molecular adsorption on the anatase (101) surface and dissociative adsorption on the (001) surface [13].

There is still controversy for water adsorption on the TiO_2 surface based only on the DFT total energy calculations, e.g., on rutile (110). Despite numerous studies on this topic in literature, the mechanism of water dissociation on TiO_2 surfaces remains to be clarified. Moreover, there is still a lack of a generic understanding of different surfaces including surfaces with steps and vacancies.

Due to a recent experiment, a large percentage of exposed (211) facet has been prepared and it is found that the (211) surface can effectively improve the photocatalytic activity of TiO_2 for water dissociation reactions [45, 46]. Since the surface exposes both Ti_4 and Ti_5 atoms, the anatase (211) surface is studied in this work as an ideal model surface to illustrate the mechanism of water dissociation on TiO_2 surface [47]. The Ti_4 atom plays a critical role for splitting water molecule. Furthermore, we introduce a bond-charge counting model according to TiO_2 structure and conclude that two unsaturated Ti bonds are the necessary conditions for splitting water on the surface. Then, we examine as much as possible TiO_2 surfaces and find the model is applicable to all surfaces.

The paper is organized as follows: the reactivity of anatase TiO_2 (211) surface is briefly reviewed in Section 2; then, we propose the bond-charge counting model and give some typical examples in Section 3; finally, we present conclusion remakes in Section 4.

2. The reactivity of anatase TiO_2 (211) surface

2.1. Surface structure and stability

We begin with the experimental result. Recently, a large percentage of exposed (211) facet has been identified by X-ray diffraction (XRD) on N-doped TiO_2 film deposited using RF magnetron sputtering [45]. N-doped TiO_2 films were deposited on quartz glass substrates (2 cm × 4 cm) by RF reactive magnetron sputtering. The crystalline structure of the as-deposited N-doped TiO_2 films was identified by X-ray diffraction (XRD). **Figure 1** shows the XRD patterns of N-doped TiO_2 film. Diffraction peaks observed at 2θ = 25.28°, 36.95°, 37.88°, 38.55°, 48.05°, 54.09°, 54.88°, 62.67°, and 68.76°correspond well with (101), (103), (004), (112), (200), (105), (211), (204), and (116) planes of anatase phase of TiO_2. It can be seen that N-doping can greatly influence the growth orientation of anatase TiO_2 particles. As shown in **Figure 1**, the intensities of the (004), (112), (200), and especially (211) peaks become stronger, while (101) peak become weaker for the N-doped TiO_2 film, compared with those of the undoped TiO_2 film. Especially, the increase of exposed (211) facets can effectively improve the photocatalytic activity of TiO_2 for water dissociation reactions [46], showing the high reactivity of exposed (211) facets.

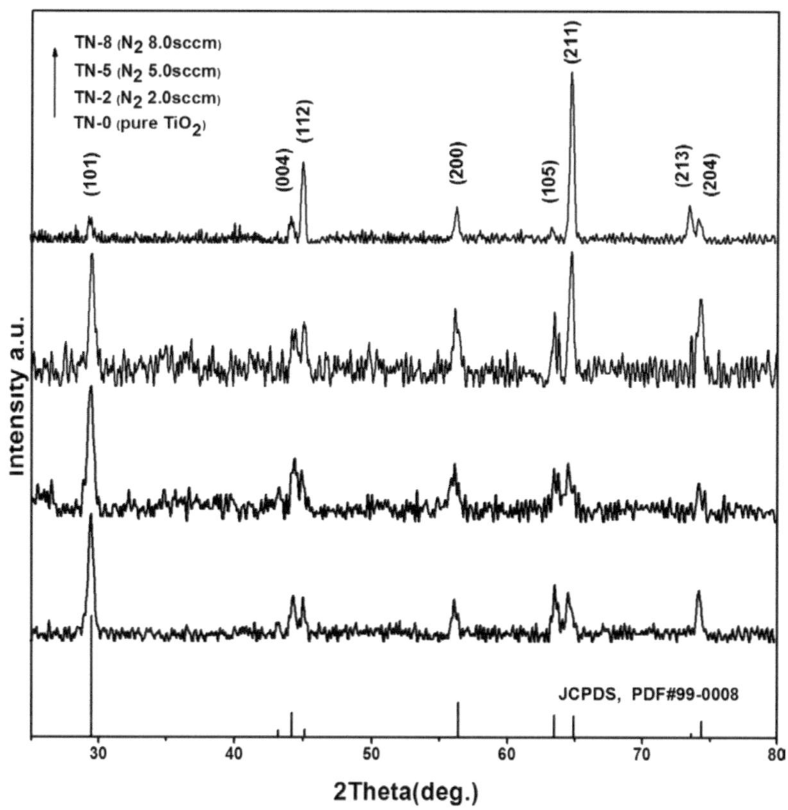

Figure 1. The X-ray diffraction (XRD) on N-doped TiO_2 film.

Motivated by the above experimental findings, in a recent theoretical work [47], we have made a systematic study of the surface reactivity and water adsorption on anatase (211) using *ab initio* calculations. Calculations have been performed by the Vienna ab initio simulation package (VASP) [48–50] with all-electron projector augmented wave (PAW) method [51]. The generalized gradient approximation (GGA-PW91) [52] is set as the exchange and correlation functional. The valence states $3d^24s^2$ for Ti, $2s^22p^4$ for O, and $1s^1$ for H are used with an energy cutoff of 500 eV for the plane wave basis set. The calculated lattice constants for bulk anatase TiO_2 are used to construct the diverse facets listed in **Table 1**. The anatase (211) surface was modeled by a slab of six layers with a unit surface cell of 7.706 Å × 5.501Å × 27.677Å (γ = 101.44°) comprising a total of 54 atoms separated by a vacuum region of 12 Å. The Monkhorst-Pack scheme [53] was adopted for the Brillouin zone integration with a 6 × 6 × 1 k-point mesh. All atoms are relaxed during geometry optimizations with the given surface cell. Convergence criteria employed for both the electronic self-consistent relaxation and the ionic relaxation were set to 10^{-6} eV and 0.01 eV/Å for the total energy and Hellmann-Feynman force, respectively.

We firstly discuss the surface stability. The stoichiometric unrelaxed termination of the (211) surface is shown in **Figure 2(a)** There are five under-coordinated and four fully-coordinated atoms exposed to the vacuum. The five under-coordinated atoms include three inequivalent twofold-coordinated oxygen atoms denoted by O_{21}, O_{22}, and O_{23}, a fourfold-coordinated Ti_4 and a fivefold-coordinated Ti_5, respectively. The four fully-coordinated atoms are O_{31}, O_{32}, O_{33}, and Ti_6 [see **Figure 2(a)**]. Different from (001) and (101) surfaces [12], fourfold-coordinated Ti_4 atoms are present on the (211) surface. **Figure 2(b)** shows the optimized anatase (211) surface. After relaxation, the (211) surface shows a very corrugated structure, with a characteristic, saw tooth-like profile along the [1–31] direction. All under-coordinated oxygen (O_{2i}) atoms are displaced outward, while the under-coordinated Ti_4 and Ti_5 atoms are relaxed inward. Both angles $\angle Ti_4–O_{22}–Ti_5$ and $\angle Ti_5–O_{31}–Ti_6$ become smaller, 100.3° and 147.7°, respectively. The largest relaxations are those of the fully-coordinated oxygen O_{31}, which relaxes outward by approximately 0.35 Å, and O_{32}, which relaxes inward by approximately 0.34 Å. Meanwhile, the surface atoms form four-membered-ring (O-Ti-O-Ti) structures on the surface. These O-Ti-O-Ti rings are slightly deformed and the distances between oxygen atoms in these rings increased from the bulk value of 2.458 Å to 2.492–2.515 Å.

Facet	N_{at}	E (J/m²)	$n(Ti_5)$ (10^{-2}Å^{-2})	$n(Ti_4)$ (10^{-2}Å^{-2})	$n(O_2)$ (10^{-2}Å^{-2})
(101)	24	0.52	5.1		5.1
(001)	18	1.08	6.9		6.9
(211)	54	0.97	2.4	2.4	7.2
$(103)_s$[14]	48	0.99		3.5	
(110)[14]	42	1.15		3.8	

Table 1. Calculated surface energies (E in J/m²) and surface densities Ti_5, Ti_4, and oxygen O_2 atoms [$n(Ti_5)$, $n(Ti_4)$, and $n(O_2)$]. N_{at} is the total number of atoms in the slab.

http://dx.doi.org/10.5772/intechopen.69141

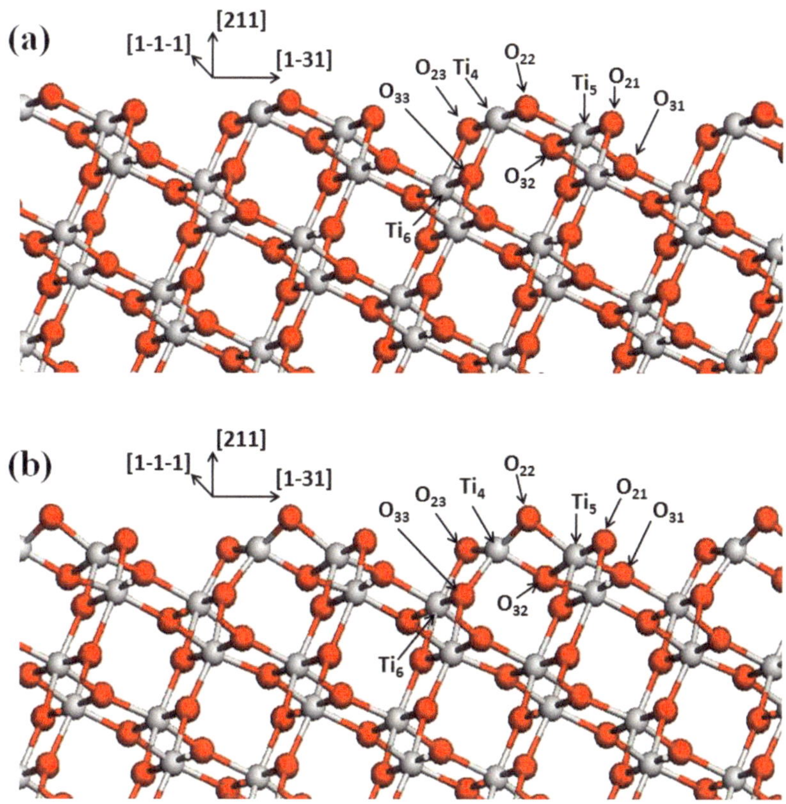

Figure 2. Structures of the unrelaxed (a) and relaxed (b) anatase (211) surfaces. The O and Ti atoms are in red and grey spheres, respectively, indicated with different coordination numbers.

To investigate the surface stability, we calculated the electronic density of states (DOS) for bare anatase (211) surface after relaxation. The result is shown in **Figure 3** in comparison with the bulk TiO_2. There is a big gap, as large as in bulk, in the DOS of relaxed bare anatase (211) surface, which indicates the chemical stability of the (211) surface.

2.2. Surface energetics

The surface energies for (101), (001), and (211) are estimated using the expression, $E=(E_{tot}-nE_{bulk})/A$, where E_{tot} is the total energy of the slab and E_{bulk} is the energy of TiO_2 unit in the bulk, n is the number of TiO_2 units in the slab, A is the total surface area of the slab, including both sides of the slab. The calculated surface energies are listed in **Table 1**. The surface energy of the (001) surface is estimated to be 1.08 J/m^2, which is nearly twice that of the most stable anatase (101) surface (0.52 J/m^2), in agreement with previous theoretical studies [14]. Similarly, the (211) surface has a high surface energy of 0.97 J/m^2, close to that of the (001) surface.

The value of the surface energy is known to be strongly correlated to the presence of under-coordinated Ti atoms on the surface [14]. The (001) surface energy is large because of the high surface density of Ti_5 (see **Table 1**). However, the surface energy of anatase (211) is large even

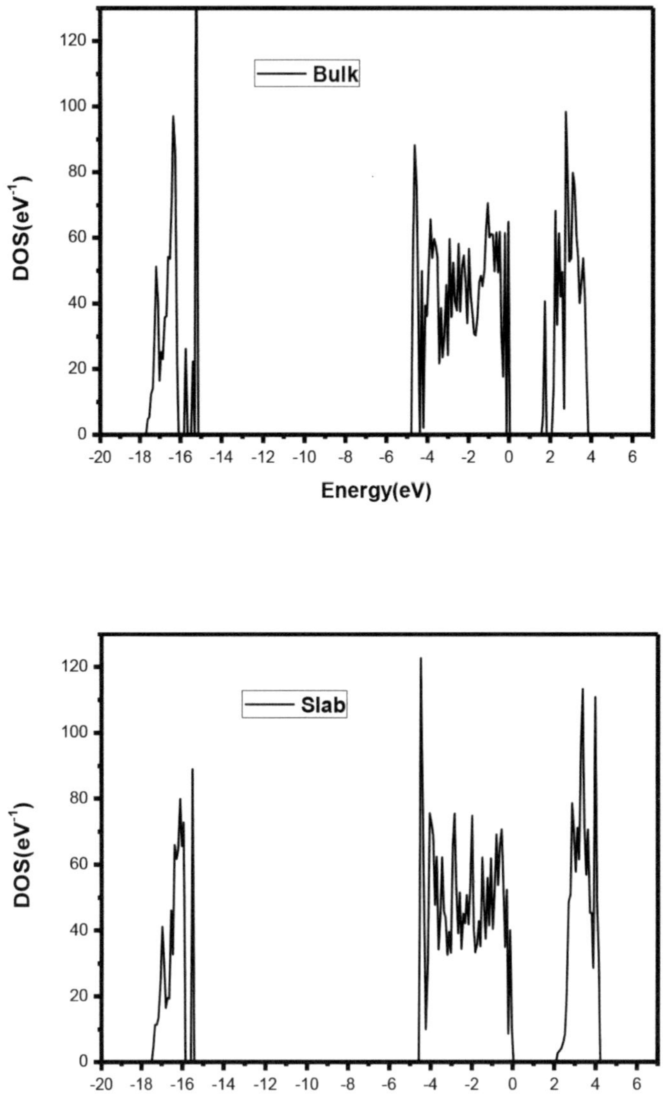

Figure 3. Electronic density of states for the bulk TiO_2 and bare anatase (211) surface after relaxation. The Fermi level is set at 0 eV.

though the total density of under-coordinated Ti_4 and Ti_5 atoms is smaller than that on the (001) surface [even smaller than on (101)]. This result already suggests that Ti_4 atoms, with two unsaturated bonds, have a higher reactivity than Ti_5 atoms with one unsaturated bond. A similar behavior was found also for the anatase (110) and $(103)_s$ surfaces [14].

From the above calculation results in subsections 2.1 and 2.2, we can see that the (211) surface has a large electronic band gap and high surface energy. It shows those two surface properties, stability and reactivity, seem contradictory, could uniformly hold on the (211) surface.

2.3. Water adsorption

We next present a detailed picture for the adsorption of water on the TiO_2 (211) surface by considering one, two and three adsorbed water molecules corresponding to various coverages $\theta = 1/3$, 2/3 and 1 ML per surface unit cell, respectively.

For a single water molecule (1/3 ML), there are four possible adsorption modes, corresponding to different adsorption positions (Ti_4 or Ti_5) and different (molecular or dissociative) adsorption conformations. *For molecular water adsorption on Ti_5 site* [see **Figure 4(a)**], the oxygen of water bonds to Ti_5 with bond length of 2.226 Å, and two surface oxygen atoms via

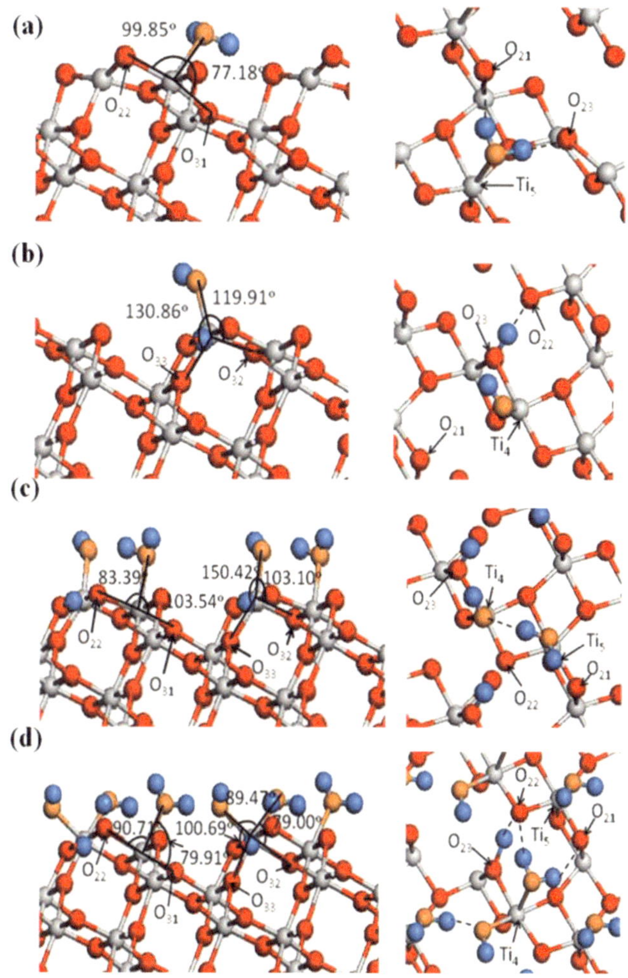

Figure 4. Side (left) and top (right) views of the structures for water adsorption on the anatase TiO_2 (211) surface. (a) Molecular water on a Ti_5 site. (b) Dissociative water on a Ti_4 site. (c) Mixed state with one dissociative H_2O on Ti_4 and one molecular H_2O on Ti_5 sites at 2/3 ML. (d) Mixed state on Ti_4 and Ti_5 sites at 1 ML coverage. The O atom of water is plotted in orange, the H atom is in blue, and the H-bond is indicated by a dashed line.

Ti_5 form two bond angles $\angle O_{water}–Ti_5–O_{22}$ = 99.85° and $\angle O_{water}–Ti_5–O_{31}$ = 77.18°, close to the bulk angles of 101.90° and 78.10°, respectively. Upon water adsorption, the Ti_5 site becomes sixfold-coordinated: the Ti atom has six Ti-O bonds with their orientations similar to those in the bulk. At the same time, the two hydrogen atoms of water form H-bonds (HBs) with two neighboring surface under-coordinated oxygen atoms, O_{21} and O_{23}, with bond lengths 2.339 and 1.873 Å, respectively. As a result, the computed molecular adsorption energy on the Ti_5 site is 0.784 eV. For dissociative water adsorption on Ti_5 site, the water molecule is dissociated into hydroxyl (OH) and H fragments. The OH group bonds to Ti_5 with a length of 1.857 Å and then further bonds to O_{23} with a weak HB. The H fragment forms a new OH moiety with a nearby O_{21}. As a result, the adsorption energy for a dissociated water on Ti_5 site is 0.77 eV, which is slightly smaller than that of molecular adsorption. Thus, molecular adsorption is preferred on Ti_5 site.

On the other hand, for molecular water on a Ti_4 site, the oxygen atom of water binds to Ti_4 with a bond length of 2.207 Å, and Ti_4 is located at the position of one of the bulk Ti-O bonds indicating that there is one Ti bond left. The Ti_4 adsorption site becomes fivefold-coordinated and the adsorption energy for molecular water on Ti_4 site is estimated to be 0.99 eV. Finally, *for dissociative water on* Ti_4 *site* [see **Figure 4(b)**], the O atom of the OH group is strongly bonded to the Ti_4 atom with a short bond length of 1.847 Å (as compared to the Ti-O bond length of 2.207 Å in the molecular adsorption case) so that the Ti_4 adsorption site becomes fivefold-coordinated. It is worthwhile to point out that the adsorption position of the O atom of the OH group does not correspond to the position of a bulk Ti-O bond, but is in the middle of the two missing bulk Ti-O bonds, and the orientation of Ti_4-O_{OH} bond clearly deviates from its direction in the bulk, as shown by the two bond angles $\angle O_{OH}–Ti_4–O_{33}$ = 130.86° and $\angle O_{OH}–Ti_4–O_{32}$ = 119.91°. This adsorption geometry with short bond length and a middle position indicates that the dissociated water interacts with two unsaturated Ti_4 bonds indeed. Furthermore, the hydrogen atom of OH forms a weak HB with a neighboring O_{23} atom of length 2.534 Å. The dissociated H from water interacts with a surface oxygen O_{23} forming a new OH moiety with a bond length of 1.013 Å and further forms an HB of 1.598 Å with O_{22}. As a result, the adsorption energy for dissociated water on Ti_4 site is estimated to be 1.28 eV, which is significantly larger than the value of 0.99 eV obtained for molecular adsorption.

To obtain further insight, we show the projected densities of states of the surface with a dissociative water on a Ti_4 site and molecular water on Ti_5 site in **Figure 5**. We can see that the O-2p orbitals in the OH group are extended to a wide range between −3.2 and −0.9 eV, indicating that the O atom of OH is strongly interacting with the substrate. On the other hand, for the case with a molecular water adsorption on a Ti_5 site, all peaks from the water molecule are sharp and are simply superimposed on those of the bare surface, indicating that they interact weakly with the surface.

Combined together the above four calculation results: the adsorption energy, bond length, bond angle, and DOS, we can conclude that dissociative adsorption can *easily* happen at the Ti_4 site while *hardly* happens on Ti_5. In fact, these different behaviors can be understood in terms of a simple model based on the bond-charge distribution, which is the key issue in this paper and will be discussed in the next section.

http://dx.doi.org/10.5772/intechopen.69141

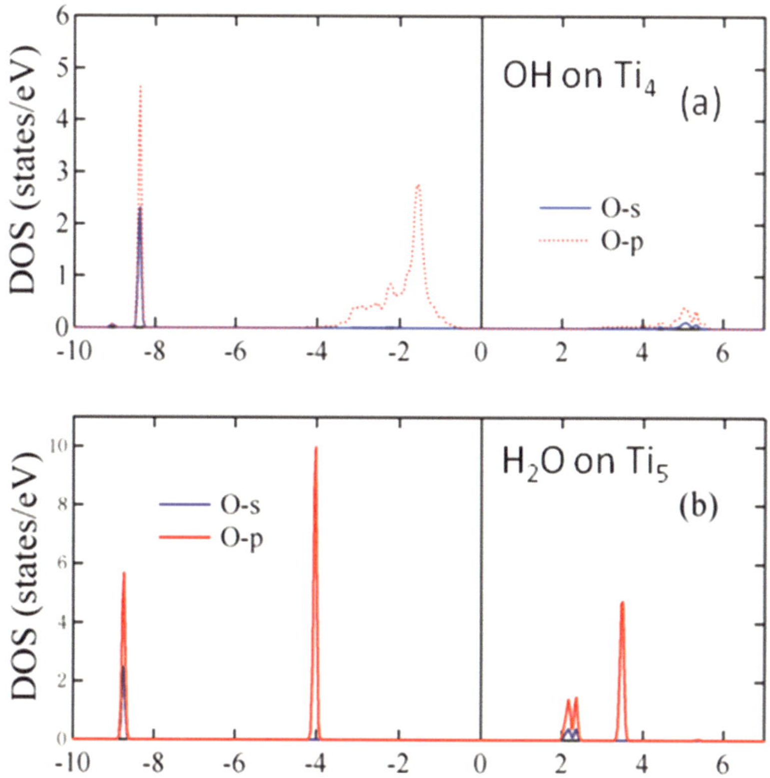

Figure 5. Projected electronic density of states on anatase (211) surface (a) for O-2*s,p* in the OH group on Ti_4 site; (b) for O-2*s, p* in water on Ti_5 site; The Fermi level is set at 0 eV.

Now, let us consider more water adsorption case. For two adsorbed H_2O molecules (2/3 ML coverage), the first H_2O prefers to adsorb at a Ti_4 site in dissociative form according to the single water-adsorption results; next, another water molecule should adsorb on a Ti_4 or Ti_5 site. *For the structure with one dissociated H_2O on a Ti_4 site and one molecular H_2O on a Ti_5 site* [see **Figure 4(c)**], the O-Ti_5 bond is 2.230 Å, while the O-Ti_4 bond length is much shorter, 1.992 Å. The dissociated H combines with an O_{23} atom forming a new OH moiety and further forms a strong HB (1.590 Å) with an O_{22} atom. An additional HB between two water molecules forms with a bond length of 1.775 Å, which makes two water molecules closer and changes the values of the bond angles. In this mixed structure (c), the adsorption energy is 1.045 eV/mol, which is a little larger than the averaged value of 1.034 eV [(1.284 + 0.784)/2] for the single water adsorption on Ti_4 and Ti_5 sites due to the contribution of the new HB. On the contrary, for the structure with two dissociated H_2O on Ti_4 and Ti_5 sites, the adsorption energy is estimated to be 0.991 eV/mol, which is about 0.036 eV smaller than the averaged value of 1.027 eV [(1.284 + 0.770)/2] for the single water adsorption and also 0.054 eV lower than that of the mixed configuration. These results confirm that molecular adsorption is preferred on Ti_5 in the mixed structure. Moreover, for a structure with one dissociated H_2O and one associated H_2O on Ti_4 sites and structure with two dissociated H_2O on Ti_4 sites, the adsorption energy is estimated to be 0.960 and 0.916 eV/mol, respectively. Those are clearly lower than 1.045 eV/mol for the mixed structure. Therefore, with increasing coverage, water molecules prefer to be adsorbed in a mixed form with one dissociated H_2O on a Ti_4 site and one molecular H_2O on a Ti_5 site.

Finally, we discuss the monolayer coverage where three water molecules are adsorbed per surface unit cell. Following the 2/3 ML result with one dissociated H_2O on a Ti_4 site and one molecular H_2O on a Ti_5 site, the third molecule would adsorb molecularly on a Ti_4 site [see **Figure 4(d)**]. It is an O atom that binds to Ti_4 with a bond length of 2.259 Å, while the two H atoms form HBs with nearby O atoms, where H-O_{21} is 2.156 Å and H-O_{22} is 1.907 Å. The bond length for molecular water on a Ti_5 atom is 2.221 Å, while for dissociated water on Ti4 it is 1.934 Å. Two bond angles $\angle O_{OH}–Ti_4–O_{33}$ and $\angle O_{water}–Ti_4–O_{32}$ are 100.69° and 79.00°, respectively. Thus, all Ti atoms are sixfold-coordinated with orientations similar to those of a bulk Ti atom. The dissociated H is captured by an O_{23} atom to form an OH moiety and further interacts with an O_{22} atom forming an HB of 1.835 Å. An additional HB of 1.652 Å also exists between these two adsorbed H_2O. All surface atoms become saturated. The adsorption energy has a larger value of 0.946 eV/mol for this mixed configuration on Ti_4 sites. For comparison, we also consider configuration with two dissociated water molecules on Ti_4 sites and one molecular H_2O on Ti_5. Though all surface atoms are also saturated, its adsorption energy is 0.909 eV/mol, which is lower than that of the mixed configuration with one dissociative and one molecular adsorption on Ti_4. Thus, molecular adsorption is favored on Ti_4 for the second water molecule. Meanwhile, for structure with one intact water on Ti_4 and two dissociated water molecules on Ti_4 and Ti_5 sites, the computed adsorption energy is 0.902 eV/mol; for structure with three dissociated water molecules on Ti_4, Ti_4, and Ti_5 sites, the adsorption energy is 0.833 eV/mol (see **Table 2**). These results suggest that a mixed water configuration is formed at monolayer coverage, with one dissociated water on Ti_4, one molecular water on Ti_4, and one molecular water on Ti_5.

Our results show that Ti_4 is an only active site which can dissociate water. Once Ti_4 is saturated with a water, there will be no more water can be dissociated. It corresponds well with the experimental observations that the dissociation does occur only at low coverages and the probability of H_2O dissociation is decreased with increasing surface coverage. That is the purpose that we describe adsorption in detail from low water coverage to full coverage in this section.

3. The bond-charge counting model for TiO_2 surfaces

3.1. The mechanism and a simple phenomenological model

According to the above results, the adsorption energies of water on a Ti_4 site are always larger than those on a Ti_5 site (see **Table 2**). Moreover, a water molecule can be easily dissociated on a Ti_4 site while it hardly dissociates on Ti_5. We use a word "hardly" here because there is still controversy for the adsorption of H_2O on Ti_5 site. In our case, the adsorption energy for a molecular adsorption on Ti_5 site is only slightly higher than the dissociated one, which is not adequately to convince that water cannot be dissociated on Ti_5. Actually, there are a lot of arguments on this issue in literature. Lindan et al. [12] suggested that dissociative adsorption happened on the rutile

θ (ML)	Structure	H_2O	H, OH	ΔH (eV)
1/3	**Figure 4(a)**	Ti_5		0.784
			Ti_5	0.770
		Ti_4		0.994
	Figure 4(b)		Ti_4	1.284
2/3	**Figure 4(c)**	Ti_5	Ti_4	1.045
			$Ti_{4,}Ti_5$	0.991
		Ti_4	Ti_4	0.960
			$Ti_{4,}$ Ti_4	0.916
3/3	**Figure 4(d)**	$Ti_{4,,}Ti_5$	Ti_4	0.946
		Ti_5	$Ti_{4,}$ Ti_4	0.909
		Ti_4	$Ti_{4,}Ti_5$	0.902
			$Ti_{4,}$ $Ti_{4,}Ti_5$	0.883

The adsorption states are shown as H_2O = molecular and H, OH = dissociative on a Ti_4 or Ti_5 site.

Table 2. Adsorption energy (ΔH in eV) per H_2O molecule on anatase (211) at various water coverages θ = 1/3, 2/3, and 1 ML.

(110) surface, while Schaub's result [43] is in contrast to that. However, there is an important case that water is indeed dissociated on anatase (001) surface with only Ti_5 atom [13]. Therefore, whether Ti_5 atom can dissociate water is still a matter for controversy on TiO_2 surfaces.

Many research works are only concentrated on the total energy calculations in literature. DFT total energy calculations are a definitely powerful tool. But here there is a shortcoming that total energy calculations just tell the total energy of the system, not the local interaction energy. For example, in dissociative adsorption, the total energy includes the adsorption site of the dissociated H as well as H-bond energy at high coverage case. Thus, it is difficult to extract the onsite interaction energy and it is hard to obtain a clear conclusion just from the total energy. All those controversies come from the information of total energy and less considerations for the reaction mechanism. Therefore, it is necessary to think it over from the origin of physics and chemistry beyond total energy calculation.

The water dissociation on surface is a chemical adsorption, which can be regarded as a chemical reaction analogue to a chemical displacement reaction, e.g., $2Na + 2H_2O = 2NaOH + H_2\uparrow$. In this displacement reaction, the necessary condition is that more active metal atom can substitute the less active metal atom or hydrogen. Also, from a physical point of view, the O atom in water must gain more electrons that H atom provided so that such a dissociation process happens by losing an H atom. From these considerations, we propose a simple bond-charge counting model based on the charge distribution on Ti bond in TiO_2.

In bulk TiO_2, each Ti atom has six nearest neighboring O atoms and Ti atom has four outmost electrons, i.e., each Ti^{4+} ion is surrounded by an octahedron of six O^{2-} ions. Thus, on an average, each Ti atom can offer 4/6 electron charge at each Ti-O bond. It holds for both rutile and anatase since they have the same TiO_6 octahedra structure. Therefore, we could make a simple bond-charge counting for this system. When a Ti_4 atom interacts with the oxygen atom of H_2O, two unsaturated Ti bonds participate in the interaction and offer about 4/3 electron charge to this O atom by forming a strong bond [see **Figure 4(b)**]. Thus, it satisfies the reaction requirement and Ti_4 is more active than H atom. Then, one H atom can be released from the water molecule on a Ti_4 site. In fact, one H atom dissociates spontaneously from the water molecule as this adsorbs on a Ti_4 atom. On the other hand, Ti_5 can only provide about 2/3 electron charge to the water O atom, less than the charge contribution from an H atom. Therefore, Ti_5 is less active than H atom and hardly causes dissociation of an H atom. Although there is very little difference in energy (0.014 eV) between molecular and dissociated structures, we can clearly judge that water favors molecular adsorption and is unfavorable on Ti_5 site.

The essence of the model is qualitatively taking account of average charge on each Ti bond in TiO_2 material. The model is phenomenological and does not intend to provide the precise value of charge transferred during interaction due to the complex of 3d orbitals of Ti atom in TiO_6 octahedra. Nevertheless, the bond-charge counting model clarifies the intrinsic charge difference between Ti_4 and Ti_5 atoms on surfaces where Ti_4 can provide more than one electron and Ti_5 much less one electron. The necessary condition for water splitting is that the surface Ti atoms must provide more than one electron to O atom of water. The charge contributed from a single Ti bond is not sufficient for water dissociation. Therefore, two unsaturated Ti bonds satisfy the condition corresponding to the four-coordinated Ti_4 atom that has chemical reactivity, while Ti_5 atom not.

3.2. Typical examples

In order to verify this model, we have made an intensive investigation for TiO_2 surface as much as we could find in literature including steps and vacancies. We found that all reactive surfaces splitting water are associated with Ti_4 atom or equivalent Ti_4 atom without any exception. We are not able to exhaust all surfaces here, rather list typical and important surfaces in different geometric categories as follows.

3.2.1. Surfaces with Ti_4 atom

We start the survey from the TiO_2 surfaces with Ti_4 atoms. At anatase surfaces, the (110) and (103)*s* contain Ti_4 atoms. Therefore, their surface energies are 1.15 and 0.99 J/m^2, respectively [14]. Those rather high values indicate that the surfaces have a very high reactivity contributed from Ti_4 atoms.

3.2.2. Surface with only Ti_5 atom

The surfaces with only Ti5 atom are more interesting. Whether the surfaces can dissociate water is controversy. We will see it strongly depends on its surface energy. As anatase (001) surface, the surface energy has a rather high value of 0.98 J/m^2 indicating a high reactivity. Thus, it is clearly pointed out in Ref. [13] that the structure of the dissociated state (in **Figure 6(a)** same

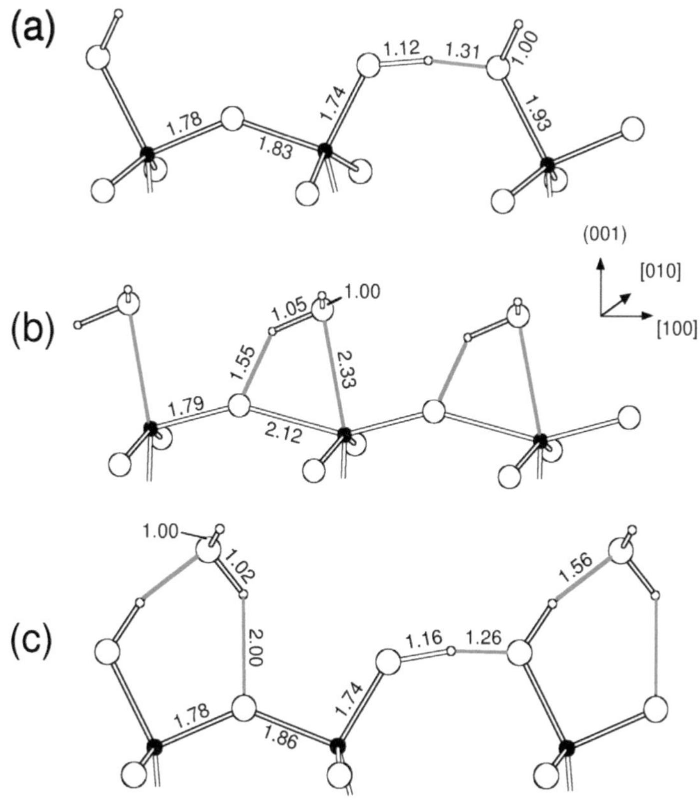

Figure 6. (**Figure 3** in Ref. [13]) Atomic structure (side view) for adsorbed water molecule(s) on anatase (001). (a) Dissociated state. (b) Molecular state. (c) Mixed state. Gray lines in (b) and (c) indicate hydrogen bonds. Bond lengths are in Å.

as in **Figure 3(a)** in Ref. [13]) is characterized by the breaking of the bond between the bridging 2c oxygen and the Ti_5 atoms involved in the adsorption, i.e., the Ti_5 atom actually becomes fourfold-coordinated after breaking its bond to a bridging O atom. Therefore, the dissociative water is absorbed on an equivalent Ti_4 site. It is expected that such phenomenon of Ti_5 atom breaking bond with neighboring oxygen becomes an effective Ti_4 atom which would be held for the anatase $(103)_f$ surface with the high surface energy of 0.9 J/m^2.

However, for the other surfaces only containing Ti_5 atoms, their surface energies are very small. The values are 0.49 and 0.58, and 0.35 J/m^2 for anatase (101) and (100), and rutile (110), respectively. Note that the rutile (110) surface has smallest surface energy. Thus, the anatase (101) [13] and rutile (110) [11] are thermodynamically most stable structures. We can rule out the possibility of water dissociation on those surfaces [10].

Combined with the surface energy, we may estimate which surface with only Ti_5 atom can dissociate water by breaking a Ti bond. According to calculations, the surface energy should approach to ~1 J/m^2. For the high reactive surfaces with only Ti_5, the Ti_5 atoms eventually become too effective Ti_4 atoms during the interaction.

3.2.3. *Surfaces with steps*

Step edges are the most common intrinsic defects on the surface. In this subsection, we give two examples of steps on two most stable surfaces, anatase (101) and rutile (110) surface.

Gong et al. [40] have made an intensive investigation on anatase (101) surface. The structure models of step edges they studied are shown in **Figure 7**. We can divide those surface step structures into two categories with Ti_4 (in **Figure 7(a)**, **(d)**, **(e)**, **(f)**, and **(h)**) and without Ti_4 atoms (in **Figure 7(b)**, **(c)**, and **(g)**). Then, we recapitulate their surface energy calculation results γ (θ) in unit 10^{-2}eV/Å^2 and rearrange in an order of containing Ti_4 and Ti_5 atoms as follows:

AI(**Figure 7(a)**)	5.36	4.68	4.34	Ti_4
CI(**Figure 7(e)**)	5.49	4.81	4.46	Ti_4
CII(**Figure 7(f)**)	6.22	5.22	4.78	Ti_4
E(**Figure 7(h)**)	6.82	5.86	5.34	Ti_4
AII(**Figure 7(b)**)	4.60	4.24	4.06	Ti_5
BI(**Figure 7(c)**)		3.66	3.59	Ti_5
D(**Figure 7(g)**)	4.52	4.11	3.94	Ti_5

We omit the labels of vicinal surfaces here for simplicity. The surface energies with Ti_4 are larger than that with Ti_5 in each column. Gong et al. further studied water adsorption on D, BI, and AII with Ti_5 atom (see Supplementary Information in Ref. [40]). On D-type step edge, both molecular and dissociative H_2O adsorption can occur, but energy difference is much smaller than on flat TiO_2 (101). For BI step, the adsorption energies are smaller than that on the

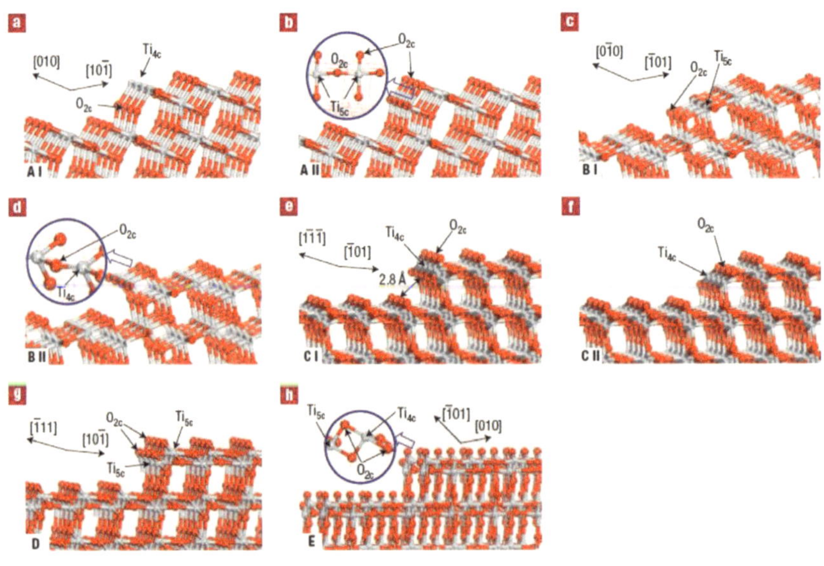

Figure 7. Structure models of step edges A–E on anatase (101) (**Figure 2** in Ref. [40]).

(101) surface. Note BI has the smallest surface energy among all step structures. On the least stable AII with the highest surface energy among steps containing Ti_5 atom, the water adsorption is found to be similar to the anatase (001) surface where water is dissociatively adsorbed with adsorption energy 1.28 eV. Again, here surface Ti_5 atom becomes Ti_4 by breaking a bond with neighboring oxygen.

Hong et al. [41] investigated water adsorption behavior step edges on rutile TiO_2 (110) surface using DFT calculations. They found that the < 1-10 > edge exhibits significantly enhanced water adsorption, especially dissociative adsorption, as compared to the pristine (110) surface and < 001 > step edge due to the existence of fourfold coordinated Ti_4 atoms at the < 1-10 > step edge, which lead to charge transfer to adsorbates more easily than fivefold coordinated Ti_5 atoms on the (110) surface and < 001 > step edge.

Later, Zheng et al. [42] studied the associative and dissociative adsorption of water molecules on rutile TiO_2 (110) surface with step defects by DFT calculations. The step structures were created by removing the TiO_2 unit along the < 1-11 > direction and exposing the Ti_4 atoms (terminating the Ti rows of the upper terrace) and the Ti_5 atoms (see **Figure 1(a)**, **(b)** in Ref. [42]). They only considered the case of Ti bonds fully saturated. Their results show that the molecular and dissociative adsorptions of H_2O can both be observed on Ti_4 sites and the molecularly absorbed water is more favorable on the Ti_5 sites. The lowest energy corresponds to the configuration where water molecule on Ti_5 site and one dissociated and one molecular water on Ti_4 site, same as our result for anatase (211) surface [47].

Thus, dissociative adsorption is also attributed to the existence of Ti_4 atoms and/or equivalent Ti_4 atoms exposing at step edges on anatase (101) and rutile (110) surfaces.

3.2.4. *Surfaces with O vacancy*

Oxygen vacancy is also one of the most common defects on the surface. Again, in this subsection, we give two examples of O vacancy on two most stable surfaces, rutile (110) and anatase (101).

Schaub et al. [43] investigated the O vacancy on rutile (110) through both experiments and DFT calculation and determined the O vacancies as active sites responsible for the dissociation of water molecules. Their DFT calculation results show that the dissociation of water is only at oxygen vacancies with an adsorption energy of 0.94, while on the perfect (100) surface, water molecule binds to the surface by 0.56 eV and the dissociation of water is even endothermic by 0.23 eV. The dissociative adsorption is unfavorable on the perfect (100) surface. They explained that the large reactivity of the vacancies is associated with the high-energy defect and water dissociation is simply to refill the coordination shell of Ti underneath the vacancy. However, there will be something more than that. Note that the *two* Ti_6 atoms underneath the vacancy become Ti_5 atoms when O leaves the surface. Thus, when O atom of water is to refill the O vacancy, the *two* Ti_5 atoms interact with O of water simultaneously. Then, in this case, we should count two Ti bonds participating in interaction. Therefore, the dissociation of water is due to the combined contribution of two unsaturated Ti atoms.

On the anatase (101) surface, dissociation of water close to the oxygen vacancy is energetically favored compared to molecular adsorption. Tilocca and Selloni [44] have done a detailed calculation of energy barrier between the molecular and dissociated states. When a surface O atom is removed, the Ti_6 and Ti_5 atoms bridging this O atom tune into Ti_5 and Ti_4, respectively. Water oxygen bonds to Ti_4 and then dissociates through a dissociation pathway. Thus, the dissociation of water on O vacancy of anatase (101) is also attributed to that Ti_4 atom.

3.2.5. Surfaces with Ti_3 atom

Recently, the three-coordinated Ti atom (Ti_3) are found on the (111) anatase TiO_2 surfaces. Xu et al [34] reported that they prepared anatase TiO_2 single crystals exposed (111) facet. Their DFT calculations showed that the (111) facet has a much higher surface energy of 1.61 J/m^2, which is attributed to the large percentage of under-coordinated atoms. They also found that this material has much higher photocatalytic activity than other surfaces. Here, there are Ti_3 and Ti_5 atoms with the ratio 1:3 on the (111) surface. According to our model, the Ti_3 atom could contribute two electrons participating interaction. Thus, it is expected that two water molecules could be dissociated on a single Ti_3 site. It could explain the (111) surface has much higher photocatalytic activity than other surfaces.

4. Conclusions

In conclusion, we use the anatase (211) surface as an idea model surface, containing one Ti_4 and one Ti_5 under-coordinated atoms in unit cell, to investigate their distinct properties. Our *ab initio* calculations show that the (211) surface is indeed a high reactivity surface with a high surface energy of 0.97 J/m^2. In addition, the four-coordinated Ti_4 atoms with two unsaturated bonds have a stronger chemical reactivity in comparison to the Ti_5 atoms with one unsaturated bond. Studies of water adsorption suggest two distinct states of adsorbed water on the (211) surface, one related to molecular water on Ti_5 sites and the other to dissociated water on Ti_4 sites. These results indicate that the Ti_4 atoms will play a critical role in water decomposition. According to TiO_2 structure, we propose a simple bond-charge counting model where each unsaturated Ti bond contributes 2/3 charge in average. As a necessary requirement of chemical reaction, the dissociation of water only occurs when Ti atoms provide more electrons to oxygen in water than H atom. Then, we reach to a conclusion that only Ti_4 atom or equivalent Ti_4 can dissociate water. The controversy about whether Ti_5 can dissociate water is resolved that Ti_5 atom will eventually become Ti_4 by breaking bond to neighboring O atom at surface with high surface energy while the dissociation will not happen for surface with low surface energy. Besides traditional DFT total energy calculation, this model is considered in a fundamental way. We can also declare that the bond charge offered by surface Ti atoms is the mechanism for water dissociation on TiO_2 surface. Furthermore, the model is generic and applicable to both rutile and anatase surfaces including defects, e.g., step edges and vacancies.

Acknowledgements

We acknowledge the useful discussions with Y.G. Cao. This work was supported by the National Science Foundation of China (Grants No. 10774177, No. 10634070, and No.11174297), and the National Basic Research Project (Grants No. 2010CB923002 and No. 2011CB922204), and the Fundamental Research Funds for the Central Universities, and the Research Funds of Renmin University of China.

Author details

Jing Xu[1], Li-Fang Xu[2]*, Jian-Tao Wang[2,3] and Annabella Selloni[4]

*Address all correspondence to: lfxu@iphy.ac.cn

1 Beijing Key Laboratory of Optoelectronic Functional Materials and Micro-nano Devices, Department of Physics, Renmin University of China, Beijing, China

2 Beijing National Laboratory for Condensed Matter Physics, Institute of Physics, Chinese Academy of Sciences, Beijing, China

3 School of Physics, University of Chinese Academy of Sciences, Beijing, China

4 Department of Chemistry, Princeton University, Princeton, New Jersey, USA

References

[1] Fujishima A, Honda K. Electrochemical photolysis of water at a semiconductor electrode. Nature (London). 1972;**238**: 37-38. DOI: 10.1038/238037a0

[2] Regan BO, Grätzel M. A low-cast, high-efficiency solar-cell based on dye-sensitized colloidal TiO_2 films. Nature. 1991;**353**: 737-740. DOI: 10.1038/353737a0

[3] Linsebigler AL, Lu G, Yates JT. Photocatalysis on TiO_2 surface principles, mechanisms, and selected results. Chemical Reviews. 1995;**95**: 735-758. DOI: 10.1021/cr00035a013

[4] Grätzel M. Photoelectrochemical cells. Nature. 2001;**414**: 338-344. DOI: 10.1038/35104607

[5] Fujishima A, Zhang X, Tryk DA. TiO_2 photocatalysis and related surface phenomena—Review, Surface Science Reports. 2008;**63**: 515-582. DOI: 10.1016/j.surfrep.2008.10.001

[6] Li Y, Somorjai GA. Nanoscale advances in catalysis and energy applications, Nano Letters. 2010;**10**: 2289-2295. DOI: 10.1021/nl101807g

[7] Liu G, Yang HG, Pan J, Yang YQ, Lu GQ, Cheng HM. Titanium dioxide crystals with tailored facets, Chemical Reviews. 2014;**114**: 9559-9612. DOI: 10.1021/cr400621z

[8] Fang WQ, Gong XQ, Yang HG. On the unusual properties of anatase TiO_2 exposed by highly reactive facets, The Journal of Physical Chemistry Letters. 2011;**2**: 25-734. DOI: 10.1021/jz200117r

[9] Hagfeldt A, Boschloo G, Sun L, Kloo L, Pettersson H. Dye-sensitized solar cells, Chemical Reviews. 2010;**110**: 6595-6663. DOI: 10.1021/cr900356p

[10] Henderson MA. Structural sensitivity in the dissociation of water on TiO_2 single-crystal surfaces, Langmuir. 1996;**12**: 5093-5098. DOI: 10.1021/la960360t

[11] Charlton G, Howes PB, Nicklin CL, Steadman P, Taylor JSG, Muryn CA, Harte SP, Mercer J, McGrath R, Norman D. Relaxation of TiO2(110)-(1x1) using surface X-ray diffraction, Physical Review Letters. 1997;**78**: 495-498. DOI: 10.1103/PhysRevLett.78.495

[12] Lindan, PJD, Harrison NM, Gillan MJ. Mixed dissociative and molecular adsorption of water on the rutile (110) surface, Physical Review Letters. 1998;**80**: 762-765. DOI: 10. 1103/PhysRevLett.80.762

[13] Vittadini A, Selloni A, Rotzinger FP, Grätzel M. Structure and energetics of water adsorbed at TiO_2 Anatase (101) and (001) surfaces, Physical Review Letters. 1998;**81**: 954-2957. DOI: 10.1103/PhysRevLett.81.2954

[14] Lazzeri M, Vittadini A, Selloni A. Structure and energetics of stoichiometric TiO_2 anatase surfaces, Physical Review B. 2001;**63**: 155409. DOI: 10.1103/PhysRevB.63.155409

[15] Diebold U. The surface science of titanium dioxide. Surface Science Reports. 2003;**48**: 53-229. DOI: 10.1016/S0167-5729(02)00100-0

[16] Gong XQ, Selloni A. Reactivity of anatase TiO_2 nanoparticles: The role of the minority (001) surface, The Journal of Physical Chemistry B. 2005;**109**: 19560-19562. DOI: 10.1021/ jp055311g

[17] Gong XQ, Selloni A, Vittadini A. Density functional theory study of formic acid adsorption on anatase TiO_2 (001): Geometries, energetics, and effects of coverage, hydration, and reconstruction, The Journal of Physical Chemistry B. 2006;**110**: 2804-2811. DOI: 10.1021/ jp056572t

[18] Yang HG, Sun CH, Qiao SZ, Zou J, Liu G, Smith SC, Cheng HM, Lu GQ. Anatase TiO_2 single crystals with a large percentage of reactive facets, Nature (London). 2008;**453**: 638-641. DOI: 10.1038/nature06964

[19] Selloni A. Crystal growth—Anatase shows its reactive side, Nature Materials. 2008;**7**: 613-615. DOI: 10.1038/nmat2241

[20] Angelis FD, Valentin CD, Fantacci S, Vittadini A, Selloni A. Theoretical studies on anatase and less common TiO_2 phases: Bulk, surfaces, and nanomaterials, Chemical Reviews. 2014;**114**: 9708-9753. DOI: 10.1021/cr500055q

[21] Li JM, Cao K, Li Q, Xu DS. Tetragonal faceted-nanorods of anatase TiO_2 with a large percentage of active {100} facets and their hierarchical structure. CrystEngComm. 2012;**14**: 83-85. DOI: 10.1039/c1ce06229h

[22] Wu L, Yang BX, Yang XH, Chen ZG, Li Z, Zhao HJ, Gong XQ, Yang HG. On the synergistic effect of hydrohalic acids in the shape-controlled synthesis of anatase TiO_2 single crystals, CrystEngComm 2013;**15**: 3252-3255. DOI: 10.1039/c2ce26744f

[23] Lai ZC, Peng F, Wang Y, Wang HJ, Yu H, Liu P, Zhao HI. Low temperature solvothermal synthesis of anatase TiO_2 single crystals with wholly {100} and {001} faceted surfaces, Journal of Materials Chemistry. 2012;**22**: 23906-23912. DOI: 10.1039/c2jm34880b

[24] Li JM, Xu DS. Tetragonal faceted-nanorods of anatase TiO_2 single crystals with a large percentage of active {100} facets, Chemical Communications (Cambridge). 2010;**46**: 2301-2303. DOI: 10.1039/b923755k

[25] Ding XK, Ruan HC, Zheng C, Yang J, Wei MD. Synthesis of TiO_2 nanoparticles with tunable dominant exposed facets (010), (001) and (106), CrystEngComm. 2013;**15**: 3040-3044 DOI: 10.1039/c3ce27097a

[26] Wu BH, Guo CY, Zheng NF, Xie ZX, Stucky GD. Nonaqueous production of nanostructured anatase with high-energy facets {001} or {010} facets, Journal of the American Chemical Society. 2008;**130**: 17563-17567. DOI: 10.1021/ja8069715

[27] Wu D, Gao Z, Xu F, Chang J, Gao S, Jiang K. Anatase TiO_2 nanocrystals enclosed by well-defined crystal facets and their application in dye-sensitized solar cell, CrystEngComm. 2013;**15**: 516-523. DOI: 10.1039/c2ce26454d

[28] Gordon TR, Cargnello M, Paik T, Mangolini F, Weber RT, Fornasiero P, Murray CB. Nonaqueous synthesis of TiO_2 nanocrystals using TiF_4 to engineer morphology, oxygen vacancy concentration, and photocatalytic activity. Journal of the American Chemical Society. 2012;**134**: 6751-6761. DOI: 10.1021/ja300823a

[29] Wu N, Wang J, Tafen DN, Wang H, Zheng J-G, Lewis JP, Liu X, Leonard SS, Manivannan A. Shape enhanced photocatalytic activity of single-crystalline anatase TiO_2 (101) nanobelts, Journal of the American Chemical Society. 2010;**132**: 6679-6685. DOI: 10.1021/ja909456f

[30] Chen C, Hu R, Mai K, Ren Z, Wang H, Qian G, Wang Z. Shape evolution of highly crystalline anatase TiO_2 nanobipyramids {001} and {101} facets, Crystal Growth & Design. 2011;**11**: 5221-5226. DOI: 10.1021/cg200457g

[31] Pan JH, Han G, Zhou R, Zhao XS. Hierarchical N-doped TiO_2 hollow microspheres consisting of nanothorns with exposed anatase {101} facets, Chemical Communications (Cambridge). 2011;**47**: 6942-6944. DOI: 10.1039/c1cc11796c

[32] Horvat B, Rečnik A, Dražić G. The growth of anatase bipyramidal crystals during hydrothermal synthesis, Journal of Crystal Growth. 2012;**347**: 19-24. DOI: 10.1016/j.jcrysgro.2012.03.027

[33] Liu M, Piao L, Zhao L, Ju S, Yan Z, He T, Zhou C, Wang W. J. Anatase TiO_2 single crystals with exposed {001} and {110} facets: Facile synthesis and enhanced photocatalysis, Chemical Communications (Cambridge). 2010;**46**: 1664-1666. DOI: 10.1039/b924172h

[34] Xu H, Reunchan P, Ouyang SX, Tong H, Umezawa N, Kako T, Ye JH. Anatase TiO_2 single crystals exposed with high-reactive {111} facets toward efficient H_2 evolution, Chemistry of Materials. 2013;**25**: 405-411. DOI: 10.1021/cm303502b

[35] Yang WG, Xu YY, Tang Y, Wang C, Hu YJ, Huang L, Liu J, Luo J, Guo HB, Chen YG, Shi WM, Wang YL. Three-dimensional self-branching anatase TiO_2 nanorods: Morphology control, growth mechanism and dye-sensitized solar cell application, Journal of Materials Chemistry A. 2014;**2**: 16030-16038. DOI: 10.1039/c4ta02761b.

[36] Jiang HB, Cuan Q, Wen CZ, Xing J, Wu D, Gong X-Q, Li C, Yang HG. Anatase TiO_2 crystals with exposed high-index facets (105), Angewandte Chemie International Edition. Ed. 2011;**50**: 3764-3768. DOI: 10.1002/anie.201007771.

[37] Wu L, Jiang HB, Tian F, Chen Z, Sun C, Yang HG. $Ti_{0.89}Si_{0.11}O_2$ single crystals bound by high-index {201} facets showing enhanced visible-light photocatalytic hydrogen evolution, Chemical Communications (Cambridge). 2013;**49**: 2016-2018. DOI: 10.1039/c3cc38105f

[38] Yang MH, Chen PC, Tsai MC, Chen TT, Chang IC, Chiu HT, Lee CY. Alkali metal ion assisted synthesis of faceted anatase TiO_2 {101} and {301} exposed were synthesized, CrystEngComm. 2013;**15**: 2966-2971. DOI: 10.1039/c3ce26965e

[39] Wu HB, Chen JS, Lou XW, Hng HH. Asymmetric anatase TiO_2 nanocrystals with exposed high-index facets and their excellent lithium storage properties. Nanoscale. 2011;**3**: 4082-4084. DOI: 10.1039/c1nr10854a

[40] Gong XQ, Selloni A, Batzill M, Diebold U. Steps on anatase TiO_2 (101), Nature Materials. 2006;**5**: 665-670. DOI: 10.1038/nmat1695

[41] Hong F, Ni YH, Xu WJ, Yan YF. Origin of enhanced water adsorption at <1-10> step edge on rutile TiO_2 (110) surface, The Journal of Chemical Physics. 2012;**137**: 114707. DOI: 10.1063/1.4753951

[42] Zheng T, Wu C, Chen MJ, Zhang Y, Chummings PT. A DFT study of water adsorption on rutile TiO_2 (110) surface: The effects of surface steps, The Journal of Chemical Physics. 2016;**145**: 044702. DOI: 10.1063/1.4958969

[43] Schaub R, Thostrup P, Lopez N, Lægsgaard E, Stensgaard I, Nørskov JK, Besenbacher F. Oxygen vacancies as active sites for water dissociation on rutile TiO_2 (110), Physical Review Letters. 2001;**87**: 266104. DOI: 10.1103/PhysRevLett. 87. 266104

[44] Tilocca A, Selloni A. Reaction pathway and free energy barrier for defect-induced water dissociation on the (101) surface of TiO_2-anatase, The Journal of Chemical Physics. 2003;**119**: 7445. DOI: 10.1063/1.1607306

[45] Wang C, Hu QQ, Huang JQ, Wu L, Deng ZH, Liu ZG, Liu Y, Cao YG. Efficient hydrogen production by photocatalytic water splitting using N-doped TiO_2 film. Applied Surface Science. 2013;**283**: 188-192. DOI: 10.1016/j.apsusc.2013.06.080

[46] Wang C, Hu QQ, Huang JQ, Deng ZH, Shi HL, Wu L, Liu ZG, Cao YG. Effective water splitting using N-doped TiO_2 films: Role of preferred orientation on hydrogen production, International Journal of Hydrogen Energy. 2014;**39**: 1967-1971. DOI: 10.1016/j.ijhydene.2013.11.097

[47] Xu J, Xu LF, Li ZZ, Wang JT, Lin ZS, Liu K, Cao YG, Selloni A. *Ab Initio* study of water adsorption and reactivity on the (211) surface of anatase TiO_2, Physical Review Applied. 2016;**5**: 064001. DOI: 10.1103/PhysRevApplied.5.064001

[48] Kresse G, Hafner J. *Ab initio* molecular dynamics for liquid metals, Physical Review B. 1993;**47**: 558-561. DOI: 10.1103/PhysRevB.47.558

[49] Kresse G, Furthmüller J. Efficiency of *ab-initio* total energy calculations for metals and semiconductors using a plane-wave basis set, Computational Materials Science. 1996;**6**: 15-50. DOI: 10.1016/0927-0256(96)00008-0

[50] Kresse G, Furthmüller J. Efficient iterative schemes for *ab initio* total-energy calculations using a plane-wave basis set, Physical Review B. 1996;**54**: 11169-11186. DOI: 10.1103/PhysRevB.54.11169

[51] Blöchl PE. Projector augmented-wave method. Physical Review B. 1994;**50**: 17953-17979. DOI: 10.1103/PhysRevB.50.17953

[52] Wang Y, Perdew JP. Correlation hole of the spin polarized electron gas, with exact small-wave-vector and high-density scaling, Physical Review B. 1991;**44**: 13298-13307. DOI: 10.1103/PhysRevB.44.13298

[53] Monkhorst HJ, Pack JD. Special points for Brillouin zone integrations, Physical Review B. 1976;**13**: 5188-5192. DOI: 10.1103/PhysRevB.13.5188

Chapter 2

Rare Earth-Doped Anatase TiO_2 Nanoparticles

Vesna Đorđević, Bojana Milićević and

Miroslav D. Dramićanin

Additional information is available at the end of the chapter

http://dx.doi.org/10.5772/intechopen.68882

Abstract

Titanium dioxide is a wide band-gap semiconductor of high chemical stability, nontoxicity and large refractive index. Because of the high photocatalytic activity, anatase is a preferred TiO_2 form in many applications such as for air and water splitting and purification. Doping of TiO_2 with various ions can increase the photocatalytic activity by enhancing light absorption in visible region and can alter structure, surface area and morphology. Also, by doping TiO_2 with optically active ions, visible light via up- or downconversion luminescence can be produced. It is a challenge to optimize the synthesis procedure to incorporate rare earth RE^{3+} ions into the TiO_2 structure due to large mismatch in ionic radii between the Ti^{4+} and RE^{3+} and because of the charge imbalance. Visible (VIS) and ultraviolet (UV) luminescence of several RE^{3+} ions can be obtained when incorporated into anatase TiO_2, also affecting microstructural characteristics of TiO_2. It is of great importance to summarize publications on rare earth-doped anatase TiO_2 nanoparticles to find correct TiO_2-RE combination to sensitize trivalent rare earths luminescence, as well as to predict or tune structural and morphological properties. A better understanding on these topics may progress the desired design of this kind of material towards specific applications.

Keywords: anatase, rare earth ions, photoluminescence, photocatalysis

1. Introduction

Rare earth (RE) elements are sixth period elements in the periodic table, from ^{57}La to ^{71}Lu. Because of many similarities, such as ionic +3 charges and similar ionic radius, ^{39}Y that also belongs to the III transition group and is positioned just above ^{57}La is also often considered as a part of the RE group. Even though the group is regarded as rare earth elements, they are not particularly rare. However, they are costly but highly efficient for many technological applications, mainly in lighting and display devices. With the absence of ^{57}La and ^{71}Lu, RE atoms, all have incompletely filled 4f orbitals that are positioned in the inner shell of xenon [Xe: $1s^2\,2s^2\,2p^6\,3s^2\,3p^6\,4s^2\,3d^{10}\,4p^6\,5s^2\,4d^{10}\,5p^6$] electron configuration, which are responsible for their emission properties. Since they are shielded by outer $5s^2$ and $5p^6$ orbitals, electrons from 4f orbitals do not participate in bonding and are only slightly affected by the surroundings of the ions. Ionic +3 charges are the most often, although some cases +2 and +4 can be stable as presented in **Table 1**. Electronic states are noted as $^{2S+1}L_J$, where L is the orbital angular momentum, S is the spin angular momentum, and J is the total angular momentum, and corresponding notations are also presented in **Table 1**. Lanthanide contraction makes significant decrease of ionic radii in the series with an increase in atomic number, and the values for six-coordinated RE^{3+} are also presented in **Table 1**.

Laporte's selection rule states that electron transitions between 4f states are forbidden, but they become partially allowed when RE ions are incorporated in non-symmetric sites [2, 3]. In that way, each ion has characteristic 4f energy levels with narrow-emission lines that depend on the crystalline environment of the host material in the order of few hundred cm^{-1}. The Dieke diagram is the energy-level diagram of trivalent lanthanide 4f electrons of RE^{3+}

Atomic number	Name	RE symbol	Atom	RE^{2+}	RE^{3+}	RE^{4+}	$^{2S+1}L_J$	Radii RE_{VI}^{3+} [Å]
57	Lanthanum	La	$5d^1\,6s^2$	–	[Xe]	–	$^1S_0$	1.032
58	Cerium	Ce	$4f^1\,5d^1\,6s^2$	–	$4f^1$	[Xe]	$^2F_{5/2}$	1.020
59	Praseodymium	Pr	$4f^3\,6s^2$	–	$4f^2$	$4f^1$	$^3H_4$	0.990
60	Neodymium	Nd	$4f^4\,6s^2$	$4f^4$	$4f^3$	$4f^2$	$^4I_{9/2}$	0.983
61	Promethium	Pm	$4f^4\,6s^2$	–	$4f^4$	–	$^5I_4$	0.970
62	Samarium	Sm	$4f^6\,6s^2$	$4f^6$	$4f^5$	–	$^6H_{5/2}$	0.958
63	Europium	Eu	$4f^7\,6s^2$	$4f^7$	$4f^6$	–	$^7F_0$	0.947
64	Gadolinium	Gd	$4f^7\,5d^1\,6s^2$	–	$4f^7$	–	$^8S_{7/2}$	0.938
65	Terbium	Tb	$4f^9\,6s^2$	–	$4f^8$	$4f^7$	$^7F_6$	0.923
66	Dysprosium	Dy	$4f^{10}\,6s^2$	–	$4f^9$	$4f^8$	$^{11}H_{15/2}$	0.912
67	Holmium	Ho	$4f^{11}\,6s^2$	–	$4f^{10}$	–	$^5I_8$	0.901
68	Erbium	Er	$4f^{12}\,6s^2$	–	$4f^{11}$	–	$^4I_{15/2}$	0.890
69	Thulium	Tm	$4f^{13}\,6s^2$	$4f^{13}$	$4f^{12}$		$^3H_6$	0.880
70	Ytterbium	Yb	$4f^{14}\,6s^2$	$4f^{14}$	$4f^{13}$	–	$^2F_{7/2}$	0.868
71	Lutetium	Lu	$4f^{14}\,5d^1\,6s^2$	–	$4f^{14}$	–	$^1S_0$	0.861

Table 1. Outer electronic configurations of RE atoms and ions, outside of the [Xe] shell, ground-state term of RE^{3+} and radii of 6-coordinated RE^{3+} (taken from Ref. [1]).

incorporated in $LaCl_3$ crystals, which can be found in the original or revised form, which is informative for many materials [4–7]. It schematically represents variations between ground- and excited-level energies or rare earth ions, proposing emissions of almost any colour in visible spectra by using one, or a combination of various RE ions in hosts.

Luminescent materials that absorb energy as light and do not emit it as heat, but as ultraviolet, visible or infrared (IR) light, are called phosphor materials. Typically, they are composed of insulating or semiconducting host material that is doped with activator ions. Phosphors with RE ions as activators are important materials that have found applications in artificial light, cathode-ray tubes, vacuum fluorescent and field emission displays, solid-state lasers, and so on [8]. It is now a custom to refer materials that have at least one dimension less than 100 nm as nanomaterials. The great number of atoms in top layers of nanoparticles significantly alters their optical properties; hence, it is justified to name nanostructured phosphors as a nanophosphors. Today, nanophosphors can be found in many forms, such as nanopowders, composites, coatings and thin films, giving new possibilities for application in bio-imaging and various types of physical and chemical sensing [9–11].

Photoluminescence of RE ions can be induced by the absorption of light through host lattice (host, H) that is transferred to RE ion (activator, A), directly exciting A, or energy transfer from other exited ions (sensitizer, S) that are also incorporated in matrix. A schematic diagram showing direct and indirect excitations with energy transfer resulting in the emission of light or heat is presented in **Figure 1(a)**.

When RE ions are used as activators in phosphor materials, depending on the positions of energy levels in RE ion, two main energy conversion mechanisms can lead to radiative energy transfer that results in the emission of light, one being downconversion, and the other upconversion. As it can be seen in **Figure 1(b)**, the principal difference between the two is the difference in excited and emitted energies. As schematically presented, in downconversion process electrons are excited by higher-energy photons compared to energy obtained from emission. In the process, prior to the emission of photons some energy is lost by non-radiative transitions. Oppositely, in upconversion process electrons are excited by lower-energy photons compared to energy obtained from emission. In order to preserve energy conservation rule, more than one photon is necessary for either single-ion excited-state absorption process, or in energy transfer upconversion process where the second ion is the sensitizer ion.

In order to fully understand the processes of downconversion light emission, we refer to energy-level diagram scheme presented in **Figure 2**. In honour of professor Alexander Jablonski, this type of energy diagrams is often called the Jablonski diagram. It qualitatively represents electronic energy levels as bolded horizontal lines and vibrational energy levels as a stack of horizontal lines in vertical energy diagram. Straight and wavy vertical arrows represent transitions between the states, where straight arrow represents transition associated with photon, while wavy arrows represent non-radiative transfers. A radiative decay process is a process in which electron releases some of its excitation energy as photon, while in a non-radiative decay excess energy is transferred into thermal motions, as vibration, rotation and translation processes, heat. Once an electron is excited through very quick process of absorption of photon, into, for example, some vibronic state of second excited singlet state, there are several ways that energy may be dissipated. The first is through vibrational relaxation, a non-radiative process that lowers energy of electron to the lowest excited singlet state, with or without non-radiative internal conversion process, depending on the overlap of vibrational and electronic energy of different states. Next, a radiative process of energy transfer to ground singlet state is followed by emission of photons in terms of fluorescence. There is no change in multiplicity $S_1 \rightarrow S_0$, so the transition is spin allowed and consequently fast. Since there are a large number of vibrational levels in electronic states, transitions can result in a range of emitted wavelengths. There is also a probability of non-radiative relaxations between

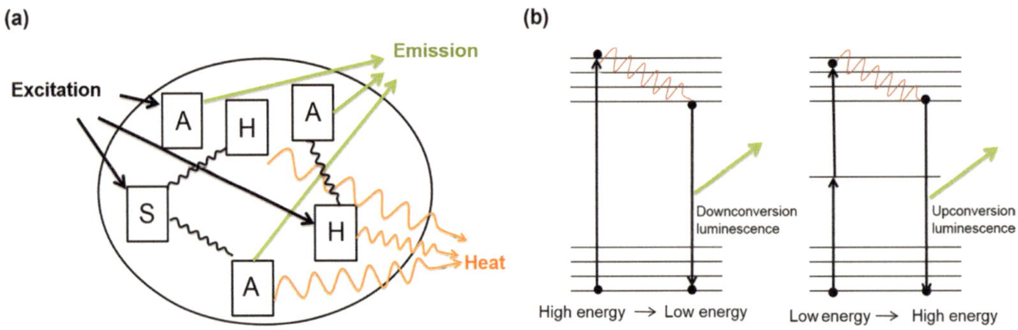

Figure 1. (a) Direct and indirect excitation with energy transfer resulting in emission of light or heat, by activators (A), hosts (H) and sensitizers (S). (b) Basic mechanisms of downconversion and upconversion luminescence.

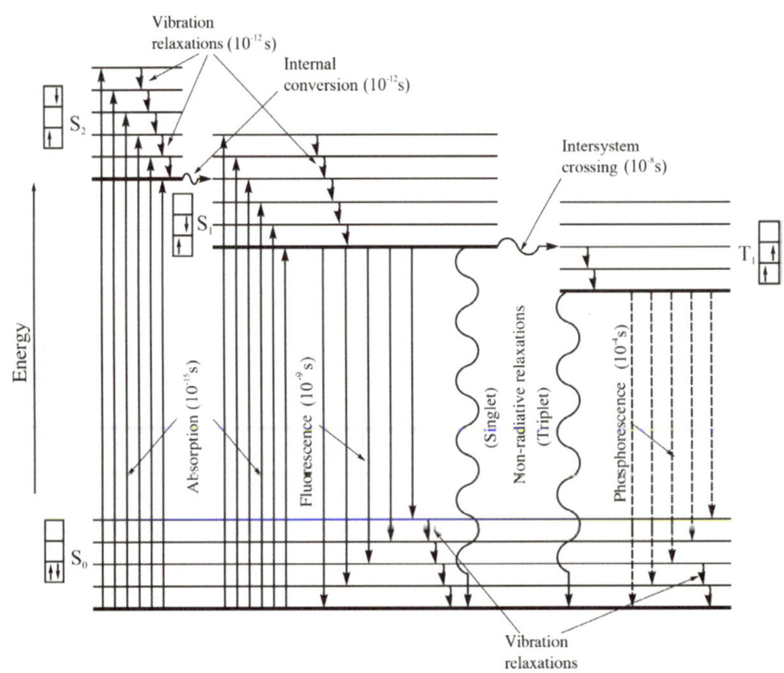

Figure 2. Radiative and non-radiative processes with corresponding approximate time interval of the processes in energy-level diagram scheme. S_0, ground singlet state; S_1, S_2, excited singlet states; T_1, excited triplet state.

the singlet states ($S_1 \rightsquigarrow S_0$). If in the process of dissipating of energy spin multiplicity changes by slower process of intersystem crossing, energy can be radiatively emitted from lowest excited triplet state to ground singlet state by phosphorescence $T_1 \rightarrow S_0$, or non-radiatively by relaxations between the triplet and singlet states ($T_1 \rightsquigarrow S_0$). Intersystem crossing and therefore phosphorescence are spin-forbidden processes; nevertheless, by coupling vibrational factors into the selection rules transitions become partially allowed, and they are consequently much slower.

2. Synthesis of rare earth-doped anatase TiO_2 nanoparticles

TiO_2 nanoparticles present several advantages for applications compared to their bulk counterparts. Their high-surface-to-volume ratio, improved charge transport and lifetime, afforded by their dimensional anisotropy, allows efficient contribution to the separation of photo-generated holes and electrons [12]. The properties of TiO_2 depend on its crystal structure, surface chemistry, dopants, doping levels, crystallization degree, size and morphology [13]. Hence, it is of great importance to control the particle size, shape and distribution of the synthetized TiO_2. To achieve desired characteristics, a variety of TiO_2 nanostructures have been prepared, such as nanoparticles, nanotubes, nanorods, nanofibres, nanosheets and nanofilms. These structures can be synthetized through various preparation methods, such as sol-gel, direct oxidation, micelle and inverse micelle techniques, sonochemical, hydrothermal/solvothermal, microwave, chemical vapour deposition, physical vapour deposition and electrospray deposition [14–17]. Significant progress has been made in the last 10 years regarding new approaches to the preparation of TiO_2. These include doping TiO_2 with optically active rare earth ions (RE). TiO_2 can be considered as an 'unusual' matrix for doping with RE^{3+} ions due to the large mismatch of both charge and ionic radius between the dopant and the host constituent cations. It is a challenge even now to optimize the synthesis procedure in the way to efficiently incorporate RE^{3+} ions into TiO_2 nanostructure and to obtain material with high crystallinity. Spectroscopic studies have showed that the RE ions can reside in the anatase in three different sites [18–20]. In nanopowders, substantial number of RE ions occupies the sites near the surface with the lowest point symmetry.

TiO_2 occurs in three most abundant crystalline phases in nature: anatase (tetragonal), rutile (tetragonal) and brookite (orthorhombic). Rutile TiO_2 is the most stable form, while anatase and brookite phases are metastable and can be transformed to rutile phase at higher temperatures. Even though rutile is denser and thermodynamically more stable than anatase, this significant temperature treatment is not favourable for the formation of nanoparticles with a diameter lower than 15 nm, which is a feature of anatase form TiO_2 [21, 22].

Sol-gel synthesis is the most common method for the preparation of RE-doped TiO_2 nanoparticles, being simple, cost-effective and low-temperature procedure, with the ability to fabricate nanostructure with high purity, homogeneity and controllable morphology. This synthesis includes the process of hydrolysis and poly-condensation of Ti–OH–Ti or Ti–O–Ti bonds forming densely three-dimensional structure that after heating changes from sol to gel, and after thermic treatment results in the form of oxide. Titanium source precursors can be alkoxides (such as titanium (IV)-isopropoxide (TTIP), titanium (IV)-butoxide (TBT)) or titanium (IV)-chloride ($TiCl_4$). RE ion precursors can be acid-soluble oxides (RE_2O_3) or water-soluble nitrates, acetates or chlorides ($RE(NO_3)_3 \cdot xH_2O$, $RE(CH_3COO)_3 \cdot x(H_2O)$, $RECl_3$).

In the method of hydrolysis of TTIP, products are characterized by low surface area, wide pore size distribution with contribution to pores of mesopores scale (<50 nm) [22]. The sol-gel synthesis with a two-step procedure of mixing precursor solutions was successfully used to obtain RE-doped TiO_2 [18, 23–36]. The gels obtained in such procedures undergo various temperature treatments, which are summarized in **Table 2**. In the method of hydrolysis of $TiCl_4$, which is another sol-gel method for the preparation of RE-doped anatase TiO_2, minor amounts of brookite phases are often present and slightly larger crystallite size compared to RE-doped TiO_2 from the titanium alkoxides is reported [13, 37].

Hydrothermal synthesis is a heterogeneous chemical reaction in the presence of an aqueous solvent, above room temperature (<200°C) in a closed system, where the pressure is elevated.

Dopant ions	Doping conc. (%)	Calcination temperature (°C)	Crystalline phase	Crystallite size* (nm)	BET surface area (m²/g)	Pore diameter (nm)	Refs.
–	–	400–700	A	8.14–79.1	25–117	3.26–6.4	[13, 18, 22–28, 38, 39]
–	–	500–800	A + R	14.1–101.8	0.59–17.94	4.68	[22–24, 29]
–	–	800–1000	R	32.7–100	0.34–16.7	–	[22–24]
Sc	2	500	A + B	16.6	–	–	[37]
Sc	2	500–550	A	16.6–26.9	–	–	[13]
Sc	2	600	A + R	45.0	–	–	[13]
Sc	2	650–800	R	51.7–65.2	–	–	[13]
Y	0.25–2	400–500	A	8.5–9.4	89.68–151	–	[28, 30, 31]
La	0.1–10	500	A	8.57–13.40	46.51–105.66	4.90–12.34	[25]
La	–	600	A + R	17.2	36.7	–	[32]
Ce	0.1–10	500	A	8.68–13.79	53.31–94.49	5.46–12.52	[24, 25]
Ce	5	800	A + CeO_2	–	–	–	[24]
Pr	0.25–1	400–650	A	9–20	77.5–134	–	[28, 33, 40]
Nd	0.05–4	400–700	A	10–20	7.5–75	–	[24, 34, 40, 41]
Nd	0.1–5	800	A + R	25	<1.0	–	[24, 41]
Nd	0.1–5	900–1000	A + R + $Nd_4Ti_9O_{24}$	–	–	–	[24]
Sm	0.3–3	420–700	A	5.8–12	50.78–95.9	5.20	[18, 29, 34, 35, 38, 42]
Sm	0.3–0.5	800	A + R	–	16.1–24.7	–	[38]

Dopant ions	Doping conc. (%)	Calcination temperature (°C)	Crystalline phase	Crystallite size* (nm)	BET surface area (m^2/g)	Pore diameter (nm)	Refs.
Eu	0.25–3	400–500	A	6–12	88.55–178.3	3.6–7.5	[18, 27–30, 39, 42]
Eu	5	800	A + R	27	–	–	[24]
Gd	1–2	500–700	A	6.9–15.1	32.8–97.7	–	[22]
Gd	2	800	A + $Gd_2Ti_2O_7$	–	–	–	[22]
Gd	5	600–800	A + R	26–27	–	–	[24, 32]
Gd	5–10	800–900	A + R + $Gd_2Ti_2O_7$	7.2–14.7	15.3–51.5	–	[22]
Tb	0.7–3	420–500	A	8.69–9	88.34	5.43	[18, 29]
Tb	5	800	A + R	25.5	–	–	[24]
Dy	0.3	450–650	A	9–31	60.4–80.6	–	[33]
Dy	5	800	A + R	24	–	–	[24]
Ho	0.3–2	500–800	A	12.5–20.5	76.76–98.81	–	[23, 36]
Er	0.25–5	400–700	A	8.5–21.9	18–132	–	[24, 26, 28, 42]
Er	5	800	A + $Er_2Ti_2O_7$	23.8	–	–	[24]
Yb	0.21–1.13	500	A	–	–	–	[43]
Yb	5	600–800	A + R	19–23	–	–	[24, 32]

*Anatase phase.

Table 2. The sol-gel synthesis conditions and major physicochemical properties of RE-doped TiO_2 nanostructures; A-anatase, B-brookite, R-rutile.

The synthesis has been used to produce homogeneous, high-purity, crystalline nanostructured RE-doped TiO_2 with different morphologies: nanotubes, nanobelts, nanowires or spherical nanoparticles. Alkoxide Ti precursors and water-soluble RE precursors are activated by acids or bases prior to the temperature treatments in Teflon-liners autoclave up to several days [28, 44–53]. Obtained precipitates should be washed to neutral pH [47] before calcination in order to gain well-defined TiO_2 nanoparticles. Also, hydrothermal route can use the synthesized or commercial available TiO_2 nanoparticle without the post-calcination treatment [44, 46]. The main difference of *solvothermal* synthesis is using other solvents than water. The obtained samples are spherical nanoparticles with an average diameter of 16 nm and the doping process can be easily achieved without significant loss of dopants [54]. The main characteristics and major physicochemical properties of RE-doped TiO_2 nanostructures synthetized by hydrothermal and solvothermal are summarized in **Table 3**.

Electrospinning method can be employed to produce nanostructure RE-doped TiO_2 with fibre morphology and the average fibre diameter in the range of 35–80 nm. Typically, RE-doped TiO_2

Dopant ions	Doping conc. (%)	Hydrothermal treatment (°C)	Calcination temperature (°C)	Crystalline phase	Crystallite size (nm)	BET surface area (m^2/g)	Morphology	Refs.
–	–	140–160	≤400	A	9.3–30	102–312.5	Spherical particle (d^* = 10–30 nm)	[28, 47–49, 54]
–	–	200	500	A + R	22.8	53–165	Spherical particle	[44, 51]
Y	0.25	150–160	≤400	A	9.8	120–157	Spherical particle (d = 5–15 nm)	[28, 47]
Y	0.3	80	–	A + R	–	–	–	[51]
La	0.11–0.53	200	500	A + R	22.32–24.38	69–86	Spherical particle	[44]
La	0.3	80	–	A + R	–	–	–	[51]
Pr	0.25–2.0	100	400	A	5.04–6.22	155–170	Spherical particle (d < 10 nm)	[55]
Pr	0.25	160	400	A	9.0	127	Spherical particle (d = 5–15 nm)	[28]
Pr	0.3	80	–	A + R	–	200	–	[51]
Nd	0.3	80	–	A + R	–	220	–	[51]
Sm	1	150	500	A	16	–	Spherical particles (d ~ 16 nm)	[54]
Eu	0.25–0.5	130–200	400–500	A	8.6	133	Spherical particle (d = 5–15 nm)	[28, 56]
							Sub-microspheres (d = 300 nm)	[52]
							Spindle particles (d = 50–100 nm, l^* = up to several μm)	[53]
							Nanorods (d = 10–20 nm, l = up to several μm)	[53]
							Nanobelts (w^* = 200–400 nm, l = several μm)	[45]

http://dx.doi.org/10.5772/intechopen.68882

Dopant ions	Doping conc. (%)	Hydrothermal treatment (°C)	Calcination temperature (°C)	Crystalline phase	Crystallite size (nm)	BET surface area (m^2/g)	Morphology	Refs.
Eu	–	180	700	A + R	–	–	Nano-belts forming aggregates (*d* = 50–200 nm)	[45]
Eu	–	180	900	R	–	–	Nano-belts forming aggregates (*d* = 50–200 nm)	[45]
Eu	1	150	500	A	16	–	Spherical particle (*d* ~ 16 nm)	[54]
Ho	0.75	150**	–	A + R	7.6–20.4	–	Nanowires (*d* = 500 nm, *l* = 15 nm)	[46]
2% Ho + Yb	2% Yb	120**	25, 100, 280	A	–	–	Nanotube	[50]
Er	0.25–4	140–160	>400	A	8.9–16	98.1–127	Spherical particles (*d* < 16 nm)	[28, 48, 49, 54]

*d, diameter; w, weight; l, length.

**TiO_2 calcined at 550°C was used in the synthesis route.

Table 3. Hydrothermal and solvothermal synthesis conditions and major physicochemical properties of RE-doped TiO_2 nanostructures.

nanofibres are fabricated with the use of polymer solvents of polyvinyl pyrrolidone (PVP) or polyvinyl alcohol (PVA), titanium alkoxides and RE chlorides or nitrates. Starting solutions in glass syringe with stainless-steel needle are connected to a high voltage and electrospun in air at different tensions, needle-target distances and feed rates [57–61]. In order to remove the polymeric component and obtain nanocrystalline anatase, RE-doped TiO_2, as-spun nanofibres were calcined at 500°C. However, the pure phase of RE-doped rutile TiO_2 can be obtained after higher calcination temperature (>1000°C). The synthesis conditions and major physicochemical properties of RE-doped TiO_2 nanostructures reported in the literature are summarized in **Table 4**.

Precursor materials	Dopant ions	Doping conc. (%)	Calcination temperature (°C)	Crystalline phase	Crystallite size (nm)	Fibre diameter (nm)	Refs.
PVP, TTIP	–	–	400–500	A			[57, 58]
PVP, TTIP	–	–	500–900	A + R	15.71–40		[57–59]
PVP, TTIP	–	–	1000	R			[57]
PVP, TTIP, $Y(NO_3)_3$	Y	1–2	500	A + R	11.35–13.8		[59]
PVP, TTIP, $Y(NO_3)_3$	Y	3	500	A	8.8		[59]
PVP, TTIP, $La(NO_3)_3$	La	1	500–800	A		40	[57]
PVP, TTIP, $La(NO_3)_3$	La	1	900–1000	A + R			[57]
PVP, TTIP, $La(NO_3)_3$	La	1	1100	R			[57]
PVA, TTIP, $La(NO_3)_3$	La	1	500	A			[58]
PVA, TTIP, $La(NO_3)_3$	La	1	700	A + R	12.51		[58]
PVA,,TTIP, $Ce(NO_3)_3$	Ce	1	500	A			[58]
PVA, TTIP, $Ce(NO_3)_3$	Ce	1	700	A + R	11.49		[58]
PVA, TTIP, $Nd(NO_3)_3$	Nd	1	500	A			[58]
PVA, TTIP, $Nd(NO_3)_3$	Nd	1	700	A + R	10.2		[58]
PVP, TTIP, $Eu(NO_3)_3$	Eu	1, 3	500–800	A		60, 70	[57]
PVP, TTIP, $Eu(NO_3)_3$	Eu	1	900	A + R			[57]
PVP, TTIP, $Eu(NO_3)_3$	Eu	3	900	A + R + $Eu_2Ti_2O_7$			[57]

Precursor materials	Dopant ions	Doping conc. (%)	Calcination temperature (°C)	Crystalline phase	Crystallite size (nm)	Fibre diameter (nm)	Refs.
PVP, TTIP, $Eu(NO_3)_3$	Eu	1, 3	1000–1100	R + $Eu_2Ti_2O_7$			[57]
PVP, TTIP, $Tb(NO_3)_3$	Tb	1, 3	400–800	A		35, 80	[60]
PVP, TTIP, $Tb(NO_3)_3$	Tb	1	900	A + R			[60]
PVP, TTIP, $Tb(NO_3)_3$	Tb	3	900	A + R + $Tb_2Ti_2O_7$			[60]
PVP, TTIP, $Tb(NO_3)_3$	Tb	1, 3	1000–1100	R + $Tb_2Ti_2O_7$			[60]
PVP, TTIP, $Er(NO_3)_3$	Er	1	400–900	A		60	[57]
PVP, TTIP, $Er(NO_3)_3$	Er	1	1000–1100	A + R + $Er_2Ti_2O_7$			[57]
PVP, TTIP, $Er(NO_3)_3$	Er	3	500–800	A		77	[57]
PVP, TTIP, $Er(NO_3)_3$	Er	3	900	A + R + $Er_2Ti_2O_7$			[57]
PVP, TTIP, $Er(NO_3)_3$	Er	3	1000–1100	R + $Er_2Ti_2O_7$			[57]
PVP, TBT, $ErCl_3$	Er	0.5–1.5	500	A	11.5–8.1		[61]
PVP, TBT, $ErCl_3$	Er	0.5	600–700	A + R	17.9–23.1		[61]
PVP, TBT, $ErCl_3$	Er	0.5	800	R	27		[61]
PVP, TTIP, $Yb(NO_3)_3$	Yb	1, 3	400–800	A		55, 70	[60]
PVP, TTIP, $Yb(NO_3)_3$	Yb	1	900	A + R			[60]
PVP, TTIP, $Yb(NO_3)_3$	Yb	3	900	A + R + $Yb_2Ti_2O_7$			[60]
PVP, TTIP, $Yb(NO_3)_3$	Yb	1, 3	1000–1100	R + $Yb_2Ti_2O_7$			[60]

Table 4. The electrospinning synthesis conditions and major physicochemical properties of RE-doped TiO_2 nanostructures.

Thermal plasma pyrolysis is rarely used for the synthesis and preparation of RE-doped TiO_2 nanopowders, which enables highly crystallized and well-dispersed nanoparticles due to the processing temperature (up to 1.0×10^4 K), rapid quenching rate at the plasma tail (~10^{5-7} K/s) and very short residence time [62]. The advantage of this synthesis is that well-dispersed and highly crystalline nanoparticles in a single processing step are obtained, without post-annealing treatment. On the other hand, it promotes crystallization of several crystalline phases of TiO_2, and with small amount of RE dopants mixtures of anatase and rutile are formed, while at higher temperatures dititanate structures were also formed [62].

Electrochemical synthesis is a significant method in the preparation of TiO_2 nanotubes at substrates, providing the precise control of nanotube morphology, length and pore size, and the formation of thick walls at substrates. Electrolytes used in this procedure are fluorides, where the concentration strongly effects on the dimensions and pH on the thickness of TiO_2 nanotubes [63, 64]. With anodic potential from 10 to 30 V, nanotubes with diameters between 15 and 200 nm are formed, and by cathodic electrochemical process RE ions are incorporated into the nanotubes. Also, magnetron-sputtering method can be used to prepare RE-doped TiO_2 films [65] as well as evaporation-induced self-assembly method [66–69].

In order to investigate structural, morphological, photocatalytic and optical properties of RE-doped anatase TiO_2 nanopowders with a series of RE^{3+} ions (Pr, Nd, Sm, Eu, Dy, Tb, Ho, Er and Tm) at a fixed concentration of 1 at.%, the sol-gel method has been used. To prepare samples, titanium (IV)-isopropoxide, water, ethanol and nitric acid were mixed in 1:3:20:0.08 molar ratios and the synthesis procedure is schematically shown in **Figure 3** and given in Ref. [27].

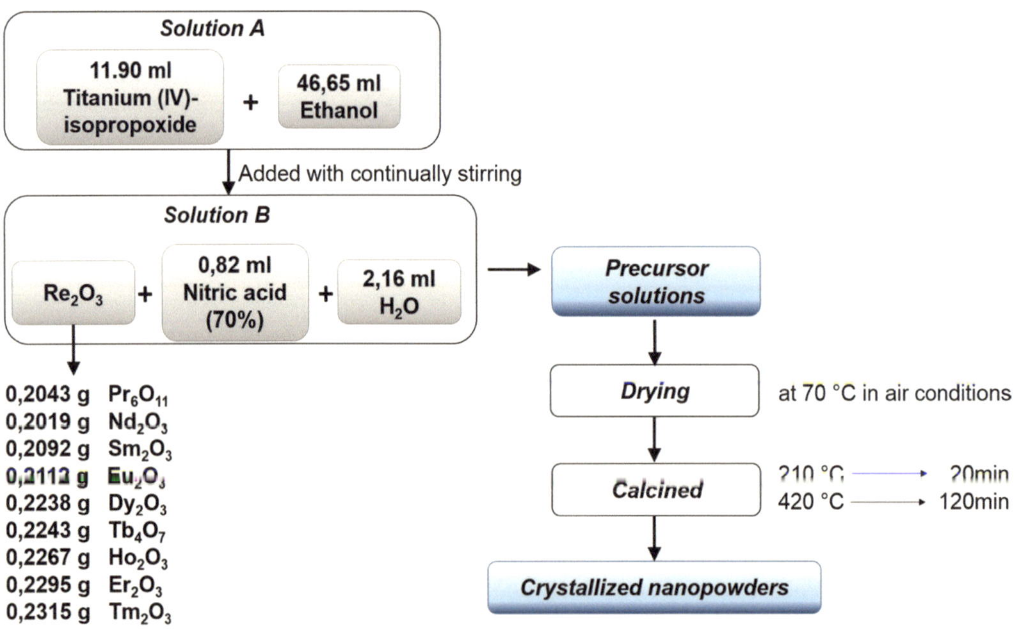

Figure 3. Schematic representation of the sol-gel synthesis with quantities of precursors used to prepare 3 g RE^{3+}-doped TiO_2 nanopowders.

http://dx.doi.org/10.5772/intechopen.68882

3. The influence of rare earth doping on the stability of phase structure, surface area and morphology of anatase TiO_2 nanoparticles

In most morphologies of calcinated TiO_2 powders, anatase phase is stabile up to temperatures below 500°C. Anatase to rutile crystalline phase transformation occurs above this temperature. In RE ions doped of anatase materials, the temperature of phase transformations shifts to higher values, suggesting the stabilization of anatase phase. As it can be seen in **Tables 2–4** in Section 2, phase transformations of RE-doped anatase to rutile crystalline phase occur in the temperature range of 500–1000°C. There are three types of dominant nucleation modes in forming rutile from anatase, bulk, interface and surface, which lead to the phase transformation. The proposed mechanisms affect the rate of grain forming and the density of rutile nucleation sites. The bulk nucleation of rutile particles is most likely to occur at temperatures above 500°C, when the grain boundary is surrounded by RE ions hindering the surface nucleation. The interface nucleation mode is dominant in the range of 550–680°C, when rutile particles with a larger crystallite size are formed on account of anatase particles, probably through aggregating of some anatase particles at the surfaces [70]. When calcination temperature increases, the phase transformation is not completed because the surface region is still in the mixed phases of anatase and rutile, with increasing percentage of rutile particles. At the same time, the formation of multiphase RE-titanate structures can also be noticed at higher temperatures, usually dititanates pyrochlore structures with a general formula of $RE_2Ti_2O_7$ [22, 57, 60, 71]. The contribution of these structures increases with RE-doping concentration [57], and it is more pronounced with RE ions with smaller ionic radius (heavier ions). When RE ions with a larger ionic radius occupy TiO_2 lattice sites, ionic mobility is hindered and the possibility of forming other titanate phases is lower. The electrospinning sol-gel route can be used to fabricate RE-doped TiO_2 with pure rutile phase at higher calcination temperature (>1000°C) without the formation of the $RE_2Ti_2O_7$ phase [57].

The influence of doping TiO_2 with RE, where larger RE ions of different charge (+3) compared to Ti ions are introduced into the anatase phase, gives rise to substitutional defects and, consequently, the large decrease in the short lattice order, thus in the reduction of the crystallite size. With increasing the concentration of RE ions, amorphization of crystalline powders is expected. The contents of RE ions used in sol-gel synthesis are usually in the 0.1–3% range, while further addition of RE ions (≥5%) effectively obstructs TiO_2 crystallinity owing to a lattice distortion, and remarkably reduces the crystallite size [22, 25]. The increase of doping concentration leads to a higher content of RE–O–Ti bonds that inhibit the growth of TiO_2 crystal grains restricting the direct contact of anatase particles, shifts diffractions to lower 2 theta angles, and as a consequence of smaller crystallites, broadening of X-ray diffraction (XRD) maxima [18, 55, 72, 73]. Even in undoped TiO_2, the anatase phase is reported to be thermodynamically stable at very low particle size. In respect to the particle size, it is reported that rutile phase can be formed when the crystallite size reaches a critical value of 12–20 nm [22]. Therefore, with the temperature increase, the crystallite size increases, which also favours anatase to rutile phase conversion. The influence of the incorporation of RE ions into the TiO_2 is reflected in the reduction of the crystallite size that inhibits the transformation of anatase to rutile phase. Taking into account all possible RE-doping effects on the stability of anatase phase, size and concentration of RE ion, applied synthesis method and calcination temperature, a number of parameters may be varied in an attempt to optimize desired TiO_2 powder structure and properties.

RE-doped TiO_2 nanopowders were prepared by the sol-gel route using a series of RE (Pr, Nd, Sm, Eu, Dy, Tb, Ho, Er, Tm) oxides and titanium(IV)-isopropoxide. The final calcination treatment is carried out at a temperature of 420°C for 2 h. XRD measurements were done on synthesized powders using Rigaku SmartLab instrument under the Cu $K_{\alpha 1,2}$ radiation, in a 2θ-range from 10° to 120° in 0.02° steps, and are shown in **Figure 4**. The XRD patterns indexed according to the ICDD card No. 00-021-1272. These patterns consist of the characteristic, intense peaks corresponding to 101, 004, 200, 105, 211 and 204 main reflections from anatase phase TiO_2 in all RE-doped TiO_2 nanopowders. There are no diffraction peaks of another crystalline phase of TiO_2 (rutile or brookite), rare earth oxide phase or dititanate pyrochlore structures. The analysis of relevant structural parameters was obtained using PDXL Integrated software, and calculated results are presented in **Table 5**. The average crystallite size of undoped TiO_2 was determined to be 149.6 Å, which is a much higher value than for the doped ones, suggesting the decrease in crystallinity with doping with RE as a result of the RE–O–Ti bonds in doped TiO_2 nanopowders. A consequence of the incorporation of larger RE ion compared to Ti ion ($r(Ti_{VI}^{4+})$ = 0.605 Å into anatase structure results in an increase in cell parameters that result in an increase of cell volume.

Mesoporous materials have important properties for potential applications, such as well-defined pore structure, uniform pores in the range between 2 and 50 nm and high surface area that provides a large number of active sites. Nevertheless, during the calcination treatment, TiO_2 nanoparticles pass through the process of crystal growth and anatase-to-rutile phase transformation causing the collapse of the mesoporous framework and a decrease of surface area. Incorporation of RE ions into the TiO_2 matrix has been presented as a potential strategy to overcome these disadvantages, with a possibility for thermal stability of the mesoporous structure and retarded decreasing of surface area of TiO_2 nanoparticles at high temperatures [25]. Also, RE ion-doped nanocrystalline TiO_2 has a significant number of active sites at anatase wall, leading to different physicochemical properties compared to undoped TiO_2 nanoparticles.

One of the problems in the synthesis of mesoporous TiO_2 is to achieve an appropriate balance between the hydrolysis and condensation processes of the titanium precursor. A slow hydrolytic condensation could lead to a small surface area in pure mesoporous TiO_2, because small quantities of water influence the reactivity of titanium precursor materials, and affects polymerization of TiO_2 [25]. On the other hand, higher reactivity of the titanium precursor towards hydrolysis and condensation leads to denser inorganic networks, which is promoted by the influence of hydrated RE precursors. In that way, relatively higher surface area and pore diameter are expected in RE-doped TiO_2 nanoparticles compared to undoped TiO_2 [25]. In sol-gel synthesis of anatase, TiO_2 nanoparticles crystallize with a pore diameter in the range of 3.26–6.4 nm and the surface area in the range of 25–117 m^2/g [13, 18, 22–28, 38]. In the low-concentration RE-doped anatase TiO_2 nanoparticles annealed at the intermediate temperatures, pores have almost the same size as in the undoped ones. However, relatively high doping concentrations of RE ions (up to 10%) induce significant change in pore size distribution, indicating the significant process of filling the pores, additionally promoted at higher

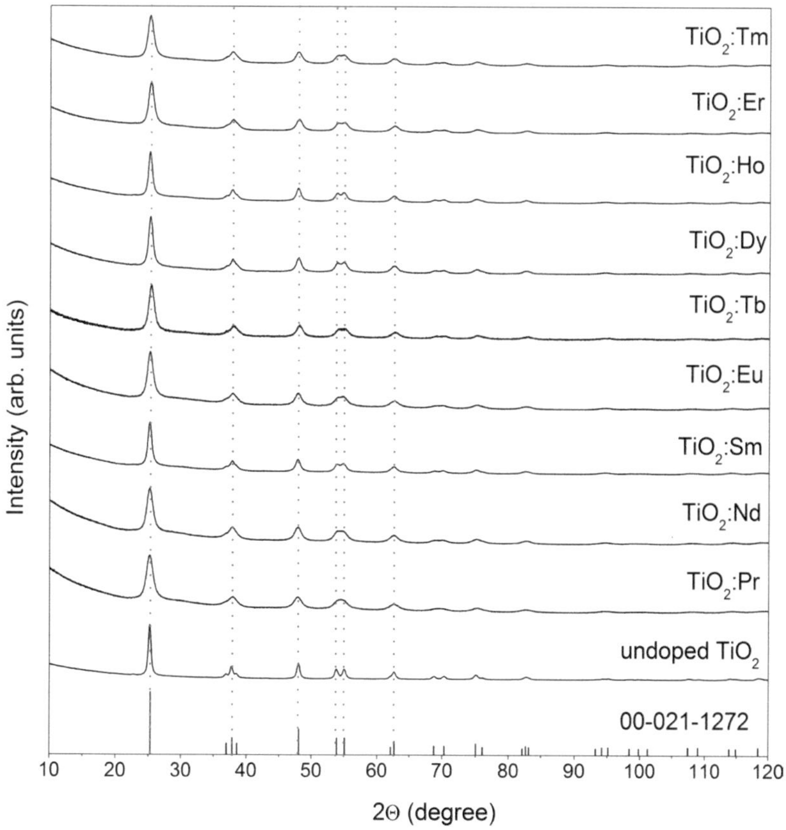

Figure 4. XRD patterns of undoped TiO_2 and TiO_2 doped with series of RE ions (RE = Pr, Nd, Sm, Eu, Dy, Tb, Ho, Er and Tm).

Sample	Crystallite size (Å)	Strain (%)	Lattice parameters $a = b$ (Å)	Lattice parameter c (Å)	Unit cell volume (Å^3)	Specific area (m^2/g)
Undoped TiO_2	149.6	0.35	3.785	9.502	136.128	9.7
TiO_2: Pr	72.1	0.85	3.803	9.508	137.512	54.4
TiO_2: Nd	68.4	0.46	3.796	9.505	136.963	101.5
TiO_2: Sm	103.1	0.48	3.804	9.521	137.643	68.2
TiO_2: Eu	81.6	0.73	3.796	9.494	136.805	52.4
TiO_2: Dy	101.3	0.56	3.794	9.505	136.189	87.4
TiO_2: Tb	83.1	0.66	3.789	9.494	136.301	–
TiO_2: Ho	102.63	0.40	3.806	9.535	138.120	81.0
TiO_2: Er	81.3	0.68	3.797	9.516	137.194	68.2
TiO_2: Tm	79.5	0.58	3.801	9.528	137.657	63.7

Table 5. XRD and BET results of undoped TiO_2 and RE doped TiO_2.

temperatures. For most of the RE ion-doped anatase TiO_2 nanoparticles, porosity can be presented by unimodal distributions, while the bimodal distribution may occur in some cases of higher doping concentration of RE ions and higher calcination treatments, when their pore diameter exceeded 100 nm [38].

The adsorption isotherms of RE-doped TiO_2 nanoparticles prepared by sol-gel route show type IV behaviour with the typical hysteresis loop. Undoped TiO_2 often show tails in their hysteresis loops at higher relative pressure, which are usually attributed to wide distribution of mesopores with some percentage of macropores (>50 nm). With the increase in calcination temperature, the crystallite size increases, also resulting in the significantly larger average pore size, but also with reduction in surface area values. The RE-doped TiO_2 are characterized by high degree of pore-size uniformity and a well-defined narrow pore size distribution without any contribution of macropores. On the contrary to the undoped TiO_2, high surface area can be retained even at relatively high temperatures [22]. Different trends are observed in samples prepared by impregnation sol-gel synthesis based on the later addition of RE metals that can lead to blockage pores and the formation of agglomerations due to low dispersion over the surface. The comparison of surface areas reveals that the specific surface area decreases by adding the metal oxides on the surface [71, 74]. The pore diameter of the RE-doped TiO_2 nanoparticles prepared with co-precipitation synthesis is larger and basically consists of some percentage of macropores (>50 nm). The formation of macroporous structure in the RE-doped TiO_2 nanoparticles was attributed to the agglomerations of TiO_2 particles and higher calcination treatment, as already known that higher calcination temperature will facilitate the growth of grains, obviously the smaller pores endured much greater stress and collapsed first during the calcination treatment [32].

RE-doped TiO_2 prepared by hydrothermal route shows higher Brunauer, Emmett and Teller (BET) surface area values when compared to undoped TiO_2. Probably, the increase in the BET surface area with increasing the doping level of RE ions is a consequence of smaller crystallite size for RE-doped TiO_2 [28]. However, the lack of linear correlation between the crystallite size of TiO_2 and the specific surface area may suggest that small amounts of RE_2O_3 were accumulated on the surface of TiO_2 nanoparticles resulting in higher surface area [28].

The specific surface area of the synthesized materials estimated by BET method is summarized in **Table 5**. The significant influence of RE^{3+} ions in doped anatase TiO_2 is obvious by the huge increase in the surface area of doped materials compared to the undoped one. The crystallite size and BET surface area have no linear correlation, suggesting a small amount of RE_2O_3 accumulated on the surface of TiO_2. The result could also be discussed regarding agglomeration of nanoparticle which is unavoidable in this kind of synthesis.

Transmission electron microscopy (TEM) was performed in order to investigate the surface morphology of the undoped TiO_2 nanopowder and nanopowders doped with the series of RE ions. RE-doped TiO_2 nanopowders were prepared by the sol-gel method using the series of RE (Pr, Nd, Sm, Eu, Dy, Tb, Ho, Er and Tm) oxides and titanium(IV)-isopropoxide, as previously discussed. The final calcination treatment is carried out at a temperature of 420°C for 2 h. As it can be seen from **Figure 5(A)**, the undoped sol-gel anatase sample consists of densely aggregated crystalline nanoparticles of irregular shapes, and variable dimensions of about 10–20 nm in size. Using selected area electron diffraction (SAED) technique, local crystal structure was confirmed to be pure anatase phase. The ring pattern was indexed by ICDD card no. 00-021-1272 with rings that correspond to 101, 004, 200, 105, 211 and 204 main reflections, presented in **Figure 5(B)**. The presence of rings suggests polycrystalline sample, and the characteristic grainy appearance of the rings suggests that crystallites have a size of 20 nm or more, suggesting only few joint unit cells per particle.

In **Figure 6(A–I)**, TEM of RE-doped TiO_2 nanopowders is collected at different magnifications, all showing a bar of 20 nm. All of the doped samples show agglomerated nanoparticles, only the estimated particles are smaller in size compared to the undoped sample.

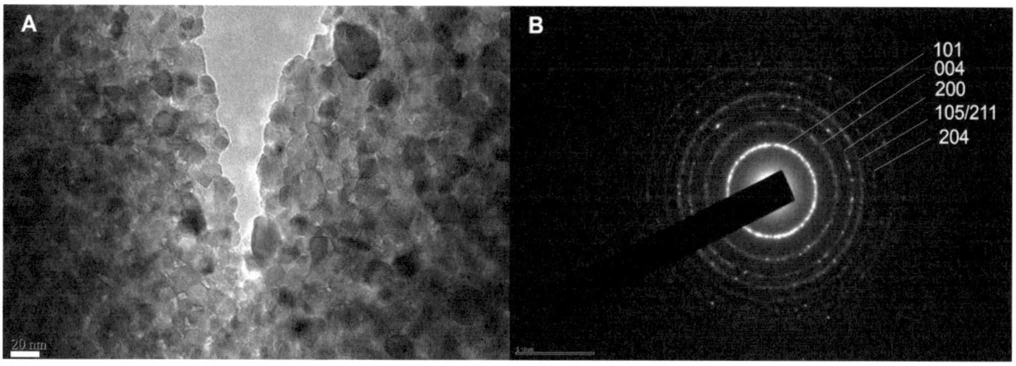

Figure 5. Transmission electron micrograph of undoped TiO_2 nanopowders recorded at magnification of ×67,000 (A), with corresponding selected area electron diffraction (B).

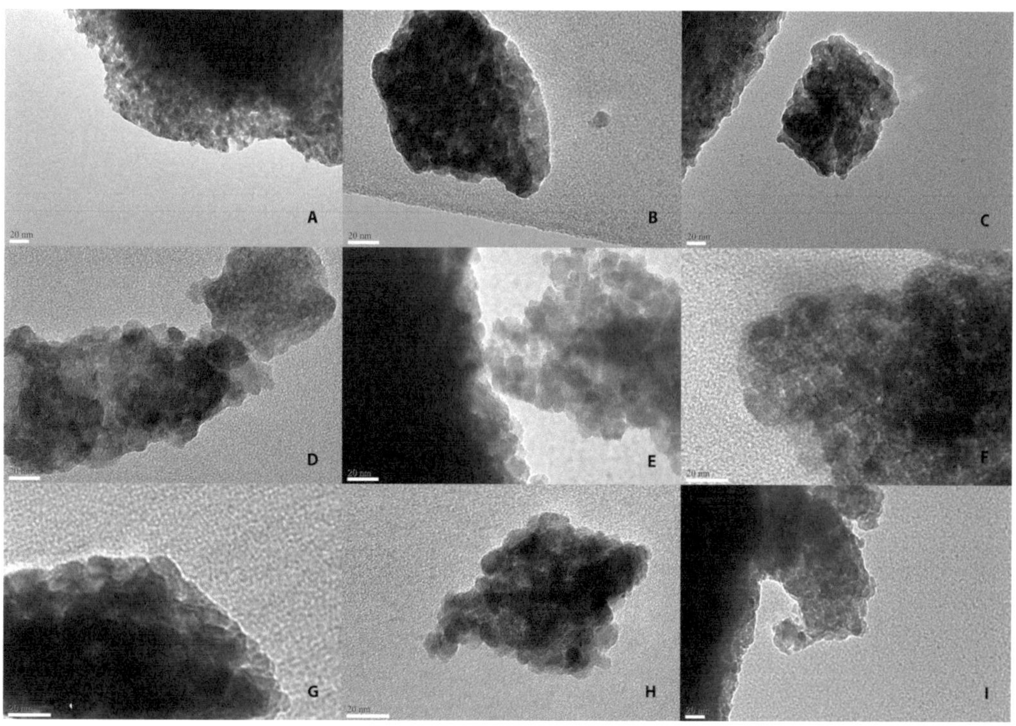

Figure 6. Transmission electron micrographs of RE-doped TiO_2 nanopowders at different magnification with bar of 20 nm: (A) TiO_2:Pr, (B) TiO_2:Nd, (C) TiO_2:Sm, (D) TiO_2:Eu, (E) TiO_2:Dy, (F) TiO_2:Tb, (G) TiO_2:Ho, (H) TiO_2:Er and (I) TiO_2:Tm.

4. The influence of rare earth doping on photocatalytic activity of anatase TiO_2 nanoparticles

One of the main challenges in photocatalytic research is the increase of spectral sensitivity of TiO_2 from ultraviolet (UV) to visible (VIS) spectrum. Incorporation of various RE ions into the anatase TiO_2 can increase the photocatalytic activity by enhancing the light absorption, adjustment of the phase structure, crystallinity, doping concentration, surface area and morphology. An overview of literature where RE-doped TiO_2 was used as a photocatalyst in respect to variables to experiments is given in **Table 6**. For detailed information about the type of artificial light source, time of illumination, as well as the percentage of dye degradation, the readers are advised to inquire the reference list provided in **Table 6**.

Dopant ion	Optimal doping conc. (%)	Synthesis method	Optimal calcination temperature (°C)	Crystalline phase	Dye	Refs.
Sc	2	Sol-gel	500	A + B	Rhodamine B	[37]
Y	1.5	Sol-gel	500	A	Methyl orange	[31]
Y	–	Hydrothermal	150	A	Methyl orange	[47]
Y	0.25	Hydrothermal	400	A	Phenol	[28]
Y	0.3	Hydrothermal	400	A + R	Phenol	[51]
La	0.3	Hydrothermal	400	A + R	Phenol	[51]
La	1	Sol-gel	550	A	Direct blue dye (DB53)	[75]
Pr	0.3	Sol-gel	450	A	Herbicide metazachlor	[33]
Pr	0.25, 0.5	Hydrothermal	400	A	Methyl orange	[55]
Pr	0.3	Hydrothermal	400	A + R	Phenol	[51]
Nd	0.3	Hydrothermal	400	A + R	Phenol	[51]
Nd	1	Sol-gel	550	A	Direct blue dye (DB53)	[75]
Sm	0.3	Sol-gel	500	A	Diuron	[38]
Sm	0.7	Sol-gel	500	A	Remazol red RB-133	[29]
Sm	1	Sol-gel	500	A	Methylene blue	[42]
Sm	1	Sol-gel	550	A	Direct blue dye (DB53)	[75]
Eu	0.5–2.0	Sol-gel	400	A	Methylene blue	[39]
Eu	1	Sol-gel	500	A	Rhodamine B	[71]
Eu	1	Sol-gel	420	A	Crystal violet	[27]

Dopant ion	Optimal doping conc. (%)	Synthesis method	Optimal calcination temperature (°C)	Crystalline phase	Dye	Refs.
Eu	1.3	Sol-gel	500	A	Remazol red RB-133	[29]
Eu	1	Sol-gel	550	A	Direct blue dye (DB53)	[75]
Eu	1.5	Sol-gel	500	A	Methylene blue	[30]
Eu	0.5	Hydrothermal	50	A	Phenol	[56]
Gd	1	Sol-gel	550	A	Direct blue dye (DB53)	[75]
Gd	5	Sol-gel	800	A + $Gd_2Ti_2O_7$	Methylene blue	[22]
Gd	0.3-0.6	Magnetron sputtering	1000	R	Methyl orange	[65]
Tb	0.7	Sol-gel	500	A	Remazol red RB-133	[29]
Ho	0.3	Sol-gel	500	A	Methyl orange	[23]
Ho	0.5	Sol-gel	600	A	Methyl orange	[23]
Ho	0.5	Sol-gel	500	A	Methyl orange	[36]
Ho	0.75	Hydrothermal	150	A + R	Methylene blue	[46]
Er	1.5	Sol-gel	500	A	Orange I	[26]
Er	2	Hydrothermal	400	A	Phenol	[48, 49]
Er	0.5	Electrospinning	500	A	Methylene blue	[61]
Yb	1	Sol-gel	550	A	Direct blue dye (DB53)	[75]

Table 6. RE-doped TiO_2 used as photocatalyst in recent photocatalytic studies.

Initially, when TiO_2 is exposed to light, it produces two types of charge carriers: electrons (e^-) in conduction band and holes (h^+) in valence band, as presented in **Figure 7(a)**. These e^-/h^+ pair generations follow the processes of charge separation and migration to the surface. At the surface, active species in valence band (h^+_{vb}) reacts with adsorbed water producing $OH^\bullet$ radical and proton (H^+). At the same time, active species in conducting band (e^-_{cb}) reacts with oxygen to produce active $O_2^{\bullet-}$ radical. The radical reacts with the proton and produces $OH_2^\bullet$ radical. When paired, the $OH_2^\bullet$ radicals produce H_2O_2 which degrades into two $OH^\bullet$ radicals. The formation of $OH^\bullet$ is crucial for the degradation of organic dye. However, the rate of recombination of photogenerated e^-/h^+ pairs is very fast (few nanoseconds) and substantial number can be recombined with just the release of heat [76]. When RE-doped TiO_2 is used as photocatalyst, incorporation of RE ions into the TiO_2 host creates charge imbalance. With increasing charge imbalance, more hydroxide ions are being adsorbed on the TiO_2 surface. Hydroxide ions (OH^-) restrain the recombination of e^- and h^+, and additionally react with holes to produce surface hydroxyl radical ($OH^\bullet$), which substantially improve the photocatalytic degradation of dye [26, 28]. The main disadvantage in the application of anatase TiO_2 as catalyst is dominant absorption in UV caused by its band gap (E_g ~3.2 eV). One approach to enhance absorption in VIS is doping. In the means of energy, doping can alter absorption threshold to lower energies. Incorporation of RE ions into the TiO_2 host modifies the band gap of TiO_2 with sub-band-gap energy levels of RE ions, as illustrated in **Figure 7(b)** [64, 77]. These energy levels offer electronic transition from the TiO_2 valence band to the empty RE ion sub-band-gap energy levels. These transitions require less energy than TiO_2 valence-to-conduction band transition and can be induced by visible light. In that way, RE ions in the TiO_2 host enhance the separation of e^- and h^+, contributing to photocatalytic degradation of organic dyes [28].

The main focus on the photocatalytic activity of RE ions incorporated into the anatase TiO_2 is the influence of RE-doping concentration [23, 26, 28, 31, 46, 56, 61, 65]. On the other hand, reports of comprehensive investigation of the type of RE ions in TiO_2 matrix, in order to predict the influence of dopants on the photocatalytic activity under UV and visible light, are scarce [51, 75, 78]. The results for photocatalytic activity of 1 at.% RE (RE = Pr, Nd, Sm, Eu, Dy, Tb, Ho, Er and Tm)-doped anatase TiO_2 nanopowders are presented in **Figure 8**. All of doped nanopowders were prepared in the same way, as presented in **Figure 3**. Methylene orange (MO) aqueous solution with a concentration of 5 mg/l was used in all experiments. Solutions were photocatalytically treated up to 4 h with 0.1 g of undoped- and RE-doped TiO_2 nanopowders. UV-VIS light irradiation Ultra-Vitalux 300 W, Osram lamp was used in all experiments in order to simulate the solar radiation. Absorptions of MO solution aliquots were measured after 0, 5, 10, 20, 30, 60, 90, 180 and 240 min of illumination. The results of photodegradation of MO, observed at a maximum absorbance of MO at 464 nm, for Ho-doped TiO_2 nanopowder, are presented in **Figure 8(a)**. The results of MO degradation for all samples were calculated by $Degradation\ (\%) = \left[\frac{(C_0 - C)}{C_0}\right] \times 100\%$, where C_0 is the initial concentration of MO solution and C is the concentration of MO solution after 4 h, and is given in **Figure 8(b)**. These results show that the incorporation of the RE ions into the TiO_2

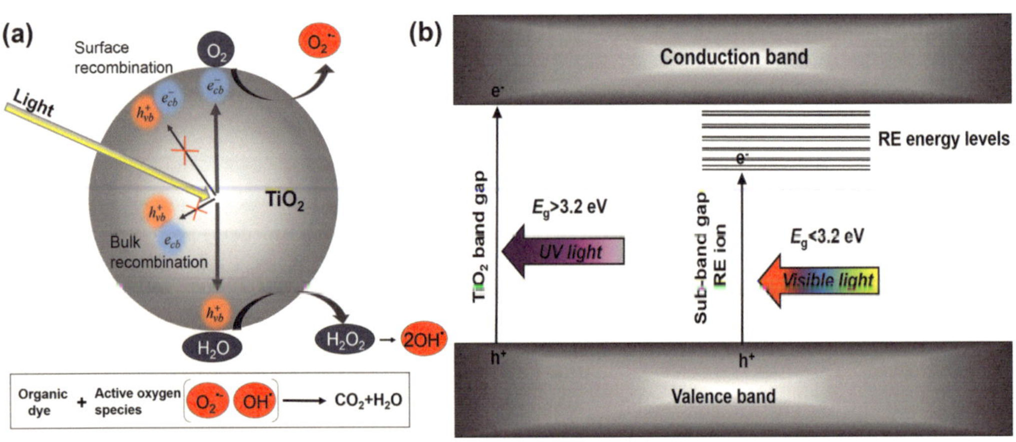

Figure 7. (a) Basic photocatalytic mechanism under UV or visible light irradiation. (b) Modification of band gap with sub-band-gap energy levels of RE ions.

matrix may bring a positive effect on the photocatalytic activity of TiO_2, as presented in **Figure 8(b)**. The reasons could be attributed to the synergetic effects of anatase phase stability, reduced crystallite size, relatively large surface area, significant improvement of the separation rate of photogenerated e^-/h^+ pairs and efficient absorption of visible light due to sub-energy levels of RE ions into the band gap of TiO_2.

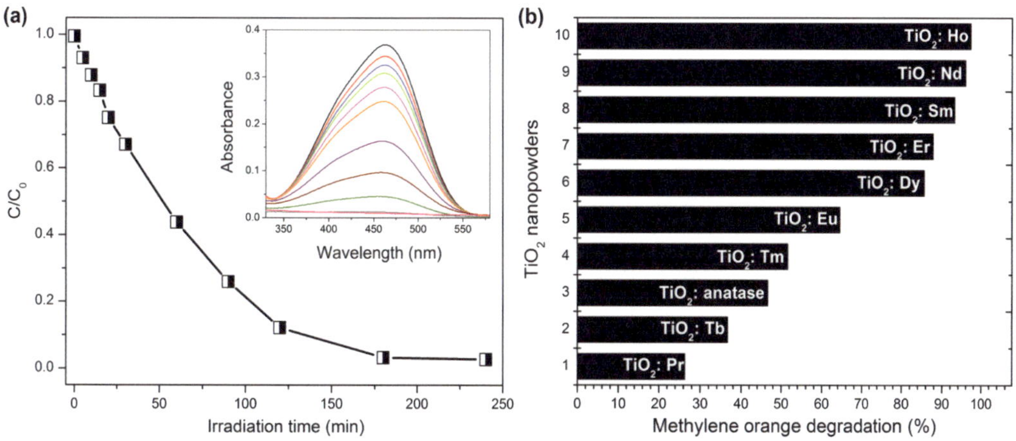

Figure 8. (a) The concentration of MO solution as a function of irradiation time for Ho-doped TiO_2 used as photocatalyst, inset: the absorption spectra of MO after different illumination times. (b) Photocatalytic degradation of MO after 4 h for various RE-doped TiO_2, with the fixed concentration of RE ions.

5. Optical properties of rare earth-doped anatase

When light interacts with matter, the material can absorb, transmit or reflect some part of the light. Absorption spectroscopy is a method to measure absorption as a function of wavelength or frequency. Since light cannot penetrate opaque samples such as powders and other solids, it is reflected on the surface of the samples. Spectrometers with integrating spheres measure the change of reflected light of a surface and compare it to a standard, most often barium sulphate, which is taken to be 100% of reflected light. Then, the obtained value is relative reflectance, and the reflectance spectrum provides the information of interaction of light in the sample as a function of wavelength. In that manner, reflectance can be directly correlated with absorption. Nowadays, research-grade spectrophotometers can combine detectors and extend detected light up to the near-infrared region of 1400 nm.

Some of the absorbed light can subsequently be emitted as light, as was already discussed in Section 1. Then, the radiative processes can be observed by photoluminescence spectroscopy (PL). In steady-state PL spectroscopy, we primarily refer to excitation and emission spectroscopy measurements obtained by a continual light source which emits a constant number of photons in time. Since exciting of electrons takes about 10–15 s^{-1}, following energy dissipation, whether radiative of non-radiative, is a much slower process so the number of excited electrons could be considered as constant. Absorption spectroscopy could suggest the wavelength that could be used to gain luminescence, but not all absorption result in emission. When we refer to the Jablonski diagram, it is obvious that absorption can occur to several excited singlet states, such as S_1, S_2, and so on, and expected emission normally occurs only from the lowest excited singlet or triplet states, S_1 and T_1. In excitation spectrum, a single emission detection wavelength is chosen that corresponds to an expected band in the emission spectrum. The excitation source is then scanned through wavelength region, and the intensity of the emission at the single selected wavelength is scanned in a function of excitation wavelength. The output of absorption and excitation spectrum is not the same, although detected maxima (or minima) at the same wavelength suggest the same excited energy levels. In luminescence emission spectroscopy, a wavelength of exciting light is selected, and emission spectrum is obtained by detecting the intensity of the emitted light as a function of wavelength. In downconversion emission spectroscopy, emitted luminescence is recorded in the spectral range above the excitation wavelength to longer wavelengths, up to the region where luminescence is expected. It was then of interest to study the influence of rare earth doping on anatase nanoparticles by the interpretation of absorption (reflectance), excitation and emission spectroscopy methods.

Samples of RE-doped anatase materials are in literature most often characterized by a positioning of the threshold of absorption of doped samples and compared to the undoped ones. Even with the reduction of nanoparticles size after rare earth ions incorporation, the difference in extrapolated slopes after Kubelka–Munk transformations in doped and undoped nanopowder samples should not be ascribed to quantum confinement effect, since particle sizes exceed the Bohr radius several times [18, 79]. Some modifications of materials density of states after the incorporation of trivalent rare earth ions are the most probable reason for small differences in observed band gaps, which is highly dependent on the synthesis procedure and the RE dopant. Kubelka-Munk transformation of reflectance spectra of RE^{3+}-doped anatase TiO_2 measured over the 360–440 nm spectral range is presented in **Figure 9**.

5.1. Praseodymium

The absorption of praseodymium ion in TiO_2 hosts is reported in Refs. [28, 55, 80, 81]. From reflectance spectrum of TiO_2:Pr presented in **Figure 10(a)**, absorptions of Pr^{3+} ions in TiO_2 absorption edge are observed at approximately 445, 480 and 595 nm that could be attributed to the transition from $^3H_4$ ground state to the $^3P_{2\text{-}0}$ and $^1D_2$ excited states of the Pr^{3+} ions. Low wide absorption at around 1000 nm could be assigned to $^1G_4$ excited state. Excitation spectrum is recorded at a fixed emission wavelength of 493 nm in the range of 260–460 nm, presented in **Figure 10(b)**. Two wide excitations are observed at 325 and 447 nm. The excitation of 447 nm was used to obtain emission spectrum in the range of 475–780 nm. Even though the room temperature emission maxima are wide, several transitions can be assigned as follows: $^3P_0 \rightarrow {}^3H_5$ (493 and 536 nm), $^3P_0 \rightarrow {}^3H_6$ (620 nm) and $^3P_0 \rightarrow {}^3F_2$ (650 nm), as can be seen in **Figure 10(c)**. $^1D_2$–$^3H_4$ transition is not observed, suggesting high concentration of Pr^{3+} ions in TiO_2 matrix, where cross-relaxation between neighbouring Pr^{3+} ions occurs [82].

5.2. Neodymium

The absorption of neodymium ion in TiO_2 hosts is reported in a spectral range up to 700 nm [41] and up to 1200 nm [34, 83]. From reflectance spectrum of TiO_2:Nd presented in

http://dx.doi.org/10.5772/intechopen.68882

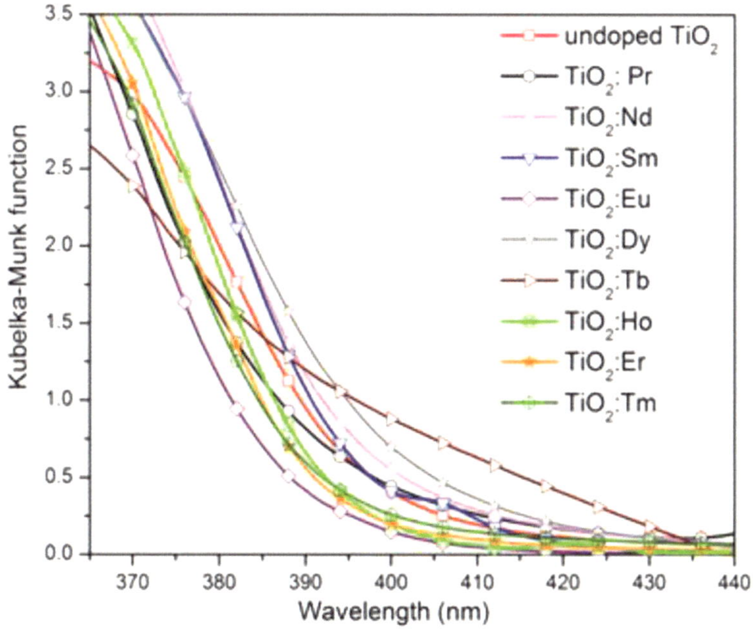

Figure 9. Kubelka-Munk transformation of reflectance spectra of RE^{3+}-doped anatase TiO_2 measured over the 360–440 nm spectral range.

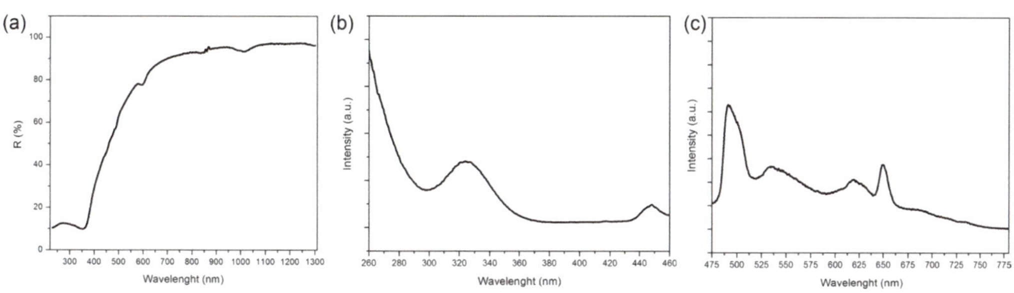

Figure 10. (a) Reflectance, (b) excitation and (c) emission spectra of anatase TiO_2:Pr nanopowders.

Figure 11(a), eight absorptions from ground $^4I_{9/2}$ to excited energy levels of Nd^{3+} ions in TiO_2 are observed and assigned in energy-level diagram in **Figure 11(b)**. Intense emission of Nd^{3+} can be obtained in the IR spectral range above 850 nm, **Figure 11(c)**. Three transitions from $^4F_{3/2}$ to its lower $^4I_{9/2}$, $^4I_{11/2}$ and $^4I_{13/2}$ are obtained with an excitation of 752 nm. The transitions correspond well with the reported data of Nd^{3+} in anatase matrix [34, 40, 84]. The position and shape of $^4F_{3/2}\rightarrow{}^4I_{9/2}$ strongly suggest Nd-doped TiO_2 anatase sample, without the presence of other compositions of segregated neodymium oxide and neodymium titanate phases [34].

5.3. Samarium

In reflectance measurements presented in **Figure 12(a)**, significant absorptions of Sm^{3+} ion can be observed, with maxima positioned at around 480 nm, which corresponds to absorption into $^4G_{5/2}$, and several strong absorptions positioned at around 947, 1080 and 1230 nm. Room temperature excitation spectrum is in the range of 310–550 nm at a fixed emission at 585 nm shown in **Figure 12(b)**. Strong wide band below 400 nm, with maximum at about 365 nm, is characteristic for Sm^{3+} in TiO_2 matrix that is assigned to charge transfer from the oxygen ligands in TiO_2 to Sm^{3+} ion [18, 29, 34, 35]. Several smaller and combined excitations at around 411 and 476 nm could be assigned to $^6G_{7/2}$ or $^6P_{5/2}$ and $^4I_{13/2,}$ respectively [18, 34]. In **Figure 12(c)**, room temperature emission spectrum in the range of 400–700 nm obtained after excitation into charge transfer at 365 nm showed only characteristic emissions from $^4G_{5/2}\rightarrow{}^6H_{5,7,9/2}$ energy levels. It is worth mentioning that the same spectral features are obtained also with exciting directly into Sm^{3+} ion by excitation with 411 nm, with all the intensities decreased as expected from the excitation spectrum. No complete splitting of Stark components caused by ligand field that are obvious at room temperatures and are in correspondence with the literature is attributed to the large number of defect at the surface [18, 29, 34, 35, 42]. When directly excited, the enhancement of Sm^{3+} emission in TiO_2 by codoping with silver dopant, caused by combined influence of plasmonic effects and sensitizing of Sm^{3+} emission by silver ions, is reported in TiO_2 films [85].

5.4. Europium

The lowest excited level ($^5D_0$) of Eu^{3+} ion is a non-degenerate ($J = 0$) singlet level, along with crystal field non-sensitive $^5D_0\rightarrow{}^7F_1$ transition and hypersensitive $^5D_0\rightarrow{}^7F_2$ emissions simplify the interpretation of emission spectra. Consequently, europium ion incorporated in

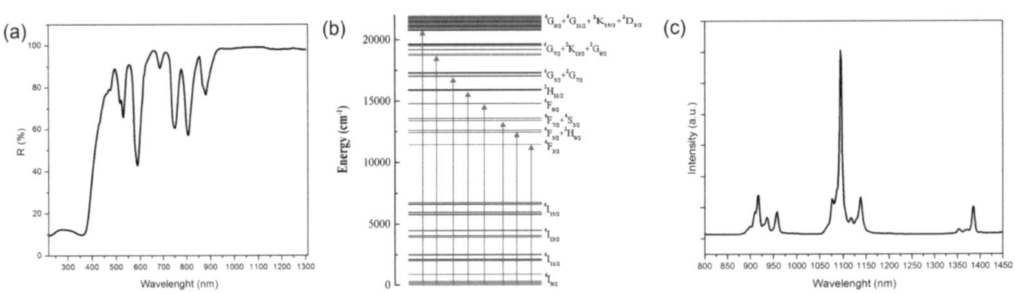

Figure 11. (a) Reflectance, (b) energy-level diagram and (c) emission spectra of anatase TiO_2:Nd nanopowders.

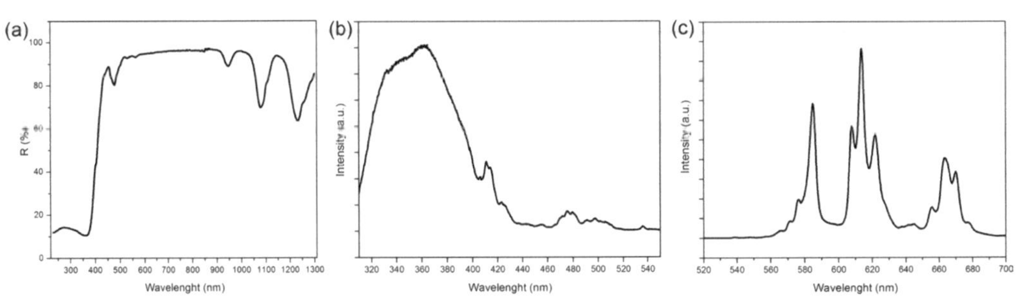

Figure 12. (a) Reflectance, (b) excitation and (c) emission spectra of anatase TiO_2:Sm nanopowders.

various matrices is often used as a luminescent probe ion in photoluminescence spectroscopy [86–90]. In **Figure 13(a)**, after a sharp rise of absorption in UV spectral range below 400 nm, low-intensity Eu^{3+} absorptions from $^7F_0 \rightarrow ^5D_2$ at around 465 nm and $^7F_0 \rightarrow ^5D_1$ at around 535 nm transitions are clearly observed. Those transitions are also present in excitation spectrum (**Figure 13(b)**) obtained with an emission fixed at 613 nm. Four dominant excitation bands originate from direct excitation of Eu^{3+} ions from ground $^7F_0$ level to $^5L_6$ (394 nm), $^5D_3$ (414 nm), $^5D_2$ (464 nm) and $^5D_1$ (532 nm) levels. By excitation into $^5L_6$ level, room temperature emission spectrum presented in **Figure 13(c)** clearly shows that emissions from $^5D_0 \rightarrow ^7F_J$ (J = 0–4) transitions are centred at around 580, 593, 613, 653 and 702 nm, respectively. A small emission observed at 540 nm is emission from higher excited $^5D_1$ level. The positions and relative intensities of wide emissions are in correspondence with extensive literature data [18, 29, 42, 45, 52, 53, 63, 67, 69, 91]. In some presented results of low-temperature site-selective spectroscopy of the materials, three possible positions of Eu ion in TiO_2 can be distinguished: Eu^{3+} can occupy Ti^{4+} site, it could enter into the interstitial site in the chain structure, or a third possible site for dopant cation is low-symmetry-distorted sites near nanoparticle's surface [18, 19, 91].

5.5. Terbium

Terbium ions often show a tendency to be stabilized by matrices in two valence states, +3 and +4. Only lower valence state is optically active in visible spectrum. The mixture of valences can additionally disturb crystallinity of matrices and introduce additional vacancies, and hence perturbations in energy states. In absorption spectra presented in **Figure 14(a)**, no clear absorption of Tb^{3+} ion can be resolved, but significant difference in absorption threshold of TiO_2 is obvious, suggesting possible weak absorption of energy in the range below 500 nm. Some reports state no or very weak emission of Tb^{3+} ion in TiO_2 matrix attributed to the mismatch of the energy levels of the $^5D_4$-emitting state of Tb^{3+} with band gap of TiO_2 [18, 29, 60, 69]. Nevertheless, as presented in **Figure 14(b, c)**, excitation and emission spectra are actually obtained. At an emission wavelength of 545 nm, excitation spectrum was measured in the range of 300–500 nm. Wide charge transfer band can be seen below 350 nm, and excitations of Tb^{3+} ion from $^7F_6$ ground level to $^5D_4$ excited level are observed at 484 nm, two excitations to $^5D_2$ 368 nm and $^5D_3$ at 377 nm. When excited into $^5D_4$ excited energy level with 484 nm, emission spectrum in the range of 510–780 nm

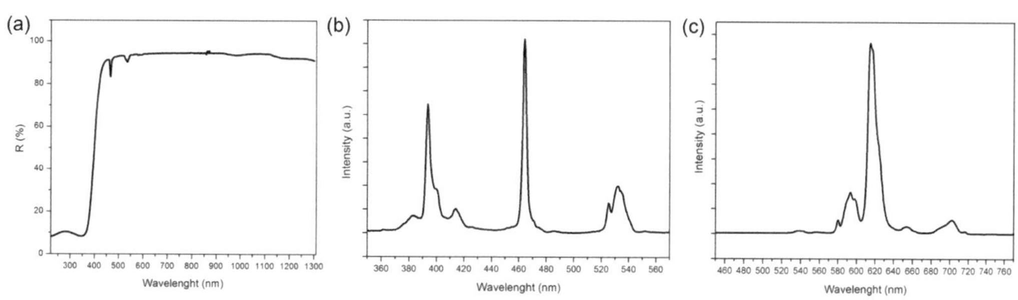

Figure 13. (a) Reflectance, (b) excitation and (c) emission spectra of anatase TiO_2:Eu nanopowders.

shows emission from $^5D_4$ to $^7F_5$ at 546 nm, $^5D_4$ to $^7F_4$ at 585 nm and $^5D_4$ to $^7F_3$ at 622 nm. The green emission at 546 nm is the dominant one. The findings are in good agreement with the literature [29, 60, 64, 69].

5.6. Dysprosium

Reflectance spectrum of Dy^{3+} ions into TiO_2 presented in **Figure 15(a)** shows low-wavelength bands of Dy^{3+} that overlaps with the absorption threshold of anatase at 450 and 470 nm and intense longer wavelength bands in the range of 700–1300 nm. Excitation spectrum of TiO_2:Dy^{3+} sample recorded at room temperature in the 300–500 nm range with a fixed emission wavelength of 577 nm showed excitations corresponding to electron transitions from the Dy^{3+} ground states to the excited states: $^4K_{17/2}$ at 391 nm, $^4G_{11/2}$ at 425 nm, $^4I_{15/2}$ at 452 nm and $^4F_{9/2}$ at 472nm, **Figure 15(b)**. When excited with 425 nm, dominant luminescence is observed with two bands observed in the blue spectral region at 483 nm, which correspond to magnetic-dipole $^4F_{9/2}\rightarrow{}^6H_{15/2}$ transition and in yellow spectral region at 580 nm, which correspond to electric-dipole $^4F_{9/2}\rightarrow{}^6H_{13/2}$ transition, **Figure 15(c)**. A low-intensity emission is observed in the red region at 674 nm that corresponds to $^4F_{9/2}\rightarrow{}^6H_{11/2}$ transition. With literature proposing no luminescence from Dy^{3+} ion in anatase host [92], this finding shows that nanocrystalline anatase powders can actually host this ion that can successfully be excited and luminescence can be observed.

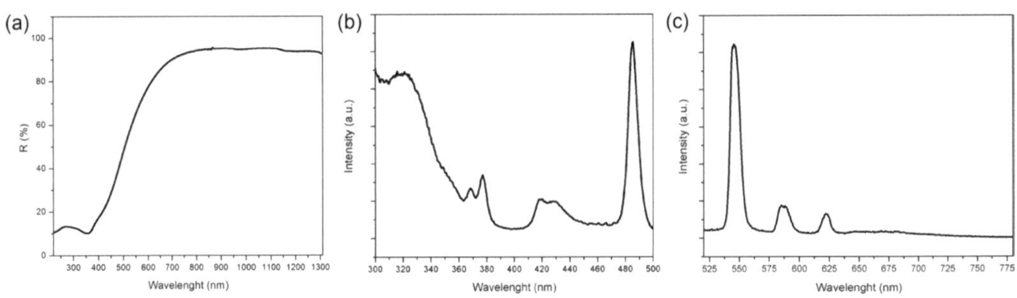

Figure 14. (a) Reflectance, (b) excitation and (c) emission spectra of anatase TiO_2:Tb nanopowders.

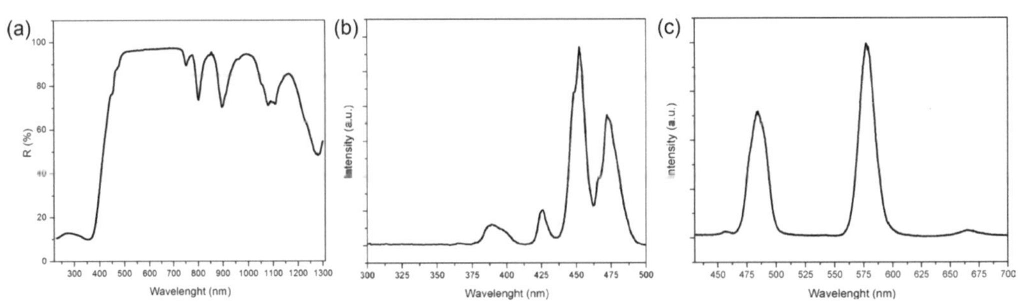

Figure 15. (a) Reflectance, (b) excitation and (c) emission spectra of anatase TiO_2:Dy nanopowders.

5.7. Holmium

Among all RE^{3+} ions doped in nanocrystalline anatase TiO_2 powders in this work, Ho^{3+} has the most pronounced absorptions in VIS. As can be seen from **Figure 16(a)**, intense bands can be observed at 420, 456, 490, 542 and 645 nm and smaller intensity bands are observed at 890, 1150 and 1200 nm. In excitation spectrum at fixed emission wavelength of 554 nm presented in **Figure 16(b)**, several excitations centred at around 422, 452, 468 and 493 nm show several possible energies for potential emission. As can be seen in **Figure 16(c)**, when excited with 452 nm, emission spectra in the range of 500–700 nm show dominant emissions from ${}^5F_4/{}^5S_2 \rightarrow {}^5I_8$ transitions at about 545, 554, and 559 nm, and emission from ${}^5F_5 \rightarrow {}^5I_8$ transition with maximum centred at 665 nm. Emissions from the same transitions can also be observed in samples sensitized with Yb^{3+} ions, when excitation wavelength was 980 nm that corresponds to the absorption of Yb^{3+} ions, and the mechanism of obtaining luminescence is upconversion [50].

5.8. Erbium

Absorptions of Er^{3+} ions in TiO_2 matrices are reported in spectral range from UV up to 700 nm [26, 28], up to 800 nm [49], and when sensitized with Yb^{3+} ions up to 1200 nm [48]. All of the reported data correspond well with results presented in **Figure 17(a)**. Absorptions located at 452, 477, 491, 525, 655, 795 and 980 nm correspond to the transitions from ${}^4I_{15/2}$ to ${}^4F_{3/2}$, ${}^4F_{5/2}$, ${}^4F_{7/2}$, ${}^2H_{11/2}$ and ${}^4S_{3/2}$, ${}^4F_{9/2}$, ${}^4G_{9/2}$, ${}^4I_{11/2}$, respectively. In excitation spectrum shown in **Figure 17(b)**, with fixed emission of 565 nm, some low-intensity excitations can be noticed at around 378, 410 and 453 nm. More pronounced excitations can be observed at 488 and 525 nm. In order to characterize emissions in the range of 520–700 nm, excitation wavelength of 488 nm was used, and the spectrum is presented in **Figure 17(c)**. From the combination of ${}^2H_{11/2} \rightarrow {}^4I_{15/2}$ and ${}^4S_{3/2} \rightarrow {}^4I_{15/2}$ transitions, wide emissions can be observed in the range of 540–575 nm, as also reported in Refs. [42, 92].

5.9. Thulium

Absorption of thulium ion in the sample presented in **Figure 18(a)** shows small absorption at 470 nm, as well as stronger absorptions at 690, 795 and 1210 nm. Excitation spectrum with

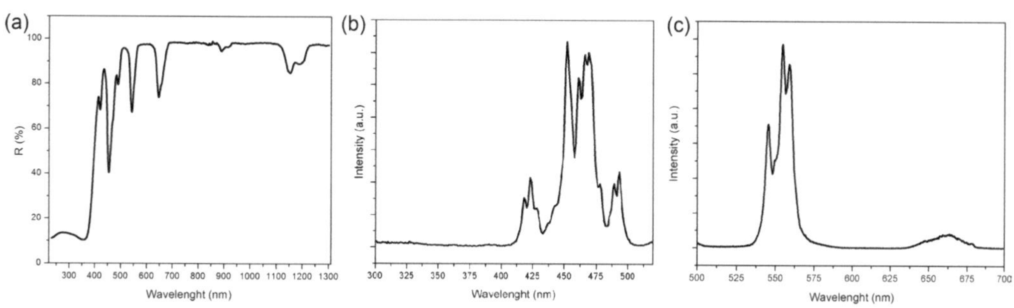

Figure 16. (a) Reflectance, (b) excitation and (c) emission spectra of anatase TiO_2:Ho nanopowders.

a fixed emission at 495 nm showed poor optical answer with some picks that most probably originate from defect, **Figure 18(b)**. In order to directly excite Tm^{3+} ion 470 nm excitation was used. Emission spectrum in the range of 490–780nm presented in **Figure 18(c)** shows shoulder of maximum at 495 nm originating from $^1G_4 \rightarrow {}^5H_6$ transition and very low intensity of group of lines in the range of 650–670 nm that could be attributed to the $^1G_4 \rightarrow {}^3F_4$ transition.

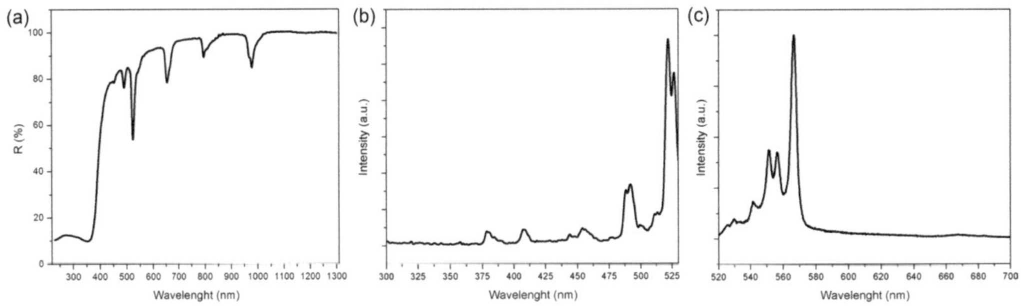

Figure 17. (a) Reflectance, (b) excitation and (c) emission spectra of anatase TiO_2:Er nanopowders.

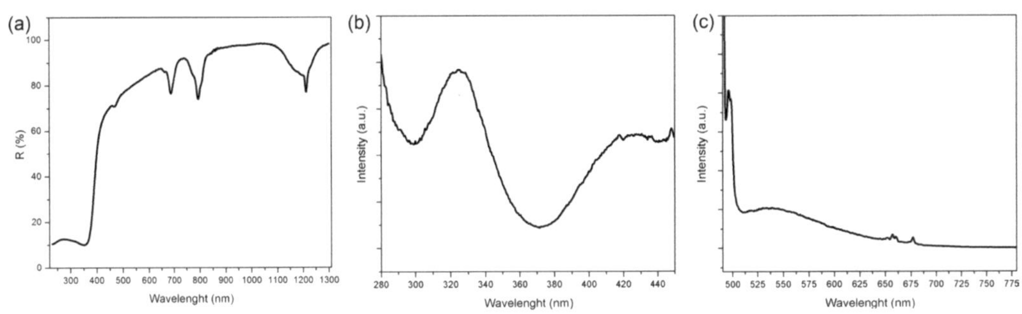

Figure 18. (a) Reflectance, (b) excitation and (c) emission spectra of anatase TiO_2:Tm nanopowders.

6. Conclusion

To conclude, the structure, morphology and optical properties of TiO_2 nanoparticles may be substantially swayed by the addition of small quantities of RE^{3+} ions. Such nanostructures deliver new options to the already broad range of important TiO_2 uses. In RE ion-doped TiO_2, anatase phase is stabilized at medium temperatures since the temperature of phase transformations shifts to higher values. The reduction of the crystallite size is readily observed and doping induces mesoporous structure with enlarged specific surface in respect to one of undoped anatase TiO_2. Thus, the photocatalytic performance of nanopowder improves with the addition of RE^{3+} in small concentrations except for Pr^{3+} and Tb^{3+}. Different rare earth ions cause TiO_2 property changes of different magnitudes. Optical properties are altered too. The modification of materials density of states after incorporation of RE^{3+} ions in TiO_2 causes changes in materials absorption which can be clearly evidenced from optical absorption spectra. Rare earth ions may be incorporated at three different sites in TiO_2 structure: they can substitute Ti^{4+} in the bulk of particle, enter vacancy site, but they at large reside near surface in low-symmetry sites. In such cases, the characteristic RE^{3+} luminescence is observed in the case of doping with the following ions: Nd^{3+}, Sm^{3+}, Eu^{3+}, Dy^{3+}, Ho^{3+} and Er^{3+}, while luminescence of low intensity is detected for Pr^{3+}, Tb^{3+} and Tm^{3+}.

Acknowledgements

The authors thank Prof. Damien Bregiroux and Alexandre Bahezre from Université Pierre et Marie Curie—LCMCP for BET and TEM measurements. This work was done as a French-Serbian collaboration under Bilateral project no. 451-03-39/2016/09/03. The financial support for this work was provided by the Ministry of Education, Science and Technological Development of Republic of Serbia (Project 172056).

Author details

Vesna Đorđević*, Bojana Milićević and Miroslav D. Dramićanin

*Address all correspondence to: vesipka@vinca.rs

Vinča Institute of Nuclear Sciences, University of Belgrade, Belgrade, Serbia

References

[1] Shannon RD. Revised effective ionic radii and systematic studies of interatomic distances in halides and chalcogenides. Acta Crystallographica. 1976;**A32**:751-767. DOI: 10.1107/S0567739476001551

[2] Laporte O, Meggers WF. Some rules of spectral structure. Journal of the Optical Society of America. 1925;**11**(5):459-463. DOI: 10.1364/JOSA.11.000459

[3] Bünzli J-CG, Eliseeva SV. Basics of lanthanide photophysics. In: Hänninen P, Härmä H, editors. Lanthanide Luminescence: Photophysical, Analytical and Biological Aspects. Berlin Heidelberg: Springer-Verlag; 2010. pp. 1-45. DOI: 10.1007/4243_2010_3

[4] Dieke GH, Crosswhite HM. The spectra of the doubly and triply ionized rare earths. Applied Optics. 1963;**2**(7):675-686. DOI: 10.1364/AO.2.000675

[5] Dieke GH. Spectra and Energy Levels of Rare Earth Ions in Crystals. New York, NY: Interscience Publishers; 1968. 401 p.

[6] Peijzel PS, Meijerink A, Wegh RT, Reid MF, Burdick GW. A complete 4fn energy level diagram for all trivalent lanthanide ions. Journal of Solid State Chemistry. 2005;**178**(2):448-453. DOI: 10.1016/j.jssc.2004.07.046

[7] Solé JG, Bausá LE, Jaque D. An Introduction to the Optical Spectroscopy of Inorganic Solids. 1st ed. Chichester, UK: John Wiley & Sons, Ltd; 2005. 283 p. DOI: 10.1002/0470016043

[8] Yen WM, Shionoya S, Yamamoto H, editors. Practical Applications of Phosphors. 1st ed. Boca Raton, FL: CRC Press; 2006. 528 p.

[9] Dramićanin MD. Sensing temperature via downshifting emissions of lanthanide-doped metal oxides and salts. A review. Methods and Applications in Fluorescence. 2016;**4**(4):42001. DOI: 10.1088/2050-6120/4/4/042001

[10] Eliseeva SV, Bunzli J-CG. Lanthanide luminescence for functional materials and bio-sciences. Chemical Society Reviews. 2010;**39**(1):189-227. DOI: 10.1039/B905604C

[11] Wolfbeis OS. An overview of nanoparticles commonly used in fluorescent bioimaging. Chemical Society Reviews. 2015;**44**(14):4743-4768. DOI: 10.1039/c4cs00392f

[12] Wang Y, He Y, Lai Q, Fan M. Review of the progress in preparing nano TiO_2: An important environmental engineering material. Journal of Environmental Sciences. 2014;**26**(11):2139-2177. DOI: 10.1016/j.jes.2014.09.023

[13] Zhang DR, Jin XZ, Li JH. Effects of Sc and V dopants on the anatase-to-rutile phase transition and crystallite size of TiO_2 nanoparticles. Materials Chemistry and Physics. 2016;**176**:68-74. DOI: 10.1016/j.matchemphys.2016.03.027

[14] Bavykin DV, Friedrich JM, Walsh FC. Protonated titanates and TiO_2 nanostructured materials: Synthesis, properties, and applications. Advanced Materials. 2006;**18**(21):2807-2824. DOI: 10.1002/adma.200502696

[15] Chen X, Mao SS. Titanium dioxide nanomaterials: Synthesis, properties, modifications, and applications. Chemical Reviews. 2007;**107**(7):2891-2959. DOI: 10.1021/cr0500535

[16] Diebold U. The surface science of titanium dioxide. Surface Science Reports. 2003;**48**(5): 53-229. DOI: 10.1016/S0167-5729(02)00100-0.

[17] Roy P, Berger S, Schmuki P. TiO_2 Nanotubes: Synthesis and applications. Angewandte Chemie International Edition. 2011;**50**(13):2904-2939. DOI: 10.1002/anie.201001374

[18] Antić Ž, Krsmanović RM, Nikolić MG, Marinović-Cincović M, Mitrić M, Polizzi S, Dramićanin MD. Multisite luminescence of rare earth doped TiO_2 anatase nanoparticles. Materials Chemistry and Physics. 2012;**135**(2-3):1064-1069. DOI: 10.1016/j.matchemphys.2012.06.016

[19] Luo W, Li R, Liu G, Antonio MR, Chen X. Evidence of trivalent europium incorporated in anatase TiO_2 nanocrystals with multiple sites. The Journal of Physical Chemistry C. 2008;**112**:10370-10377. DOI: 10.1021/jp801563k

[20] Liu Y, Luo W, Zhu H, Chen X. Optical spectroscopy of lanthanides doped in wide band-gap semiconductor nanocrystals. Journal of Luminescence. 2011;**131**(3):415-422. DOI: 10.1016/j.jlumin.2010.07.018

[21] Hanaor DAH, Sorrell CC. Review of the anatase to rutile phase transformation. Journal of Materials Science. 2011;**46**(4):855-874. DOI: 10.1007/s10853-010-5113-0

[22] Baiju KV, Periyat P, Shajesh P, Wunderlich W, Manjumol KA, Smith VS, Jaimy KB, Warrier KGK. Mesoporous gadolinium doped titania photocatalyst through an aqueous sol-gel method. Journal of Alloys and Compounds. 2010;**505**(1):194-200. DOI: 10.1016/j.jallcom.2010.06.028

[23] Shi J-W, Zheng J-T, Wu P. Preparation, characterization and photocatalytic activities of holmium-doped titanium dioxide nanoparticles. Journal of Hazardous Materials. 2009;**161**(1):416-422. DOI: 10.1016/j.jhazmat.2008.03.114

[24] Yurtsever HA, Çiftçioğlu M. The effect of rare earth element doping on the microstructural evolution of sol-gel titania powders. Journal of Alloys and Compounds. 2017;**695**:1336-1353. DOI: 10.1016/j.jallcom.2016.10.275

[25] Nguyen-Phan T-D, Song MB, Kim EJ, Shin EW. The role of rare earth metals in lanthanide-incorporated mesoporous titania. Microporous and Mesoporous Materials. 2009;**119**(1-3):290-298. DOI: 10.1016/j.micromeso.2008.10.039

[26] Liang C-H, Hou M-F, Zhou S-G, Li F-B, Liu C-S, Liu T-X, Gao Y-X, Wang X-G, Lü J-L. The effect of erbium on the adsorption and photodegradation of orange I in aqueous Er^{3+}-TiO_2 suspension. Journal of Hazardous Materials. 2006;**138**(3):471-478. DOI: 10.1016/j.jhazmat.2006.05.066

[27] Milićević B, Đorđević V, Lončarević D, Ahrenkiel SP, Dramićanin MD, Nedeljković JM. Visible light absorption of surface modified TiO_2 powders with bidentate benzene derivatives. Microporous and Mesoporous Materials. 2015;**217**:184-189. DOI: 10.1016/j.micromeso.2015.06.028

[28] Reszczynska J, Grzyb T, Wei Z, Klein M, Kowalska E, Ohtani B, Zaleska-Medynska A. Photocatalytic activity and luminescence properties of RE^{3+}–TiO_2 nanocrystals prepared by sol-gel and hydrothermal methods. Applied Catalysis B: Environmental. 2016;**181**:825-837. DOI: 10.1016/j.apcatb.2015.09.001

[29] Saif M, Abdel-Mottaleb MSA. Titanium dioxide nanomaterial doped with trivalent lanthanide ions of Tb, Eu and Sm: Preparation, characterization and potential applications. Inorganica Chimica Acta. 2007;**360**(9):2863-2874. DOI: 10.1016/j.ica.2006.12.052

[30] Wang R, Wang F, An S, Song J, Zhang Y. Y/Eu co-doped TiO_2: synthesis and photocatalytic activities under UV-light. Journal of Rare Earths. 2015;**33**(2):154-159. DOI: 10.1016/S1002-0721(14)60396-3

[31] Niu X, Li S, Chu H, Zhou J. Preparation, characterization of Y^{3+}-doped TiO_2 nanoparticles and their photocatalytic activities for methyl orange degradation. Journal of Rare Earths. 2011;**29**(3):225-229. DOI: 10.1016/S1002-0721(10)60435-8

[32] Quan X, Zhao Q, Tan H, Sang X, Wang F, Dai Y. Comparative study of lanthanide oxide doped titanium dioxide photocatalysts prepared by coprecipitation and sol-gel process. Materials Chemistry and Physics. 2009;**114**(1):90-98. DOI: 10.1016/j.matchemphys.2008.08.051

[33] Kralova M, Levchuk I, Kasparek V, Sillanpa M, Cihlar J. Influence of synthesis conditions on physical properties of lanthanide-doped titania for photocatalytic decomposition of metazachlor. Chinese Journal of Catalysis. 2015;**36**:1679-1685. DOI: 10.1016/S1872-2067(15)60943-3

[34] Luo W, Li R, Chen X. Host-sensitized luminescence of Nd^{3+} and Sm^{3+} ions incorporated in anatase titania nanocrystals. The Journal of Physical Chemistry C. 2009;**113**(20):8772-8777. DOI: 10.1021/jp901862k

[35] Hu L, Song H, Pan G, Yan B, Qin R, Dai Q, Fan L, Li S, Bai X. Photoluminescence properties of samarium-doped TiO_2 semiconductor nanocrystalline powders. Journal of Luminescence. 2007;**127**(2):371-376. DOI: 10.1016/j.jlumin.2007.01.020

[36] Shi J-W, Zheng J-T, Hu Y, Zhao Y-C. Influence of Fe^{3+} and Ho^{3+} co-doping on the photocatalytic activity of TiO_2. Materials Chemistry and Physics. 2007;**106**(2-3):247-249. DOI: 10.1016/j.matchemphys.2007.05.042

[37] Zhang DR, Liu HL, Han SY, Piao WX. Synthesis of Sc and V-doped TiO_2 nanoparticles and photodegradation of rhodamine-B. Journal of Industrial and Engineering Chemistry. 2013;**19**(6):1838-1844. DOI: 10.1016/j.jiec.2013.02.029

[38] de la Cruz D, Arévalo JC, Torres G, Bautista Margulis RG, Ornelas C, Aguilar-Elguézabal A. TiO_2 doped with Sm^{3+} by sol-gel: Synthesis, characterization and photocatalytic activity of diuron under solar light. Catalysis Today. 2011;**166**(1):152-158. DOI: 10.1016/j.cattod.2010.08.023

[39] Chen D, Zhu Q, Lv Z, Deng X, Zhou F, Deng Y. Microstructural and photocatalytic properties of Eu-doped mesoporous titanium dioxide nanoparticles by sol-gel method. Materials Research Bulletin. 2012;**47**:3129-3134. DOI: 10.1016/j.materresbull.2012.08.021

[40] Ghigna P, Speghini A, Bettinelli M. "Unusual Ln^{3+} substitutional defects": The local chemical environment of Pr^{3+} and Nd^{3+} in nanocrystalline TiO_2 by Ln–K edge EXAFS. Journal of Solid State Chemistry. 2007;**180**(11):3296-3301. DOI: 10.1016/j.jssc.2007.09.025

[41] Huang C., You W, Dang L, Lei Z, Sun Z, Zhang L. Effect of Nd^{3+} doping on photocatalytic activity of TiO_2 nanoparticles for water decomposition to hydrogen. Chinese Journal of Catalysis. 2006;**27**(3):203-209. DOI: 10.1016/S1872-2067(06)60013-2

[42] Bettinelli M, Speghini A, Falcomer D, Daldosso M, Dallacasa V, Romanò L. Photocatalytic, spectroscopic and transport properties of lanthanide-doped TiO_2 nanocrystals. Journal of Physics: Condensed Matter. 2006;**18**:S2149–S2160. DOI: 10.1088/0953-8984/18/33/S30

[43] Hamza MA, Khalil ASh, Yaseen HM. Synthesis of Yb^{3+} doped TiO_2 nano particles powder as IR filter via sol-gel. Advances in Materials Physics and Chemistry. 2013;**3**:214-216. DOI: 10.4236/ampc.2013.34032

[44] Zhang Z, Li G, Cui Z, Zhang K, Feng Y, Meng S. Influence of difference quantity La-doped TiO_2 photoanodes on the performance of dye-sensitized solar cells: A strategy for choosing an appropriate doping quantity. Journal of Solid State Chemistry. 2016;**237**:242-247. DOI: 10.1016/j.jssc.2016.02.005

[45] Li H, Zheng K, Sheng Y, Song Y, Zhang H, Huang J, Huo Q, Zou H. Facile synthesis and luminescence properties of $TiO_2:Eu^{3+}$ nanobelts. Optics and Laser Technology. 2012;**49**:33-37. DOI: 10.1016/j.optlastec.2012.12.007

[46] Zhou W, He Y. Ho/TiO_2 nanowires heterogeneous catalyst with enhanced photocatalytic properties by hydrothermal synthesis method. Chemical Engineering Journal. 2012;**179**:412-416. DOI: 10.1016/j.cej.2011.10.094

[47] Khan M, Cao W. Cationic (V, Y)-codoped TiO_2 with enhanced visible light induced photocatalytic activity: A combined experimental and theoretical study. Journal of Applied Physics. 2013;**114**:183514-1 (7pp). DOI: 10.1063/1.4831658

[48] Bhethanabotla VC, Russell DR, Kuhn JN. Assessment of mechanisms for enhanced performance of Yb/Er/titania photocatalysts for organic degradation: Part 2. Role of rare earth elements in the titania phase. Applied Catalysis B: Environmental. 2017;**202**:156-164. DOI: 10.1016/j.apcatb.2016.09.008

[49] Obregón S, Kubacka A, Fernández-García M, Colón G. High-performance Er^{3+}–TiO_2 system: Dual up-conversion and electronic role of the lanthanide. Journal of Catalysis. 2013;**299**:298-306. DOI: 10.1016/j.jcat.2012.12.021

[50] Pedroni M, Piccinelli F, Polizzi S, Speghini A, Bettinelli M, Haro-González P. Upconverting Ho–Yb doped titanate nanotubes. Materials Letters. 2012;**80**:81-83. DOI: 10.1016/j.matlet.2012.04.096

[51] Yan X, He J, Evans DG, Duan X, Zhu Y. Preparation, characterization and photocatalytic activity of Si-doped and rare earth-doped TiO_2 from mesoporous precursors. Applied Catalysis B: Environmental. 2005;**55**(4):243-252. DOI: 10.1016/j.apcatb.2004.08.014

[52] Qi X, Zou H, Song Y, Zhang H, Zhao H, Shi Z, Sheng Y. Hydrothermal synthesis and luminescence properties of TiO_2: Eu^{3+} submicrospheres. Ceramics International. 2014;**40**:12993-12997. DOI: 10.1016/j.ceramint.2014.04.162

[53] Li H, Sheng Y, Zhao H, Song Y, Gao F, Huo Q, Zou H. Facile synthesis and luminescent properties of $TiO_2:Eu^{3+}$ nanorods and spindle-shaped nanoparticles from titanate nanotubes precursors. Materials Research Bulletin. 2012;**47**(12):4322-4328. DOI: 10.1016/j.materresbull.2012.09.010

[54] Falcomer D, Daldosso M, Cannas C, Musinu A, Lasio B, Enzo S, Speghini A, Bettinelli M. A one-step solvothermal route for the synthesis of nanocrystalline anatase TiO_2 doped with lanthanide ions. Journal of Solid State Chemistry. 2006;**179**(8):2452-2457. DOI: 10.1016/j.jssc.2006.04.043

[55] Wang Y, Chen G, Shen Q, Yang H, Li L, Song Y. A hydrothermal synthesis of Pr^{3+} doped mesoporous TiO_2 for UV light photocatalysis. Journal of Nanoscience and Nanotechnology. 2014;**14**(7):4988-4994. DOI: 10.1166/jnn.2014.9487

[56] Diamandescu L, Vasiliu F, Tarabasanu-Mihaila D, Feder M, Vlaicu AM, Teodorescu CM, et al. Structural and photocatalytic properties of iron- and europium-doped TiO_2 nanoparticles obtained under hydrothermal conditions. Materials Chemistry and Physics. 2008;**112**(1):146-153. DOI: 10.1016/j.matchemphys.2008.05.023

[57] Cacciotti I, Bianco A, Pezzotti G, Gusmano G. Synthesis, thermal behaviour and luminescence properties of rare earth-doped titania nanofibers. Chemical Engineering Journal. 2011;**166**(2):751-764. DOI: 10.1016/j.cej.2010.07.008

[58] Hassan MS, Amna T, Yang O-B, Kim H-C, Khil M-S. TiO_2 nanofibers doped with rare earth elements and their photocatalytic activity. Ceramics International. 2012;**38**(7):5925-5930. DOI: 10.1016/j.ceramint.2012.04.043

[59] Kumar KS, Song C-G, Bak GM, Heo G, Seong M-J, Yoon J-W. Phase control of yttrium (Y)-doped TiO_2 nanofibers and intensive visible photoluminescence. Journal of Alloys and Compounds. 2014;**617**:683-687. DOI: 10.1016/j.jallcom.2014.08.067

[60] Cacciotti I, Bianco A, Pezzotti G, Gusmano G. Terbium and ytterbium-doped titania luminescent nanofibers by means of electrospinning technique. Materials Chemistry and Physics. 2011;**126**(3):532-541. DOI: 10.1016/j.matchemphys.2011.01.034

[61] Yang Y, Zhang C, Xu Y, Wang H, Li X, Wang C. Electrospun Er:TiO_2 nanofibrous films as efficient photocatalysts under solar simulated light. Materials Letters. 2010;**64**(2):147-150. DOI: 10.1016/j.matlet.2009.10.028

[62] Ikeda M, Li J-G, Kobayashi N, Moriyoshi Y, Hamanaka H, Ishigaki T. Phase formation and luminescence properties in Eu^{3+}-doped TiO_2 nanoparticles prepared by thermal plasma pyrolysis of aqueous solutions. Thin Solid Films. 2008;**516**(19):6640-6644. DOI: 10.1016/j.tsf.2007.11.037

[63] Liu H, Yu L, Chen W, Li Y. The progress of TiO_2 nanocrystals doped with rare earth ions. Journal of Nanomaterials. 2012;**2012**:235879. DOI: 10.1155/2012/235879

[64] Mazierski P, Lisowski W, Grzyb T, Winiarski MJ, Klimczuk T, Mikołajczyk A, et al. Enhanced photocatalytic properties of lanthanide-TiO_2 nanotubes: An experimental and theoretical study. Applied Catalysis B: Environmental. 2017;**205**:376-385. DOI: 10.1016/j.apcatb.2016.12.044

[65] Cheng XQ, Ma CY, Yi XY, Yuan F, Xie Y, Hu JM, Hu BC, Zhang QY. Structural, morphological, optical and photocatalytic properties of Gd-doped TiO_2 films. Thin Solid Films. 2016;**615**:13-18. DOI: 10.1016/j.tsf.2016.06.049

[66] Borlaf M, Caes S, Dewalque J, Colomer MT, Moreno R, Cloots R, Boschini F. Effect of the RE (RE = Eu, Er) doping on the structural and textural properties of mesoporous TiO_2 thin films obtained by evaporation induced self-assembly method. Thin Solid Films. 2014;**558**:140-148. DOI: 10.1016/j.tsf.2014.03.002

[67] Conde-Gallardo A, Garciá-Rocha M, Hernández-Calderón I, Palomino-Merino R. Photoluminescence properties of the Eu^{3+} activator ion in the TiO_2 host matrix. Applied Physics Letters. 2001;**78**(22):3436-3438. DOI: 10.1063/1.1372338

[68] Burns A, Hayes G, Li W, Hirvonen J, Demaree JD, Shah SI. Neodymium ion dopant effects on the phase transformation in sol-gel derived titania nanostructures. Materials Science and Engineering B. 2004;**111**:150-155. DOI: 10.1016/j.mseb.2004.04.008

[69] Meng Q, Zhang H, Wang S, Fu L, Zheng Y, Yang K. Preparation and characterization of luminescent thin films doped with rare earth (Tb^{3+}, Eu^{3+}) complexes derived from a sol-gel process. Materials Letters. 2000;**45**:213-216. DOI: 10.1016/S0167-577X(00)00107-5

[70] Zhang J, Li M, Feng Z, Chen J, Li C. UV Raman spectroscopic study on TiO_2. I. Phase transformation at the surface and in the bulk. The Journal of Physical Chemistry B. 2006;**110**:927-935. DOI: 10.1021/jp0552473

[71] Zhang Y, Zhang H, Xu Y, Wang Y. Europium doped nanocrystalline titanium dioxide: Preparation, phase transformation and photocatalytic properties. Journal of Materials Chemistry. 2003;**13**:2261-2265. DOI: 10.1039/b305538h

[72] Stengl V, Bakardjieva S. Molybdenum-doped anatase and its extraordinary photocatalytic activity in the degradation of orange II in the UV and Vis regions. The Journal of Physical Chemistry C. 2010;**114**(45):19308-19317. DOI: 10.1021/jp104271q

[73] Li F, Li X, Hou M. Photocatalytic degradation of 2-mercaptobenzothiazole in aqueous La^{3+}–TiO_2 suspension for odor control. Applied Catalysis B: Environmental. 2004;**48**(3):185-194. DOI: 10.1016/j.apcatb.2003.10.003

[74] Parida KM, Sahu N. Visible light induced photocatalytic activity of rare earth titania nanocomposites. Journal of Molecular Catalysis A: Chemical. 2008;**287**(1-2):151-158. DOI: 10.1016/j.molcata.2008.02.028

[75] El-Bahy ZM, Ismail AA, Mohamed RM. Enhancement of titania by doping rare earth for photodegradation of organic dye (Direct Blue). Journal of Hazardous Materials. 2009;**166**:138-143. DOI: 10.1016/j.jhazmat.2008.11.022

[76] Serp P, Machado B. Photocatalysis on nanostructured carbon supported catalysts. In: Nanostructured Carbon Materials for Catalysis. Great Britain:The Royal Society of Chemistry; 2015. pp. 412-444. DOI: 10.1039/9781782622567-00412

[77] Xie Y, Yuan C. Photocatalysis of neodymium ion modified TiO_2 sol under visible light irradiation. Applied Surface Science. 2004;**221**:17-24. DOI: 10.1016/S0169-4332(03)00945-0

[78] Xu A-W, Gao Y, Liu H-Q. The preparation, characterization, and their photocatalytic activities of rare-earth-doped TiO_2 nanoparticles. Journal of Catalysis. 2002;**207**:151-157. DOI: 10.1006/jcat.2002.3539

[79] Hörmann U, Kaiser U, Albrecht M, Geserick J, Hüsing N. Structure and luminescence of sol-gel synthesized anatase nanoparticles. Journal of Physics: Conference Series. 2010;**209**(1):012039 (7pp). DOI: 10.1088/1742-6596/209/1/012039

[80] Yang J, Dai J, Li J. Synthesis, characterization and degradation of bisphenol A using Pr, N co-doped TiO_2 with highly visible light activity. Applied Surface Science. 2011;**257**(21):8965-8973. DOI: 10.1016/j.apsusc.2011.05.074

[81] Su W, Chen J, Wu L, Wang X, Wang X, Fu X. Visible light photocatalysis on praseodymium(III)-nitrate-modified TiO_2 prepared by an ultrasound method. Applied Catalysis B: Environmental. 2008;**77**:264-271. DOI: 10.1016/j.apcatb.2007.04.015

[82] Li Y-C, Chang Y-H, Lin Y-F, Chang Y-S, Lin Y-J. Luminescent properties of trivalent praseodymium-doped lanthanum aluminum germanate LaAlGe2O7. Journal of Physics and Chemistry of Solids. 2007;**68**(10):1940-1945. DOI: 10.1016/j.jpcs.2007.06.013

[83] de Sousa Filho PC, Lima JF, Serra OA. From lighting to photoprotection: Fundamentals and applications of rare earth materials. Journal of the Brazilian Chemical Society. 2015;**26**(12):2471-2495. DOI: 10.5935/0103-5053.20150328

[84] Yildirim S, Yurddaskal M, Dikici T, Aritman I, Ertekin K, Celik E. Structural and luminescence properties of undoped, Nd^{3+} and Er^{3+} doped TiO_2 nanoparticles synthesized by flame spray pyrolysis method. Ceramics International. 2016;**42**(9):10579-10586. DOI: 10.1016/j.ceramint.2016.03.131

[85] Dolgov L, Kiisk V, Reedo V, Pikker S, Sildos I, Kikas J. Sensitizing of Sm^{3+} fluorescence by silver dopant in the TiO_2 films. Central European Journal of Physics. 2011;**9**(2):542-546. DOI: 10.2478/s11534-010-0069-x

[86] Buenzli JCG, Plancherel D, Pradervand GO. The europium(III) ion as luminescent probe: structural investigation of complexes between europium nitrate and two polyethers. The Journal of Physical Chemistry. 1989;**93**(2):980-984. DOI: 10.1021/j100339a083

[87] Wolfbeis OS, Dürkop A, Wu M, Lin Z. A europium-ion-based luminescent sensing probe for hydrogen peroxide. Angewandte Chemie International Edition. 2002;**41**(23):4495-4498. DOI: 10.1002/1521-3773(20021202)41:23<4495::AID-ANIE4495>3.0.CO;2-I

[88] Reisfeld R, Zigansky E, Gaft M. Europium probe for estimation of site symmetry in glass films, glasses and crystals. Molecular Physics. 2004;**102**(11-12):1319-1330. DOI: 10.1080/00268970410001728609

[89] Tanner PA. Some misconceptions concerning the electronic spectra of tri-positive europium and cerium. Chemical Society Reviews. 2013;**42**(12):5090-5101. DOI: 10.1039/C3CS60033E

[90] Binnemans K. Interpretation of europium(III) spectra. Coordination Chemistry Reviews. 2015;**295**:1-45. DOI: 10.1016/j.ccr.2015.02.015

[91] Xueyuan C, Wenqin L, Yongsheng L, Guokui L. Recent progress on spectroscopy of lanthanide ions incorporated in semiconductor nanocrystals. Journal of Rare Earths. 2007;**25**(5):515-525. DOI: 10.1016/S1002-0721(07)60555-9

[92] Chakraborty A, Debnath GH, Saha NR, Chattopadhyay D, Waldeck DH, Mukherjee P. Identifying the correct host-guest combination to sensitize trivalent lanthanide (guest) luminescence: Titanium dioxide nanoparticles as a model host system. The Journal of Physical Chemistry C. 2016;**120**(41):23870-23882. DOI: 10.1021/acs.jpcc.6b08421

Section 2

Rutile Structure of TiO2

Chapter 3

Structural Aspects of Anatase to Rutile Phase Transition in Titanium Dioxide Powders Elucidated by the Rietveld Method

Alberto Adriano Cavalheiro,

Lincoln Carlos Silva de Oliveira and

Silvanice Aparecida Lopes dos Santos

Additional information is available at the end of the chapter

http://dx.doi.org/10.5772/intechopen.68601

Abstract

Titanium dioxide has attracted much attention since a long time ago due to its versatility as advanced material. However, its performance as semiconductor devices is very much dependent on the predominant crystalline phase and defect concentrations, which can be adjusted through the synthesis methods, thermal treatments and doping processes. In this work, an accurate structural characterization of titanium dioxide was used by X-ray diffractometry supported by rietveld refinement and thermal analysis. The insertion of 5 mol% of zirconium silicate was able to stabilize anatase up to 900°C, permitting the oxygen vacancies to be significantly eliminated. It was demonstrated also that the changes in the isotropic thermal parameters for oxygen are related to reconstructive transformation necessary to promote the anatase-to-rutile phase transition. Independently of doping process, the crystallization process of anatase phase as a function of temperature increasing occurs exclusively due the reduction of lattice microstrain up to 600°C. However, above 650°C, that crystallization process becomes dependent of the increasing in crystallite size. The anatase crystallite growth event was only possible when the titanium dioxide was doped with zirconium silicate. Otherwise, the rutile phase amount starts to rise continually. Thus, there are optimistic expectations for that new composition to be a new semiconductor matrix for additional doping processes.

Keywords: sol-gel method, isovalent doping, phase transition, XRD, rietveld refinement

1. Introduction

The titanium dioxide TiO_2 is a semiconductor ceramic material widely applied for polluted water remediation and self-cleaning surfaces. The main basis of titanium dioxide powder and thin film performance is the electron excitation from valance band to conducting one, which can be carried out by lighting the material surface with radiation energy greater than its band gap. Thus, the larger the number of electron-hole pairs, the greater the rate of degradation of organic compounds by oxy-reduction [1–5].

There are some mechanism details to be considered after the initiation of lighting process and the final mineralization of organic compounds, which can be changed as a function of the organic molecule type and its semi-decomposed by-products, organics concentration and pH of solution, distance between the radiation source and the semiconductor material, radiation intensity and losses in the optical path, as well as some characteristics of the powder or thin film morphologies. But, the characteristic of the titanium dioxide semiconductor is the most important factor to be considered, including the morphology, composition and crystalline structure [6–9].

First of all, the importance of solid morphology is an intuitive aspect due to the basis of heterogeneous catalysis, and the several papers on the literature relating to the synthesis of titanium dioxide nanoparticles provide a good idea of the importance of a large specific area for powders. Innovative semiconductor arrangements like p-n junction, titanium dioxide glass-ceramic, ion-sodium battery based on anatase titanium dioxide ceramic matrix are also investigated in order to provide new applications concerning novel structures and morphologies degrees, making the titanium dioxide much more investigated [10–12].

In sequence, at first sight, the composition aspect seems not be applicable because high purity titanium dioxide should have stoichiometry defined as TiO_2. However, the fact is titanium dioxide is an intrinsic N-type semiconductor due to oxygen vacancies generated during the heat treatment, mainly for samples with high amount of anatase phase. Nevertheless, the non-doped n-type anatase titanium dioxide powder or thin films are usually applied as photocatalyst materials in advanced oxidation process for organic compounds. The justification is based on the yield of long-lived extrinsic photoholes, which is able to promote the water oxidation [13, 14].

The non-stoichiometry characteristic of titanium dioxide is very similar to the zinc oxide, so that many investigations consider the addition of extrinsic dopants in order to change semiconductor predominant behaviour. But, when the concentration of cation dopants becomes higher or when anion dopants are inserted in oxygen site, the crystalline structure starts to play an important role. In general, metallic cations with different oxidation states than tetravalent titanium are investigated and the energy bandgap and other photonic aspects are availed [15–17]. But, tetravalent titanium vacancies can be also formed through alternative procedures in order to generate some interesting properties, like high p-type conductivity and better photocatalytic performance [18].

Several works repot related the insertion of heterovalent metallic cations in titanium dioxide lead to the decreasing in the temperature of anatase-to-rutile phase transition and also to more amounts of rutile phase if the powder samples are heat treated up to 600°C [19–22]. Most of the doping approaches aim to shift the absorption edge to lower frequencies than ultraviolet range in order to utilize the titanium dioxide in solar photocatalysis. The nitrogen insertion by replacing lattice oxygen in anatase phase of titanium dioxide has also this central objective. However, there are strong suspicions that the desired surface electronic enhancement was followed by a considerable lattice surface distortion [23, 24].

The most of the doped titanium dioxide structures present different effects for anatase and rutile polymorphism due to the not well-controlled sample preparation as well as overlapping of several different effects. The oxygen vacancies seem to occur in association with most dopant insertion, and their concentrations are very sensitive to heat treatments. In addition to the lattice distortions and the changes in cation gap states caused by doping process, the oxygen vacancies also act as hole-trapping sites [25–27].

Even for that high-ordered single-crystal rutile sample, the trivalent titanium acting as trap site in (110) surface was found [28]. Any of those events seems to be significantly responsible for the enhancing or damaging of the titanium dioxide photocatalyst performance. Thus, well-characterized structural model systems are required to understand other changes in the material bulk or surface [29].

The exclusive structural aspects deserve special attention in order to understand a little more about the interfacial charge transfer in complex surface structure of the titanium dioxide photocatalyst, because the free energy on the single crystal surface is affected by its crystallographic orientation [30]. Different crystallographic planes were investigated separately in single crystal samples, and an interesting hypothesis was proposed concerning the surface atoms alignment. A reasonable experiment set including the etching of high aligned (001) facet proved that the more aligned the oxygen and titanium atoms on the surface, the greater are the chances of recombination of the electron-hole pairs [27].

The mechanisms of chemical reaction as a function of crystalline phase are also investigated. The single crystalline rutile (110) and anatase (101) samples submitted to the oxygen and water adsorptions at low temperatures, during the reheating up to room temperature, can be very useful to understand several aspects of photocatalysis mechanisms. Both rutile (110) and anatase (101) surfaces experience the reaction between oxygen and water molecules in order to form terminal hydroxyl groups on the phase surfaces. While the hydroxyl groups formed on the rutile (110) surface are highly bridged, on the anatase (101) surface the hydroxyl groups remain isolated and stable even after annealing above 130°C. The water molecules desorbs readily at room temperature from anatase (101) surfaces, whereas the oxygen ones remain undissociated and adsorbed in the vicinity of extrinsic P-type defects, and those differences have important consequences for photocatalysis application [31].

It is well understood that the anatase-to-rutile phase transition in polycrystalline titanium dioxide powder samples is strongly dependent on the anatase phase characteristics, mainly the impurity amount and the extrinsic dopant type. But, some structural surface aspects for non-doped and high purity powder samples are also indicative signals, such as the nucleation of the rutile phase by anatase (112) surface. A different intermediate metastable phase dissimilar to the prior anatase or the final rutile phases was observed, which showed to be sensitive to the compressive strain and the phase anisotropy [32, 33].

The photonic behaviour is strongly influenced by predominant phase characteristics, but some results attributed to the partial anatase-to-rutile phase transition through the controlled heat treatment leading to an enhancement in the photocatalytic performance. However, this hypothesis is controversial, because there are serious evidences that the anatase-to-rutile phase transition can start only the previous ordering in anatase phase. Thus, the relationship between well-crystallized anatase structure and photocatalytic activity can aid to understand some aspects not yet clarified about this material [34, 35].

The irreversible anatase-to-rutile phase transition occurring in temperatures above 600°C is a well-known fact for any experienced researcher working with titanium dioxide samples. It is also easy to verify that increasing the rutile phase amount is proportional to the temperature of heat treatment beyond 600°C. The anatase phase in titanium dioxide powder samples can be obtained even at room temperatures if no high amounts of organics are present in precursor materials, such as that provided by conventional sol-gel method. No structural differences seem to occur between the fresh hydrolysed samples and the heat treated one until 150°C, because the anatase phase nuclei is a water-driven process in order to form oxy-hydroxide titanium, which is relatively stable up to 150°C [36–38].

The continuous increase in temperature of heat treatment triggers several sequential decomposition stages in fresh hydrolysed gels, concerning the elimination of several residual compounds, such as the adsorbed water, hydroxyl groups and organic by-products. Unless the organic by-product types are able to form carbonaceous solid compounds, above the 250°C, the anatase phase is already an impurity-free material. Nevertheless, the desidroxylation stage is followed by the formation of cross-linked nuclei, which creates high amount of structural defects, in special, the oxygen vacancies. Those defects only can be eliminated overcoming the energy barrier of titanium-oxygen bonds, which leads to the structural destroy-rebuilding processes. This mechanism is the basis of the reconstructive transformation observed in the anatase-to-rutile phase transition [24, 32] and that is why the effective anatase phase crystallization only occurs when the rutile phase starts to form.

It was demonstrated earlier that the oxygen vacancies in titanium dioxide material configure an N-type semiconductor, which seems to be a not good photocatalyst due to high electron-hole combination occurring in oxygen vacancies. In addition, the formation of hydroxyl radical in aqueous media is also dependent on positive holes. Consequently, several dopants with lower oxidation states than titanium (IV) are investigated in order to make that material a P-type semiconductor [23].

In parallel, the synthesis method also plays an important role on the anatase crystallization and the sol-gel method is one of the most used to synthesize titanium dioxide powder and thin film samples. That chemical route is able to synthesize well crystallized titanium dioxide samples with anatase single phase at very low temperature because the titanium dioxide formation is a water-driven process that occurs even at room temperature and leads invariably to anatase single phase. It is possible to use only volatile coadjutant reagents, such as the metallic alkoxide precursors and acetic acid complexing agent, so that very little amount of organic wastes gets retained when the fresh gels are dried at 100°C or almost nothing if dried above 150°C. There is no chance for the existence of organic solid residues, even if metallic nitrates are used as dopant precursor reagents, which makes the sol-gel method the preferential choice in order to obtain nanoparticles of titanium dioxide in anatase single phase with relative success [39].

In addition, the amounts of alcohol solvent and water as hydrolyzing agent are also important on the particle growth stage, because the cross-linked bonds are affected directly by those coadjutant reagents during the first stages of titanium oxy-hydroxide nuclei formation. It is almost a consensus that the mean particle diameter reduces as a function of water content in the jellifying process [40], due to the more separation of nuclei from each other. Nevertheless, that means a more amount of terminal hydroxyl groups in anatase phase particle surface and, consequently, a more amount of cross-linked bonds emerging during the drying stage.

The anatase thermal stability depends on the defect concentration, including the oxygen vacancies. The oxygen vacancies concentration can be dependent on the amount of the cross-linked bonds among the nanoparticles so that the titanium dioxide nanoparticles can present a significant increase in oxygen vacancies when compared to the coarse particles. Thus, the literature shows that the thermal treatment to crystallize the titanium dioxide nanoparticles is preferentially carried out in temperatures below 500°C, which is still far from the energetic barrier for anatase-to-rutile phase transition [41]. Less often, the nanoparticles can be carefully thermal treated under step-by-step process by increasing the temperature from 250 to 600°C, at most, if it is desirable to obtain anatase single phase [42], which is probably due to the lower stability for anatase phase in nanoparticle form [43, 44].

On the other hand, homovalent cation doping seems to reduce also the oxygen vacancies, which can also be associated with cross-linked metal-oxygen bonds in anatase phase. The insertion of zirconium at 10 mol% in titanium dioxide powder samples avoids the full anatase-to-rutile phase transition at higher temperatures than 600°C, leading to the calcined material to present high amounts of anatase phase when calcined up to 750°C. By increasing the zirconium content and calcining the material at same temperature, much more amounts of anatase phase are observed in powder samples. However, an orthorhombic zirconium titanate secondary phase starts to crystallize in addition to the remaining anatase phase when the zirconium content is higher than 10 mol% [45].

By comparing the ionic radii for Ti (IV) and Zr (IV) hexacoordinate cations, it is possible to infer that the zirconium cation is bigger (72 pm) than titanium one (61 pm) [46]. As consequence, if the zirconium substitution in titanium site was successful, then a proportional anatase lattice expansion must be observed as a function of zirconium content until at certain concentration limit, at least [28, 45].

Another homovalent cation dopant investigated for anatase phase stabilization is the tetravalent silicon [47]. Thermal analysis from fresh gel has shown that the presence of silicon dopant can delay the anatase crystallization, which can be visualized by an exothermic peak, towards higher temperatures than undoped titanium dioxide. The cell volume for several calcined samples decreases continually as a function of silicon content, besides the crystallinity loss, as visualized by the X-ray diffraction patterns followed by rietveld refinement. The calcined samples in wide temperature range show an increase in temperature of anatase-to-rutile phase transition, but until a certain dopant concentration limit, signalled as 5 mol%. Different from zirconium dopant, the silicon one does not seem to generate secondary phases, but only to increase the atomic disordering in anatase phase [48]. Those results imply that the silicon dioxide can have infinite solubility in anatase titanium dioxide phase.

Some researchers have believed that the silicon cation is so smaller that its perfect accommodation in interstitial of titanium and oxygen sites of low dense anatase phase is possible. Thus, the explanation for anatase stabilization is justified, because the high dense rutile phase does not have enough interstitial space to accommodate hexacoordinate tetravalent silicon cations, avoiding the phase transition. That consideration is reasonable but is not true, because the ionic radii for hexacoordinate silicon (VI) is 40 pm [46], which means the titanium and silicon cations should experience some repulsion with each other to lead a lattice expansion in anatase phase, which is not observed in the literature.

Furthermore, an important ab-initio study taking into account the alternative silicon dopant positions in anatase and rutile phases buries the idea of interstitial silicon in titanium dioxide sample once and for all. The calculated results suggest that the interstitial tetravalent silicon cation is energetically favourable neither for anatase phase nor for rutile one. Also, the calculated results are in concordance with lattice parameter contraction; thus, it is not a good idea to continue with the consideration of interstitial tetravalent silicon cations in titanium dioxide anatase phase as the cause for anatase stabilization. At least while the X-ray diffraction data are showing the consistent anatase lattice contraction for silicon-doped titanium-doped powder samples [29].

The stabilization of anatase phase caused by homovalent dopants seems to be related to the increase in cross-linked metal-oxygen bond energies still in anatase phase. In addition, the reported enhancement in photocatalytic performance for homovalent doping cannot be related exclusively to the decrease in oxygen vacancies [28]. The effect in crystal surface can play an important role in order to reduce the recombination of electron-hole pair, and the anatase phase surface presenting oxygen and cations outer of the plane seems to be crucial to avoid that auto-neutralizing phenomenon [10, 31].

The first attempt to use simultaneously silicon and zirconium dopants was published in 2006 [49] but did not contribute to the correct understanding of the anatase phase stabilization in titanium dioxide samples. First, because the anatase phase stabilization occurring as a function temperature increasing or even significant changes in anatase-to-rutile phase transition were not demonstrated, according the X-ray diffraction patterns available on that work. Except at 700°C, all of the samples presented rutile phase and the authors chose to explain the results by considering no dopant substitutions, but the effect of anatase interparticle secondary phases.

In addition, considerable confusion can occur if the dopant concentrations were referred as weight percent, due to the enormous difference in atomic weight among the metal constituents, and is not a good choice in order to permit the correct understanding of the progressive doping effects. The results provided by simultaneous and equal weight doping elements for silicon and zirconium at 5 wt% [51] correspond to 8.4 mol% for silicon and 2.6 mol% for zirconium, which means a Si/Zr molar ratio more than 3. Even so, no considerable differences were observed comparing the isolated zirconium or silicon doping under the point of view of anatase phase stabilization [47, 48, 50–52].

Other consideration about the mechanism of the anatase-to-rutile phase transition is no longer valid, specially, the affirmation that the anatase phase starts to convert into rutile one at lower temperatures through particle agglomeration in a kinetically controlled reaction. It was demonstrated already that there is a thermodynamic stability for anatase phase at standard conditions of temperature and pressure [29], which is determined probably by titanium-oxygen bond energies [24]. The kinetic component starts to play an important role only after overcoming the energetic barrier at very high temperatures, like above 900°C [28].

Only in 2016, another publication reporting the silicon-zirconium-doped titanium dioxide sample was found [53]. There are several doped titanium dioxide nanoparticles calcined at 500°C only, but unexpectedly, the authors discuss about the anatase-to-rutile phase transition as a function of dopant type for anatase single phase samples. In addition, a sample containing only silicon and zirconium dopants was also not prepared so that it was not possible to evaluate the effects of the simultaneous silicon and zirconium doping on titanium dioxide. Despite that the authors affirm that the sample simultaneously doped with copper, silicon and zirconium cations at 15 mol% in total concentration and equal parts showed a great improvement in photocatalytic performance for the degradation of methyl orange.

No other articles about the simultaneous silicon- and zirconium-doped titanium dioxide sample were found in the literature. That way, the present work aims to provide an accurate investigation about exclusive and simultaneous silicon and zirconium doping in titanium dioxide powder samples obtained through the sol-gel method in order to demonstrate the structural basis of anatase phase stabilization. It is very important to keep in mind that equal parts of tetravalent silicon and zirconium cations represent the average ionic radii close to the tetravalent titanium cation, considering all of those in hexacoordinate sites. It is also important to know that zircon silicate $ZrSiO_4$ possesses also anatase phase, which can be thermally stable until very high temperatures [46, 54].

2. Materials and methods

In order to provide a consistent discussion about the influence of the zirconium-silicon simultaneous doping, the non-modified titanium dioxide samples were prepared in the same experimental conditions used to obtaining the doped samples. It was used the true sol-gel method involving only analytical grade reagents for the preparation of the powder samples. First, titanium (IV) isopropoxide 97% (Sigma-Aldrich) was added drop wise in glacial acetic acid 99.7% (F. Maia) under stirring. The molar ratio between the acetic acid and total metallic cations was adjusted to 4 for both samples. Due to the exothermic reaction, the homogenous mixture was cooled to room temperature by maintaining stirring for 30 min before the next step. For silicon-zirconium (Si-Zr)-doped titanium dioxide sample, tetraethyl orthosilicate 98% (Sigma-Aldrich) and zirconium (IV) propoxide 70% (Sigma-Aldrich) are added in order to obtain 5 mol

% of Si-Zr doping. Then, for both samples, the isopropyl alcohol R.G. (Qhemis) was added in order to adjust the metal concentration to 0.1 molar.

After a homogenization stage through the stirring for 30 min, acidified water was added in order to promote the acid hydrolysis. For that proposes, nitric acid solution with pH 3.5 was previously prepared in order to represent a molar ratio between water and metallic cations close to 5. Thus, both composition sols were stirred for 1 hour, capped and allowed to stand at ambient conditions for 24 hours in order to complete the jellification. Both xerogels showed transparent characteristics before the drying process carried out in drying oven at 100°C overnight and ground process in porcelain mortar. Finally, both dry gel samples were divided into several aliquots in order to perform the calcination step in wide range of temperature for 2 hours in a muffle type oven under static atmosphere. The expected compositions are TiO_2 and $Si_{0.25}Zr_{0.25}Ti_{0.95}O_2$, for pure and Si-Zr-doped titanium dioxide-calcined powder samples, respectively.

Both dry gel compositions were characterized by simultaneous thermogravimetric and differential thermal analysis (TG/DTA) in order to verify basic differences between non-doped and Silicon-zirconium-doped gel sample along the temperature of thermal treatment. For that characterization, aliquots close to 100 mg were placed in alumina crucibles and compared with the same amount of alpha-alumina powder as standard material. Sample and reference material crucibles were submitted to heating rate of 10°C min^{-1} from room temperature up to 900°C under synthetic air flux of 10 mL min^{-1} by using a Netszch—Thermische Analyse equipped with TASC 414/2 controller and Pt 10 thermocouples.

The dry gel at 100°C and all of the calcined powder samples in a wide temperature set were characterized by X-ray diffractometry by using D5005 Siemens equipment operating with K-alpha nickel-filtered Cu radiation from 20 to 80° (2-theta) in step scan mode in order to collect the diffraction signal in intervals of 0.02° (2-theta) during 1 s. The obtained diffraction patterns were phase identified using the JCPDS data bank [54], and the anatase and rutile structural models were taken from ICSD data bank [55]. The raw files were refined starting from the chosen anatase and rutile structural models by rietveld method [56] performed with the last upgraded DBWS software [57, 58] in order to include the size-strain calculation.

3. Results and discussion

Figure 1(a) shows the TG/DTG/DTA curves for both dry gel samples, where it is possible to observe higher weight losses for Si-Zr-doped titanium dioxide dry gel in all of the weight loss stages from room temperature up to 460°C. The first weight loss is an endothermic event associated to the volatilisation processes of residual alcohols and acetic acid, including probable ester and nitro compounds, which possess boiling temperatures up to 100°C. In addition to the slight narrow endothermic peak observed for Si-Zr doped titanium dioxide dry gel, the volatile compounds present in that sample were 1.7%, against 1. 3% for pure one, considering the final temperature at 180°C for both curves, according to the DTA peaks.

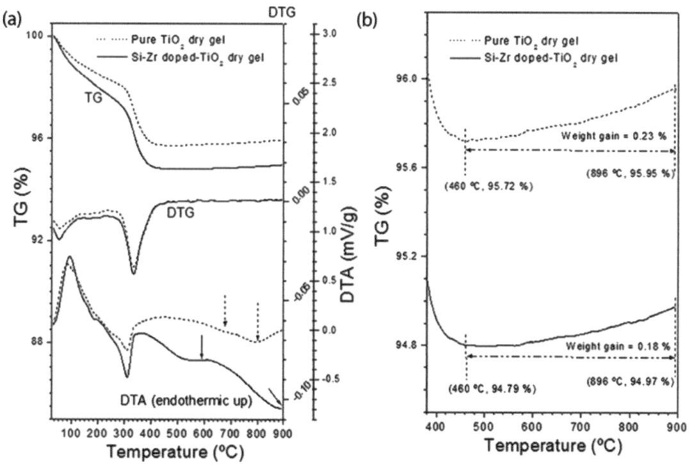

Figure 1. Thermogravimetric and differential thermal analysis for pure and silicon-zircon-doped titanium dioxide dry gel samples: (a) entire TG/DTG/DTA curves and (b) amplification of TG curves between 390 and 900°C.

Residual organic compounds and free-water are desorbed from porous matrixes between 180 and 250°C, markedly a kinetic event, as can be verified by the absence of DTG or DTA peaks in that region. That desorption event represented a weight loss of 0.7% for Si-Zr-doped titanium dioxide dry gel, against 0.4% for the same event in pure titanium dioxide sample. Thus, the total weight losses observed for Si-Zr-doped and pure titanium dioxide dry gels from 30 to 250°C increase for 2.4 and 1.7%, respectively.

The first exothermic event for both samples occurs between 250 and 460°C, which was associated to the cross-link bonding among the hydroxyl groups in nuclei surface through the dehydroxylation process. The water losses associated to that event were close to 2.8% for Si-Zr-doped titanium dioxide sample and 2.6% for pure one. It is also possible to notice that the exothermic peaks (DTA) precede the water weight loss (DTG) in almost 20°C, which prove that the water loss is slower than the dehydroxylation process.

Supposedly, this result is coherent with expected reduction of empty volume among the nuclei due to the pore constriction associated to the cross-link bonding, which harms the water desorption. It is also possible to notice that the exothermic peak for Si-Zr-doped titanium dioxide sample is considerably more intense than that observed for the pure one. Then, already during the dehydroxylation process, an important energetic difference is observed for zircon silicate titanium dioxide samples when compared to the pure one, which can justify possible changes in photocatalysis performances due to the important role the hydroxyl groups in anatase surface have in charge transference and electron-hole recombining processes.

Finally, the anatase crystallization process is observed after the last weight loss and the Si-Zr-doped titanium dioxide samples present two evidenced exothermic events. According to correlated work [59], the presence of silicon dioxide in titanium dioxide seems to favour the crystallization process as a facilitator of the oxygen vacancy elimination. In fact, it is possible to visualize that the Si-Zr-doped titanium-doped sample has both events more exothermically energetic than the pure one. But, other considerable differences are also observed concerning the temperature of maximum energy. The Si-Zr-doped titanium sample possesses the first crystallization stage starting below 400°C, already during the cross-link bonding, and ends at 550°C. For the pure titanium dioxide sample, there is only a shoulder at 650°C in DTA curve, which is very difficult to view without the graphic software help.

Taking into account that the shoulder in pure titanium dioxide sample occurs above the typical anatase-to-rutile phase transition, it is possible to infer that some amounts of rutile phase probably is already present at that 650°C. Thus, the third exothermic event for pure-doped titanium-doped sample is centred at 800°C, probably associated to the complete formation of rutile phase. On the other hand, the Si-Zr-doped titanium dioxide sample presents the third exothermic event probably above 900°C (not observed due the analysis ending). Nevertheless, there is no evidence that the rutile phase is already present above 900°C, because the zircon silicate has an increasingly stable anatase structure with increasing calcination time and temperature. Thus, 5 mol% of Si-Zr may be sufficient to prevent the rutile phase transition in titanium dioxide samples calcined until 900°C, at least.

Figure 1(b) shows the amplification of TG curves between 390 and 896°C in order to show the weight gain for both samples starting to occur immediately after the final weight loss. Then, it is acceptable that the hypothesis based on oxygen vacancies elimination starts after the ending of cross-link bonding. Both titanium dioxide samples present significant weight gain due to the oxygen incorporation. However, the pure titanium dioxide sample presents higher 0.5% more oxygen incorporation than the Si-Zr-doped one. In addition, the weight gain for Si-Zr-doped sample only becomes remarkable after 550°C.

One important relation between both the samples related to the anatase phase stabilization can be established by considering that the oxygen incorporation is not finished at 900°C due to the kinetic component. Thus, more weight gain can continue to occur until both samples reach a similar weight gain values. If it is true, then the velocity constant for oxygen incorporation in the Si-Zr-doped sample is lower because the same weight gain is reached for pure sample at 811°C, which represents 85°C lower than the Si-Zr-doped sample. An accurate investigation of that dependence would be very interesting to clarify the energetic changes provided by simultaneous silicon-zirconium doping process.

In **Figure 2**, X-ray diffraction patterns showing the phase evolution starting from 100°C for 24 hours to 900°C for 2 hours are shown. The pure titanium dioxide sample presents anatase single phase up to 600°C, which means the rutile phase starts to form at 650°C, according to DTA analysis evidence. On one hand, a continuous increase in temperature of calcination leads to much more rutile phase in the samples until the formation of rutile single phase in pure titanium dioxide powder samples above the 800°C (**Figure 2(a)**). On the other hand, the Si-Zr-doped titanium dioxide samples present no evidence of rutile phase even at 900°C (**Figure 2(b)**).

The angle degrees and the relative intensities for anatase phase peaks in all of the calcination temperatures fit with those available on JCPDS card number 21-1272, whereas the rutile phase

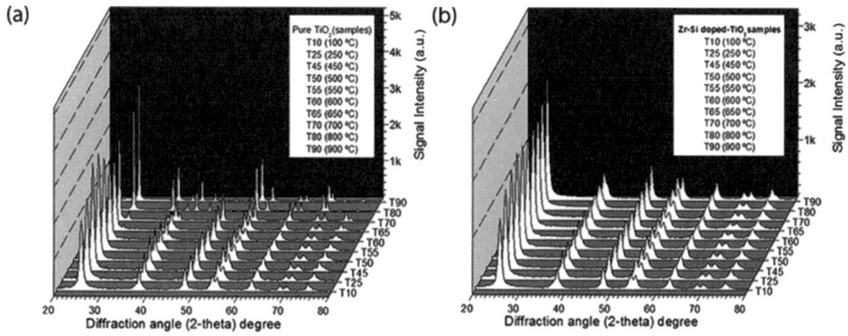

Figure 2. X-ray diffraction patterns showing the phase evolution for starting from 100°C for 24 hours up to 900°C for 2 hours: (a) pure and (b) Si-Zr titanium dioxide samples.

was perfectly identified with the angle degrees and the relative intensities available on JCPDS card number 21-1276. Nevertheless, it is easy to observe a considerable peak enlargement for Si-Zr-doped samples for all temperatures, if compared to the pure ones. However, neither the angle degrees nor the relative intensities are changed for the anatase peaks in Si-Zr-doped titanium dioxide samples along the temperature of calcination.

In order to understand the structural effects beyond the visual observation of the X-ray diffraction patterns, a structural refinement was carried out for all of the samples. That procedure starts choosing the structural models for anatase and rutile phase available on the ICSD data bank. The best adequacy was obtained by setting the lattice parameters and atomic positions according to card numbers 82084 and 53997, for anatase and rutile phase, respectively. After more than 3000 cycles, the refinement factors reach the minimum permitted for statistically expected values provided by method. More information about that methodology is available on the specific literature [56–58, 60].

The lattice parameters of each phase are shown in **Figure 3**. It is possible to observe a consistent variation for all of the refined parameters, making possible an appropriated discussion. For anatase phase, both compositions, the lattice parameters present the same variation samples and, except for 100°C, the vector a and c values are slightly higher for pure titanium dioxide samples, at least up to 600°C, where the anatase phase for pure titanium dioxide sample becomes unstable (**Figure 3(a)**). By considering tetravalent silicon and zirconium hexacoordinate dopant cations, with ionic radii of 40 pm and 72 pm, respectively, the average radii are close to 56 pm. Therefore, the substituting cations constitute as smaller than hexacoordinate tetravalent titanium substituted one (61 pm). Thus, it is a coherent result that the slight lattice contraction was observed for the Si-Zr doped samples.

Maybe it is not important to explain the basis of the anatase phase stabilization, but the inverse effect that occurred at 100°C is in consequence of the higher amount hydroxyl groups in Si-Zr-doped titanium dioxide dry gel, as was demonstrated in TG analysis. The importance of the Rietveld analysis is established just above the 650°C and remarkable difference is observed among the samples originated from both the compositions. In temperatures close to anatase-to-rutile

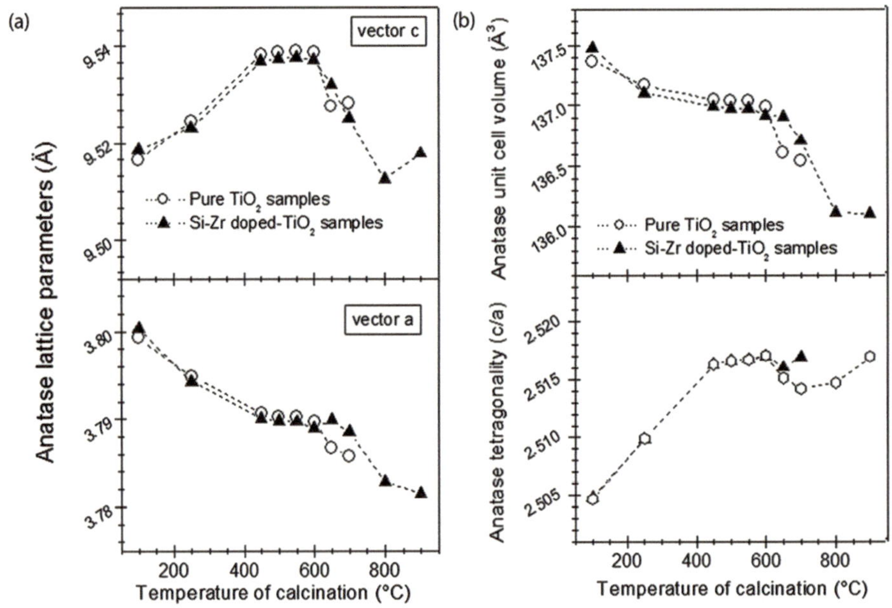

Figure 3. Refined lattice parameters for anatase phase as a function of calcination temperature for: (a) vectors values and (b) calculated unit cell volume and tetragonality (c/a).

phase transition, the both vectors a and c of anatase phase for pure titanium dioxide samples start to reduce, marking the initial collapse of anatase phase, which will no longer be present at 800°C. By observing the same variation for Si-Zr-doped samples, it is possible to observe that behaviour is characteristic of anatase phase crystallization, but only for pure titanium dioxide sample and is followed by the anatase to rutile phase transition.

By observing the variation of anatase cell volume in **Figure 3(b)** it is possible to notice an impressive variation similarity with the variation of vector a (**Figure 3(a)**), for both samples. It is possible to observe the tetragonality changes between the pure and the Si-Zr-doped titanium dioxide samples in the same way, but only displaced to higher temperature for Si-Zr-doped one. Thus, those results corroborate the previous DTA differences and the tetragonality of anatase single phase in the Si-Zr-doped titanium dioxide sample calcined at 900°C is very close to the well-crystallized anatase single phase observed in the pure titanium dioxide sample calcined at 600°C.

The reliable crystallization process was investigated through the size-strain calculation according to the methodology well-established [60] and using tungsten carbide as standard material in order to evaluate the instrumental contributions. It was noticed that the crystallite size stays in low values until the beginning of anatase-to-rutile phase formation and the peak narrowing is originated from the reduction of the lattice microstrain practically (**Figure 4(a)**). Thus, the crystallite coalescence occurs as a consequence of the destroying-rebuilt oxygen-metallic cation bonds process starting at 650°C, regardless of anatase-to rutile phase transition. Beyond that reconstructive transformation, above 700°C, the crystallite sizes considerably increase despite of

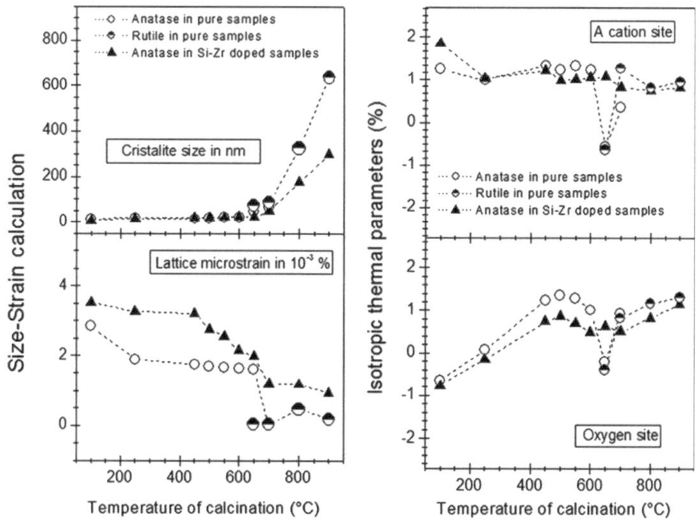

Figure 4. Calculated data through the refined lattice parameters and peak profiles: (a) crystallite size and lattice microstrain and (b) isotropic thermal parameters.

some variation on lattice microstrain. In **Figure 4(b)**, it is possible to observe that the anatase structure starts with spacious oxygen site through the negative values for isotropic thermal parameters. As the atomic scattering is associated to a specific site volume, it is probable that the hydroxyl groups push oxygen atoms away from each other. The isotropic parameters become higher after the dehydroxylation process and that results are coherent with cell volume variation and thermal analysis results. However, for pure titanium dioxide samples, the atoms experience a new localized spacing in the temperature close to the anatase-to-rutile phase transition, different from Si-Zr-doped ones. That event indicates the reconstructive transformation occurs through the destroying-rebuilt bonds, because the variation of the isotropic thermal parameters presents noticeable inflection during that process.

4. Conclusions

The accurate structural characterization was carried out by X-ray diffractometry and rietveld refinement in titanium dioxide sample synthesized through sol-gel method and calcined in wide temperature range in order to evaluate the effects on the thermal stabilization of the zirconium silicate doping at 5 mol%. The discussion of structural variations in both pure and Si-Zr-doped sample compositions was supported by corresponding thermal analysis taken from dry gels and consistent correlations were found.

The silicon and zircon dopants lead to more retention of residues coming from jellying process and also the fresh anatase formed is more hydrolyzed. All of the exothermic events associated to the phase crystallization in both samples are displaced to higher temperatures if compared to the pure titanium dioxide sample. Thus, the final crystallization associated to oxygen vacancies elimination as a function of oxygen incorporation occurs above 900°C.

By observing the X-ray diffraction patterns, it was possible to prove that the Si-Zr dopants at concentration of 5 mol% are able to prevent the rutile phase transition in titanium dioxide samples calcined up to 900°C. But, the rietveld analysis was very important to found the structural basis for that behaviour and the known reconstructive transformation was proved for the anatase-to-rutile phase transition. It was possible to show that the isotropic thermal parameters for oxygen and titanium atoms are considerably affected just during the anatase phase collapse and rutile one rising.

The crystallite coalescence responsible for anatase crystallization does not occur in pure titanium dioxide samples, and the X-ray diffraction peaks narrowing observed for calcined samples in several publications is consequence only of the microstrain reduction. Thus, the pure titanium dioxide samples cannot be crystallized as anatase phase without the parallel rutile phase conversion, different from the Si-Zr-doped titanium dioxide sample. Consequently, well-crystallized anatase single phase at 900°C can be obtained by doping the titanium dioxide samples with 5 mol% of Si-Zr. That composition can become a new semiconductor matrix for the investigation of new dopants in order to improve further the photocatalysis performance or maybe, other possible applications.

Author details

Alberto Adriano Cavalheiro[1]*, Lincoln Carlos Silva de Oliveira[2] and Silvanice Aparecida Lopes dos Santos[2]

*Address all correspondence to: albecava@gmail.com

1 State University of Mato Grosso do Sul, Naviraí, MS, Brazil

2 Federal University of Mato Grosso do Sul, Campo Grande, MS, Brazil

References

[1] Linsebigler AL, Lu G, Yates JT. Photocatalysis on TiO_2 surfaces: Principles, mechanisms, and selected results. Chemical Reviews. 1995;**95**:735–758. DOI: 10.1021/cr00035a013

[2] Nakata K, Fujishima A. TiO_2 photocatalysis: Design and applications. Journal of Photochemistry and Photobiology C: Photochemistry Reviews. 2012;**13**:169–189. DOI: 10.1016/j.jphotochemrev.2012.06.001

[3] Mills A, Davies RH, Worsley D. Water-purification by semiconductor photocatalysis. Chemical Society Reviews. 1993;**22**:417–425. DOI: 10.1039/CS9932200417

[4] Zhang J, Liu X, Gao S, Huang B, Dai Y, Xu Y, Grabstanowicz LR, Xu T. From AgI/TiO_2 to Ag/TiO_2: Effects of the annealing temperature on the compositions, porous nanostructures,

and visible-light photocatalytic properties. Ceramics International. 2013;**39**:1011–1019. DOI: 10.1016/j.ceramint.2012.07.021

[5] Ibhadon AO, Fitzpatrick P. Heterogeneous photocatalysis: Recent advances and applications. Catalysts. 2013;**3**:189–218. DOI: 10.3390/catal3010189

[6] Sclafani A, Palmisano L, Schiavello M. Influence of the preparation methods of TiO_2 on the photocatalytic degradation of phenol in aqueous dispersion. The Journal of Physical Chemistry. 1990;**94**:829–832. DOI: 10.1021/j100365a058

[7] Cavalheiro AA, Bruno JC, Saeki MJ, Valente JPS, Florentino AO. Effect of scandium on the structural and photocatalytic properties of titanium dioxide thin films. Journal of Materials Science. 2007;**43**(2):602–608. DOI: 10.1007/s10853-007-1743-2

[8] Qiu S, Starr TL. Zirconium doping in titanium oxide photocatalytic films prepared by atomic layer deposition. Journal of the Electrochemical Society. 2007;**154**(6):H472–H475. DOI: 10.1149/1.2718475

[9] Cavalheiro AA, Bruno JC, Valente JPS, Saeki MJ, Florentino AO. Photocatalytic decomposition of diclofenac potassium using silver-modified TiO_2 thin films. Thin Solid Films. 2008;**516**(18):6240–6244. DOI: 10.1016/j.tsf.2007.11.117

[10] Liu S. Liu X, Chen Y, Jiang R. A novel preparation of highly active iron-doped titania photocatalysts with a p-n junction semiconductor structure. Journal of Alloys and Compounds. 2010;**506**(2):877–882. DOI: 10.1016/j.jallcom.2010.07.103

[11] Su D, Dou S, Wang G. Anatase TiO_2: Better anode material than amorphous and rutile phases of TiO_2 for Na-Ion batteries. Chemistry of Materials. 2015;**27**(17):6022–6029. DOI: 10.1021/acs.chemmater.5b02348

[12] Orera VM, Merino RI. Ceramics with photonic and optical applications. Boletín de la Sociedad Española de Cerámica y Vidrio. 2015;**54**(1):1–10. DOI: 10.1016/j.bsecv.2015.02.002

[13] Schneider J, Matsuoka M, Takeuchi M, Zhang J, Horiuchi Y, Anpo M, Bahnemann DW. Understanding TiO_2 photocatalysis: Mechanisms and materials. Chemical Review. 2014;**114**(19):9919–9986. DOI: 10.1021/cr5001892

[14] Pesci FM, Wang G, Klug DR, Li Y, Cowan† AJ. Efficient suppression of Electron–hole recombination in oxygen-deficient hydrogen-treated TiO_2 nanowires for photoelectrochemical water splitting. The Journal of Physical Chemistry C, Nanomaterials and Interfaces. 2013;**117**(48):25837–25844. DOI: 10.1021/jp4099914

[15] Vargas S, Arroyo R, Haro E, Rodriguez R. Effect of cationic dopants on the phase transition temperature of titania prepared by the sol-gel method. Journal of Materials Research. 1999;**14**(10):3932–3937. DOI: 10.1557/JMR.1999.0532

[16] Moezzi A, McDonagh AM, Cortie MB. Zinc oxide particles: Synthesis, properties and applications. Chemical Engineering Journal 2012;**185-186**:1–22. DOI: 10.1016/j.cej.2012.01.076

[17] Gonsález MP, Tomás SA, Luna MM, Arvizu MA, Cruz MMT. Optical, structural, and morphological properties of photocatalytic TiO_2–ZnO thin films synthesized by the sol-gel process. Thin Solid Films. 2015;**594**:304–309. DOI: 10.1016/j.tsf.2015.04.073

[18] Wang S, Pan L, Song J-L, Mi W, Zou J-J, Wang L, Zhang X. Titanium-defected undoped anatase TiO_2 with p-Type conductivity, oom-temperature ferromagnetism, and remarkable photocatalytic performance. Journal of the American Chemical Society. 2015;**137** (8):2975–2983. DOI: 10.1021/ja512047k

[19] Yalçin Y, Kiliç M, Çinar Z. Fe^{+3}-doped TiO_2: A combined experimental and computational approach to the evaluation of visible light activity. Applied Catalysis B: Environmental. 2010;**99**(3-4):469–477. DOI: 10.1016/j.apcatb.2010.05.013

[20] Jamalluddin NA, Abdullah AZ. Reactive dye degradation by combined Fe(III)/TiO_2 catalyst and ultrasonic irradiation: Effect of Fe(III) loading and calcination temperature. Ultrasonics Sonochemistry. 2011;**18**(2):669–678. DOI: 10.1016/j.ultsonch.2010.09.004.

[21] Niu Y, Xing M, Zhang J, Tian B. Visible light activated sulfur and iron co-doped TiO_2 photocatalyst for the photocatalytic degradation of phenol. Catalysis Today. 2013;**201** (1):159–166, DOI: 10.1016/j.cattod.2012.04.035

[22] Harizanov O, Harizanova A. Development and investigation of sol-gel solutions for the formation of TiO_2 coatings. Solar Energy Materials and Solar Cells. 2000;**63**(2):185–195. DOI: 10.1016/S0927-0248(00)00008-8

[23] Katoh R, Furube A, Yamanaka K, Morikawa T. Charge separation and trapping in N-doped TiO_2 photocatalysts: A time-resolved microwave conductivity study. The Journal of Physical Chemistry Letters. 2010;**1**(22):3261–3265. DOI: 10.1021/jz1011548

[24] Batzill M, Morales EH, Diebold U. Influence of nitrogen doping on the defect formation and surface properties of TiO_2 rutile and anatase. Physical Review Letters. 2006;**96** (2):026103. DOI: 10.1103/PhysRevLett.96.026103

[25] Wetchakun N, Incessungvorn B, Wetchakun K, Phanichphant S. Influence of calcination temperature on anatase to rutile phase transformation in TiO_2 nanoparticles synthesized by the modified sol–gel method. Materials Letters. 2012;**82**:195–198. DOI: 10.1016/j.matlet.2012.05.092

[26] Wu S-X, Zhi M, Qin Y-N, He F, Jia L-S, Zhang Y-J. XPS study of Cooper dopping TiO_2 photocatalyst. Acta Physico-Chimica Sinica. 2003;**19**(10):967–969. DOI: 10.3866/PKU.WHXB20031017

[27] Liu X, Dong G, Li S, Lu G, Bi Y. Direct observation of charge separation on anatase TiO_2 crystals with selectively etched {001} facets. Journal of the American Chemical Society. 2016;**138**(9):2917–2920. DOI: 10.1021/jacs.5b12521

[28] Hanaor DAH, Sorrell CC. Review of the anatase to rutile phase transformation. Journal of Materials Science. 2011;**46**(4):855–874. DOI: 10.1007/s10853-010-5113-0

[29] Hanaor DAH, Assadi MHN, Li S, Yu A, Sorrell CC. Ab initio study of phase stability in doped TiO_2. Computational Mechanics. 2012;**50**(2):185–194. DOI: 10.1007/s00466-012-0728-4

[30] Selloni A. Crystal growth: Anatase shows its reactive side. Nature Materials. 2008;**7** (8):613–615. DOI: 10.1038/nmat2241

[31] Setvin M, Aschauer U, Hulva J, Simschitz T, Daniel B, Schmid M, Selloni A, Diebold U. Following the reduction of oxygen on TiO_2 anatase (101) step by step. Journal of the American Chemical Society. 2016;**138**(30):9565–9571. DOI: 10.1021/jacs.6b04004

[32] Zaleska A. Doped-TiO_2: A review. Recent Patents on Engineering. 2008;**2**(3):157–164. DOI:10.2174/187221208786306289

[33] Zhu S-C, Xie S H, Liu Z-P. Nature of rutile nuclei in anatase-to-rutile phase transition. Journal of the American Chemical Society. 2015;**137**(35):11532–11539. DOI: 10.1021/jacs.5b07734

[34] Li B, Wang X, Yan M, Li L. Preparation and characterization of nano-TiO_2 powder. Materials Chemistry and Physics. 2003;**78**(1):184–188. DOI: 10.1016/S0254-0584(02)00226-2

[35] Hu Y, Tsai HL, Huang CL. Phase transformation of precipitated TiO_2 nanoparticles. Materials Science and Engineering: A. 2003;**344**(1-2):209–214. DOI: 10.1016/S0921-5093(02)00408-2

[36] Look JL, Zukoski DF. Colloidal stability and titania precipitate morphology: Influence of short-range repulsion. Journal of the American Ceramic Society. 1995;**78**(1):21–32. DOI: 10.1111/j.1151-2916.1995.tb08356.x

[37] Ahn YU, Kim EJ, Kim HT, Hahn SH. Variation of structural and optical properties of sol-gel TiO_2 thin films with catalyst concentration and calcination temperature. Materials Letters. 2003;**57**(30):4660–4666. DOI: 10.1016/S0167-577X(03)00380-X

[38] Hemissi M, Adnani HA. Optical and structural properties of titanium oxide thin films prepared by sol-gel methods. Digest Journal of Nanomaterials and Biostructures. 2007;**2** (4):299–305. Available from: www.chalcogen.ro/index.php/journals/digest-journal-of-nanomaterials-and-biostructures. [Accessed: 15-Feburary-2017].

[39] Chaudhary V, Srivastava AK, Kumar J. On the sol-gel synthesis and characterization of titanium oxide nanoparticles. Materials Research Society Symposia Proceedings. 2001;**1352**(11):10–24. DOI: 10.1557/opl.2011.759

[40] Nagpal VJ, Davis RM, Riffle JS. In situ steric stabilization of titanium dioxide particles synthesized by a sol-gel process. Colloids and Surfaces A: Physicochemical and Engineering Aspects. 1994;**87**(1):25–31. DOI: 10.1016/0927-7757(93)02735-W

[41] Saleh A, Rasin FA, Ameen MA. TiO_2 nanoparticles prepared by sol-gel. Journal of Materials Science and Engineering. 2009;**3**(12):81–84. Serial No. 25. Available from: www.davidpublishing.com/davidpublishing/upfile/8/5/2012/2012080582702945.pdf. [Accessed: 14-Feburary-2017]

[42] Hayle ST, Gonfa GG. Synthesis and characterization of titanium oxide nanomaterials using sol-gel method. American Journal of Nanoscience and Nanotechnology. 2014;**2**(1):1–7. DOI: 10.11648/j.nano.20140201.11

[43] Nadzirah S, Foo KH, Hashim U. Morphological reaction on the different stabilizers of titanium dioxide nanoparticles. International Journal of Electrochemical Science. 2015;**10**:5498–5512. Available from: www.electrochemsci.org/papers/vol10/100705498.pdf [Accessed: 15 Feburary 2017].

[44] Vorkapic D, Matsoukas T. Effect of temperature and alcohols in the preparation of titania nanoparticles from alkoxides. Journal of the American Ceramic Society. 1998;**81**(11):2815–2820. DOI: 10.1111/j.1151-2916.1998.tb02701.x

[45] Lukáč J, Klementová M, Bezdička P, Bakardjieva S, Šubrt J, Szatmáry L, Bastl Z, Jirkovský J. Influence of Zr as TiO_2 doping ion on photocatalytic degradation of 4-chlorophenol. Applied Catalysis B-Environmental. 2007;**74**(1-2):83–91. DOI: 10.1016/j.apcatb.2007.01.014

[46] Shannon RD. Revised effective ionic radii and systematic studies of interatomic distances in halides and chalcogenides. Acta Crystallographica Section A, Foundations of Crystallographica. 1976;**A32**:751–767. DOI:10.1107/s0567739476001551

[47] Harraz FA, Salam OEA, Mostafa AA, Mohamed RM, Hanafy M. Rapid synthesis of titania-silica nanoparticles photocatalyst by a modified sol-gel method for cyanide degradation and heavy metals removal. Journal of Alloys and Compounds. 2013;**551**:1–7. DOI: 10.1016/j.jallcom.2012.10.004

[48] Okada K, Yamamoto N, Kameshima Y, Yasumori A. Effect of silica additive on the anatse to rutile phase transition. Journal of the American Ceramic Society. 2001;**84**(7):1591–1596. DOI: 10.1111/j.1151-2916.2001.tb00882.x

[49] Reidy DJ, Holmes JD, Morris MA. Preparation of a highly thermally stable titania anatase phase by addition of mixed zirconia and silica dopants. Ceramics International. 2006;**32**(3):235–239. DOI: 10.1016/j.ceramint.2005.02.009

[50] Badli NA, Ali R, Yuliati L. Influence of zirconium doped titanium oxide towards photocatalytic activity of paraquat. Advanced Materials Research. 2015;**1107**:377–382. DOI: 10.4028/www.scientific.net/AMR.1107.377

[51] Kitiyanan A, Ngamsinlapasathian S, Pavasupree S, Yoshikawa S. The preparation and characterization of nanostructured TiO_2-ZrO_2 mixed oxide electrode for efficient dye-sensitized solar cells. Journal of Solid State Chemistry. 2005;**178**(4):1044–1048. DOI: 10.1016/j.jssc.2004.12.043

[52] Ding Z, Hu X, Lu GQ, Yue P-L, Greenfield PF. Novel silica gel supported TiO_2 photocatalyst synthesized by CVD method. Langmuir. 2000;**16**(15):6216–6222. DOI: 10.1021/la000119l

[53] Ilkhechi NN, Kaleji BK. Effect of Cu2+, Si4+ and Zr4+ dopant on structural, optical and photocatalytic properties of titania nanopowders. Optical and Quantum Electronics. 2016;**48**(347):5–9. DOI: 10.1007/s11082-016-0621-z

[54] Joint Committee on Powder Diffraction Standards (JCPDS). International Center for Diffraction Data. Philadelphia; 2003.

[55] Inorganic Crystal Structure Database (ICSD). Version 1.3.1.Eggenstein-Leopoldshafen: FIZ Karlsruhe; 2003

[56] Rietveld HM. A profile refinement method for nuclear and magnetic structures. Journal of Applied Crystallography. 1969;**2**:65–71. DOI: 10.1107/S0021889869006558

[57] Young RA, Sakthive AL, Moss TS, Paiva-Santos CO. DBWS-9411—an upgrade of the DBWS programs for Rietveld Refinement with PC and mainframe computers. Journal of Applied Crystallography. 1995;**28**:366–367. DOI: 10.1107/S0021889895002160

[58] Young RA, Larson AC, Paiva-Santos CO. User's Guide to Program DBWS-9807a for Rietveld Analysis of X-ray and Neutron Powder Diffraction Patterns with a PC and Various Other Computers. Atlanta, GA: School of Physics Georgia Institute of Technology; 2000. DOI: 10.1.1.469.7262

[59] Strunk J, Vining WC, Bell AT. A study of oxygen vacancy formation and annihilation in submonolayer coverages of TiO_2 dispersed on MCM-48. Journal of Physics and Chemistry. 2000;**C**(114):16937–16945. DOI: 10.1021/jp100104d

[60] Paiva-Santos CO, Cavalheiro AA, Zaghete MA, Cilense M, Varela JA, Giotto MTS, Mascarenhas YP. An XRD study of the structure and microstructure of the laboratory synthesized crystals of $MgNb_2O_6$(NM) and PbMg1/3Nb2/3O3 (PMN). Advances in X-Ray Analysis. 2001;**44**:38–43. DOI: 10.1.1.296.1602

Chapter 4

Mechanically Activated Rutile and Ilmenite as the Starting Materials for Process of Titanium Alloys Production

Marcela Achimovičová, Christoph Vonderstein and Bernd Friedrich

Additional information is available at the end of the chapter

http://dx.doi.org/10.5772/intechopen.68747

Abstract

The consumptive conventional process of titanium alloys production needs new innovative processes. As starting materials for aluminothermic reduction, natural TiO_2 and $FeTiO_3$ concentrates can be used. The keynote of the present chapter is mechanical activation as a pre-treatment step for these concentrates, which is realized by the milling in a vibratory industrial mill. Mechanically activated rutile ore used in aluminothermic reduction saved 30% booster expenses and decreased Cl_2 emissions. Mechanical activation of ilmenite and ilmenite/aluminum mixtures was performed, and the kinetics of subsequent hydrometallurgical production of synthetic TiO_2 by pressure and normal leaching were studied. New processes with the coupling of aluminothermic production of titanium alloys were proposed.

Keywords: rutile, ilmenite, anatase, mechanical activation, aluminothermy, leaching

1. Introduction

The production of titanium alloys by using the conventional process route is tremendously cost-intensive. Therefore, there is a huge endeavor for an alternative process route. In an aluminothermic reduction process, titanium oxide can be used to achieve a titanium alloy. The energy which is needed for the autothermic reduction is released by the reduction of titanium oxides as well as by metal oxides which are needed for the titanium alloy composition. As the energy released by these oxides is not sufficient, there are boosters like $KClO_4$ or CaO_2 that are necessary. To reduce the amount of these boosters, the process of mechanical activation of titanium oxide (rutile, anatase, and ilmenite) is helpful.

1.1. Fundamentals of mechanical activation

According to Butyagin, the mechanical activation of solids is defined as an increase in reaction ability due to stable changes in solids structure [1]. Mechanical activation represents a multi-step process with changes in the energetic parameters and the amount of accumulated energy of solids in each step, and it is followed by the process of the defects accumulation, amorphization, formation of metastable polymorphous forms, and chemical reactions [2]. Depending on the number of solid phases involved, the solid-state reactions by mechanical activation are classified into single (homogeneous or inhomogeneous reactions) and multi-phase systems (heterogeneous reactions) (see **Table 1**). The homogeneous single-phase solid-state reactions are the defect reactions and inhomogeneous reactions are inter-solid diffusions with the concentration gradients. During the heterogeneous multi-phase solid-state reactions, the mass transport diffusion occurs across the phase boundaries, and one or more product phases are created [3–6].

1.2. Type of milling devices

The main problem for the technical application of the mechanical activation of minerals or ores is a suitable type of the mill. For initiation of mechanical activation, the impact load mechanism is required. This is performed in the vibratory mills that are manufactured in industrial scale. For mechanical activation examination only the lab-scale mills are commonly used. They are based on the pressure load mechanism due to centrifugal force of motion. An industrial eccentric vibratory mill (type: ESM 656–0.5 ks, Siebtechnik GmbH, Germany) for the mechanical activation of TiO_2 (rutile) and $FeTiO_3$ (ilmenite) concentrates was used (see **Figure 1**).

Type of reaction	Example
Homogeneous	$\gamma TiO_2 \xrightarrow{\beta} \alpha TiO_2$
Inhomogeneous	$\alpha TiO_2 \xrightarrow{\beta} \alpha TiO_{2-x}$
Heterogeneous	$2FeTiO_3 + 2.5O_2 \rightarrow 2TiO_2 + Fe_2O_3$
	$TiO_2 + MgO \rightarrow MgTiO_3$
	$Ti + H_2 \leftrightarrow TiH_2$

α: rutile, β: anatase, and γ: amorphous TiO_2

Table 1. Solid-state reactions of the Ti systems carried out by mechanical activation.

Figure 1. Industrial eccentric vibratory mill (type: ESM 656–0.5 ks, Siebtechnik GmbH, Germany).

2. TiO_2 concentrate, rutile

As input material, rutile concentrate of Australian origin (95% TiO_2) is used with the following composition 57% Ti, 0.7% Zr, 0.7% Fe, 0.3% Nb, 0.2% Si, 0.1% Al, 0.1% Cr, <0.03% P, and <0.03% S.

2.1. Mechanical activation

In order to identify the reaction ability of the mechanically activated rutile at the subsequent metallurgical reaction, the mechanical activation degrees, I/I_0 are assigned. The ratios I/I_0 represent the X-ray diffraction intensities at the lattice plane (110) of rutile, whereas I_0 is the measured value for the untreated rutile (defined as 100% < 10 μm), and I is value for mechanically activated rutile. This ratio I/I_0 is a parameter that gives information about grain size, lattice defects, and solid-state reactions. The results in **Figure 2** showed the activation degree of the rutile concentrate milled for 1, 2, and 3 h was dependent on the specific energy consumption of milling. The degree of crystallinity, I/I_0 of rutile crystal structure, decreased from 0.7 to 0.25, with an increase in milling time [6].

2.2. Aluminothermic reduction process for titanium alloys production

The materials with a decreased crystallinity due to mechanical activation can be very useful for the aluminothermic reduction process to reduce the amount of boosters. An aluminothermic reaction presents the reduction of a metal oxide with aluminum as a reductant. The reduction is feasible when Al shows a greater chemical affinity for the non-metal element of the compound than the desired metal which should be reduced. Regarding the redution of TiO_2 with

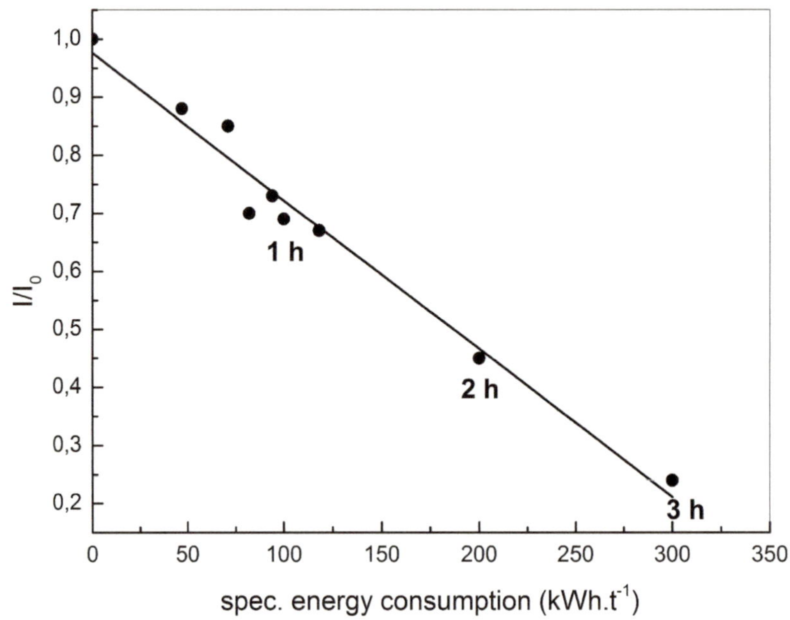

Figure 2. The mechanical activaction degree dependence on the specific energy consumption for rutile concentrate.

Al, the intrinsic heat of the reaction is not sufficient to maintain the self-sustained reaction [7]. Therefore, boosters (e.g., $KClO_4$, CaO_2) are added to reach the needed energy density. The modeling of the aluminothermic reaction mixture needs input parameters such as the required energy density, targeted adiabatic temperature, estimated heat losses, as well as aimed slag composition in order to improve the metal/slag separation. In order to reach the targeted temperature, changes in the mixture are not allowed to effect or vary the product alloy composition [7]. Because of little difference in the energy density from 50 to 100 $J.g^{-1}$ of aluminothermic reaction, the process window has to be exactly defined to supply slow propagation of reaction and early solidification. Rutile ore concentrate contains additional by-components which make the determination of the process conditions more complex. **Table 2** gives an overview of the reduction reactions of the by-components included in rutile concentrate [6].

The heats of these reactions differ significantly. Although the amount of these by-components are not very high, the released heats of each reaction have to be taken into account due to the above explained narrow process window. The reduction of several by-components is more favorable than the reduction of TiO_2 thus remaining within the metal phase. Therefore, the composition of the master alloy Ti-6Al-4V was carefully chosen, which corresponds to 60 wt.% of Ti, 24 wt.% of Al, and 16 wt.% of V. The investigations were focused on the definition of the required energy charging for the aluminothermic reaction with mechanically activated rutile to obtain a stable product. Important is the composition of the product, especially the contents of Ti, Al, and V and the by-components as well as the oxygen content in order proceed in the upcoming refining steps to achieve a valuable titanium alloy that can be produced cost-efficiently and can be used therefore in the car industry and other light-weight applications. It has to be balanced

Equations of the by-component reactions	$-\Delta H_R^0$ in kJ
$V_2O_5 + \frac{10}{3}Al \rightarrow 2V + \frac{5}{3}Al_2O_3$	1242.2
$SiO_2 + \frac{4}{3}Al \rightarrow Si + \frac{2}{3}Al_2O_3$	206.4
$Fe_2O_3 + 2Al \rightarrow 2Fe + Al_2O_3$	849.9
$Cr_2O_3 + 2Al \rightarrow 2Cr + Al_2O_3$	548.5
$ZrO_2 + \frac{4}{3}Al \rightarrow Zr + \frac{2}{3}Al_2O_3$	19.6
$Nb_2O_5 + \frac{10}{3}Al \rightarrow 2Nb + \frac{5}{3}Al_2O_3$	893.2
$MnO + \frac{2}{3}Al \rightarrow Mn + \frac{1}{3}Al_2O_3$	173.6

Table 2. Heats of reduction per mole of oxides of by-components in rutile ore.

out on how much booster can be saved by charging a reasonable mechanically activated rutile, minimizing the introduced energy for milling and maximizing the saved amount of booster [6]. Based on the promising results of the preliminary trials, further trials were conducted in small (8.4 kg of mixture) and mid-scale (18.2 kg of mixture) levels, with a variation in the reaction parameters such as time of mechanical activation of rutile concentrate (0, 1, 2, and 3 h), grain size of Al (90–300, 500–800, and 700–1200 μm), and $KClO_4$ addition to achieve energy density of 2400 and 2500 $J.g^{-1}$ for the aluminothermic reaction. The selected range for variation of each reaction parameter is shown in **Figure 3**.

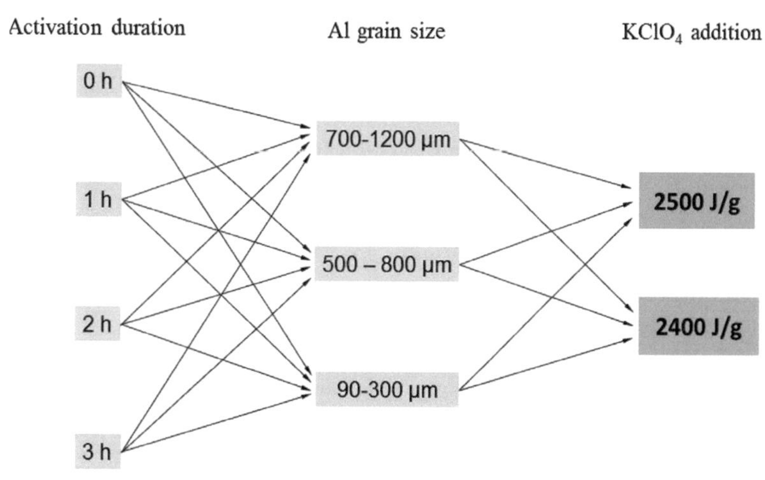

Figure 3. Variation of the trial parameters for the aluminothermic experiments [6].

The non-activated rutile (0 h) was used to compare the efficiency of the activation. The activation duration varies between 1 and 3 h. There are three types of Al grain sizes. Due to pre-trials, the two energy densities of 2400 and 2500 J.g^{-1} are used by $KClO_4$ addition [6]. This experimental setup indicates clearly that the trials with 2 h activation time and fine Al particles showed the best results according to the reaction time of the aluminothermic reduction and a good metal-slag separation. The metal composition was also the best in these trials. Small amounts of Fe and Si could be detected which would not interfere the final target alloy, Ti-Al6-V4 [6]. Because of avoiding the production of TiO_2 via chlorination, the savings were up to 50%, and because of mechanical activation, the use of $KClO_4$ decreased by 30% remarkably [6]. Due to the fact that lime is used as slag component to decrease the liquid temperature of the final slag, there are investigations to avoid the $KClO_4$ completely by using CaO_2 which will be reduced by Al to CaO and remains as slag component without producing any gas emissions. As the released heat by $KClO_4$ is nearly four times as high as the one by CaO_2, there is much more CaO_2 required to reach the needed energy density. Due to thermodynamic calculations, the amount of CaO should not exceed 50% because the liquid temperature of the slag will increase again and because of economic reasons, it should be used as sparsely as possible. As the energy density is decreased by mechanical activation, it is theoretical possible to reach a good amount of CaO in the slag and the needed energy density.

First trials were conducted with a 2 h mechanically activated rutile ore with CaO_2 as booster, varying the energy density. Besides the four trials which were conducted with an energy density from 2250 up to 2350 J.g^{-1}, the results were comparable to the experimental trials before. Due to the lower energy density in these trials, the metal-slag separation was not sufficient which resulted in high oxygen content in the alloy. The results for the oxygen content for the various trials varying the energy density can be seen in **Figure 4**.

A closer look to the trials with an energy density of 2450 J.g^{-1} shows that low oxygen contents can be reached. But there is also a large deviation for the oxygen content. This leads to the

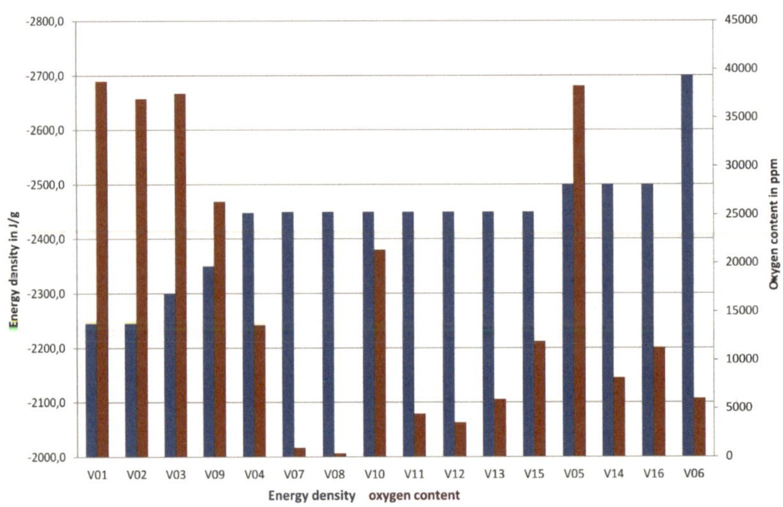

Figure 4. The various aluminothermic trials of the mechanically activated rutile for 2 h and CaO_2 as a booster.

assumption that the oxygen may not be completely diluted. Instead, there could be small oxide particles in the metal fraction which is a result of the slag-metal separation. Further investigations have to be performed. Therefore, a bigger-scale experiment to improve the separation needs to be done as well. Remelting this material in a vacuum induction furnace with a special ceramic crucible will be investigated to homogenize the material.

Mechanically activated rutile ore can be used for aluminothermic reduction to produce a titanium master alloy. $KClO_4$ can be used, as well as CaO_2 as booster. In this regard, a remarkable amount of KCl gas could be avoided. Using just CaO_2 as a booster could avoid the production of KCl completely.

3. $FeTiO_3$ concentrate, ilmenite

The beneficiation of ilmenites requires pyrometallurgical or chemical process steps to separate the iron content of approximately 30%. There are some methods of iron separation from $FeTiO_3$, which vary in their technical and energy demands. The chemical sulfate process, with H_2SO_4 at < 220°C, yielding TiO_2 in pigment quality, is difficult and complex because of the low solubility of ilmenite in H_2SO_4 [3]. Nevertheless, for the production of synthetic TiO_2 concentrates except for the pyrometallurgical processes [8, 9], the connected pyro- and hydrometallurgical processes [10–13] were developed. One possibility of the direct hydrometallurgical processing of $FeTiO_3$ is its pre-treatment by mechanical activation [14]. The studies on the solubility of $FeTiO_3$ after mechanical activation using different mills for ultrafine grinding such as vertical ball mills (attritors), planetary ball mills, and drum mills have been published, in which the structural changes of $FeTiO_3$ without technical applicability was focused [15–17]. In general, due to mechanical activation, the solids are exposed to high mechanical stress, which is responsible for their specific surface area increase, the crystalline structure defects formation and leads to the enthalpy increase. Hence, mechanically activated solids and minerals with low solubility are more leachable in subsequent hydrometallurgical process [18]. The application of mechanical activation of $FeTiO_3$ by energy-efficient milling as a pre-treatment step in the hydrometallurgical process of synthetic TiO_2 concentrates production might be a new realization to utilize such TiO_2 materials (~95%) in aluminothermic Ti alloys generation.

3.1. Mechanical activation and the kinetics of subsequent hydrometallurgical production of synthetic TiO_2 by pressure leaching

The investigations were carried out with $FeTiO_3$ concentrates (>95% $FeTiO_3$, <5% SiO_2) of Russian origin (GMD, Mineral Trade Company). The chemical composition was as follows: 34.43% Fe, 30.02% Ti, 0.76% Si, 0.47% Mg, 0.42% Al, 0.34% Mn, 0.11% Zn, 0.09% Ca, 0.07% Cr, 0.06% Co, 0.03% Ba, 31.63% O, and 1.57% insoluble rest.

For determination of the optimal milling conditions, a parameter study was done: mill feed quantities varied from 100 to 300 g/charge, activation times ranged from 15 to 60 min, the amplitude of inhomogeneous vibrations was 20 mm, and the revolutions of the motor of the mill, 960 min^{-1}, were constant. The steel balls of 30 mm diameter were applied. Activation degrees and the ratios I/I_0 at lattice plane (104) of $FeTiO_3$ after and before mechanical activation as a function of mechanical activation time are shown in **Figure 5**. It is obvious that the ilmenite structure is strongly strained by the mechanical activation.

The leaching of mechanically activated $FeTiO_3$ was performed in an autoclave, volume 2 l (Deutsch & Neumann, Germany). The following conditions were used: initial H_2SO_4 concentration of 10–30%, temperature of 100–150°C, leaching time of 15–90 min, Fe addition of 6–12%, the solid to liquid ratio of 50–200 g.l^{-1}, and stirring rate of 250 min^{-1}.

The influence of the activation time of $FeTiO_3$ on the TiO_2 extraction to the precipitation product is shown in **Figure 6**. The curve for the total Fe extraction in the product reflects the dissolution of $FeTiO_3$. The leaching tests confirm the dependence on the activation degree of $FeTiO_3$. A critical point is that, after 15 min of mechanical activation, most of $FeTiO_3$ is dissolved. This finding is extremely interesting from a technical point of view. The hydrolytic precipitation of TiO_2 is influenced by an initial dissolution accelerated with increasing activation.

At temperatures >50°C and a pH value >1.5, the hydrolysis of titanyl sulfate solution, $TiOSO_4$ to TiO_2, is triggered. In the investigated temperature range of 100–150°C, the dissolution of $FeTiO_3$ and the simultaneous precipitation of TiO_2 take place in parallel.

The leaching time was varied in a range from 15 to 90 min. A general fact is that the dissolution rate of $FeTiO_3$ is fast. **Figure 7** shows the dissolution of ~64% $FeTiO_3$ (corresponding to the dissolution of Fe) during 30 min of leaching at 120°C. By increasing the reaction temperature to 150°C, ~86% of $FeTiO_3$ dissolves during 30 min of leaching. It results from **Figure 7** that the leaching time had a relatively low effect on the hydrolysis. After 60 min of leaching, the recovery

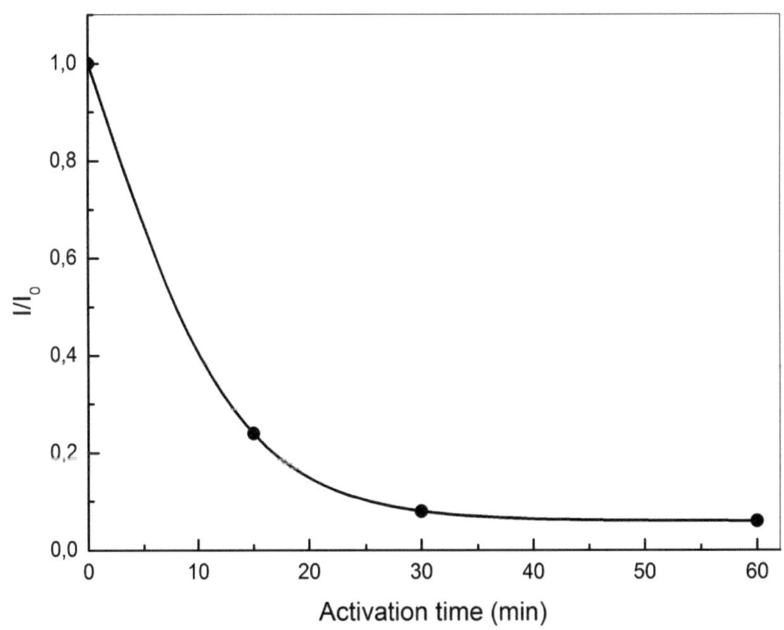

Figure 5. Activation degrees I/I_0 of $FeTiO_3$ versus mechanical activation time.

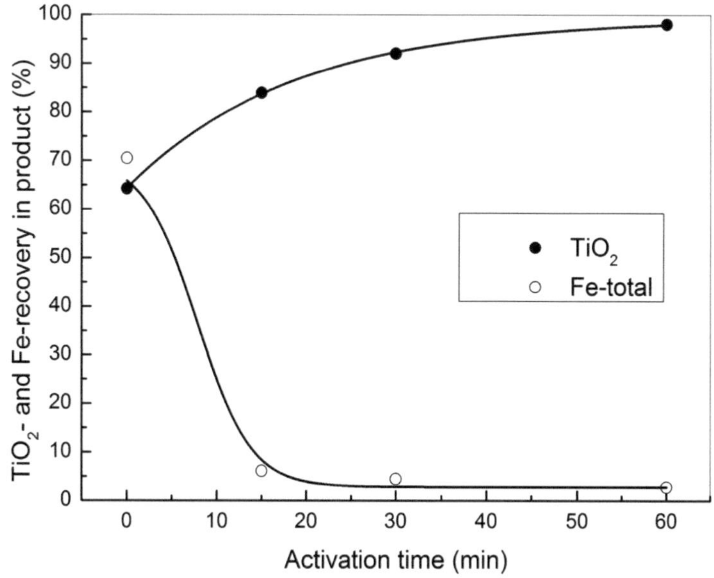

Figure 6. Influence of the mechanical activation time of $FeTiO_3$ on the precipitation of TiO_2; leaching temperature: 150°C, s/l ratio: 50 g.l^{-1}, leaching time: 60 min, H_2SO_4 : 30%, and Fe powder: 12%.

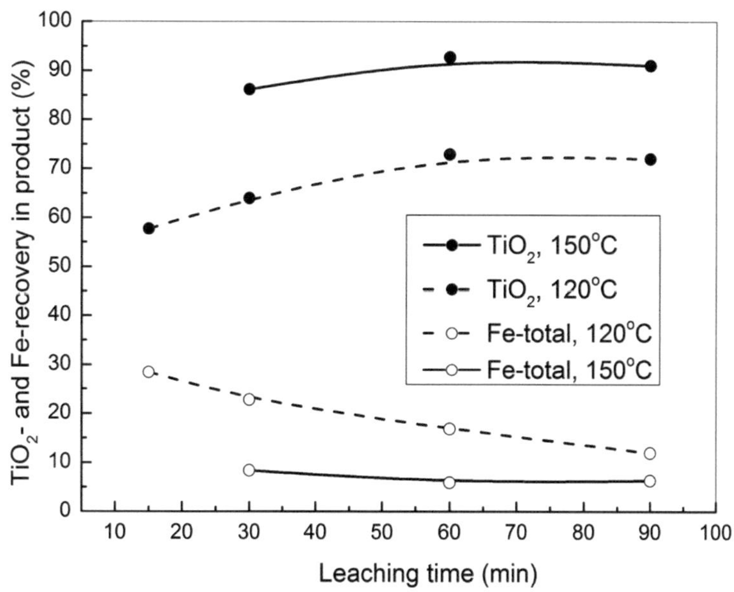

Figure 7. Influence of the leaching time and the temperature on the precipitation of TiO_2; s/l ratio: 50 g.l^{-1}, mechanical activation time: 15 min, H_2SO_4: 30%, and Fe powder: 6%.

of precipitated TiO_2 achieved ~93% and the content of Fe in TiO_2 decreased to <6%. The synthetic TiO_2 product assigned a relatively high purity. By decreasing the leaching temperature from 150 to 120°C, the obtaining of the high-quality synthetic TiO_2 concentrate is impossible [19].

It was detected that the addition of Fe powder, used as a reduction agent, influences the initial rate of $FeTiO_3$ dissolution (86% with an addition of 6% Fe and 89–92% with an addition of 12% Fe) and after a leaching time of 60 min, the influence of the Fe additive is no longer identifiable [19].

Sulfate process requires ~2 tons of concentrated H_2SO_4 per 1 ton of $FeTiO_3$ for digestion. Subsequently, the generated digestion cake is leached with H_2O by pH < 1.5. Therefore, the direct leaching of mechanically activated $FeTiO_3$ carried out by a described procedure requires a solid/acid ratio of 1:2 at least, which corresponds to 10% H_2SO_4 and 77% dissolution of $FeTiO_3$ (mechanically activated for 15 min) after 60 min of leaching, at 150°C, with addition of 10% Fe. By increasing the initial acid concentration to 20%, which corresponds to a ratio of 1:4, $FeTiO_3$ dissolves to 89%. For technical dimensioning, the ratio of ilmenite to acid should be <1:4. An excessively high amount of acid would complicate the hydrolytic conditions.

The solid contents from 50 to 200 $g.l^{-1}$ were investigated. In order to show the influence of mechanical activation on the leaching of $FeTiO_3$ with different s/l ratios, two test series with activation times of 15 and 30 min were carried out. As expected, the residual Fe content in the product increases to 7% with increasing s/l ratio to 200 $g.l^{-1}$ at a higher activation time. That means the activation time, 15 min, is sufficient to achieve 92% of dissolution of $FeTiO_3$ with s/l ratio of 200 $g.l^{-1}$ and with 30% H_2SO_4 at 150°C.

According to literature, the research on the leaching of the mechanically activated ilmenites demonstrated no technical applicability. For various types of the mills, the milling times of up to 200 h were used, and low s/l ratios (only 10 $g.l^{-1}$) for the leaching process were employed [15–17, 20, 21]. The operating conditions determined in this investigation fulfill the requirements for a technical implementation of the process for the production of a synthetic TiO_2 product (95.23% TiO_2, 3.32% Fe_2O_3, 1.7% SiO_2, 1.2% CaO, 0.34% Al_2O_3, 0.007% P_2O_5, and 0.004% ZrO_2), which is suitable for use in aluminothermic alloys production. **Figure 8** shows the process flowsheet on the coupling of hydrometallurgical processing of $FeTiO_3$ into synthetic TiO_2 concentrate (anatase) with the aluminothermic production of TiAl alloys.

The energy required per ton of synthetic anatase is expected to be 506 kWh. Based on current costs for energy, this corresponds to approximately 212 €/t of synthetic anatase. This investigated pre-treatment of ilmenite with mechanical activation and leaching enables new cost-effective production methods for titanium-based alloys. So far, only high-purity rutile pigments have been used for the aluminothermic production of TiAl alloys [7, 22]. At 95% TiO_2, the synthetic TiO_2 concentrate (anatase) that we produced meets the requirements for aluminothermy.

3.2. Mechanical activation with metal addition and the kinetics of subsequent hydrometallurgical production of synthetic TiO_2 by normal leaching

Mechanical activation of $FeTiO_3$ with Al powder in the stoichiometric ratio 1:2 caused its mechanochemical reduction already after 120 min. The product phases found on the thermodynamic calculations with 1100–1700°C (Al_2O_3, TiO, Fe_2Ti, and FeTi) have already appeared after 360 min of mechanochemical reduction. Such mechanochemical processing of $FeTiO_3$

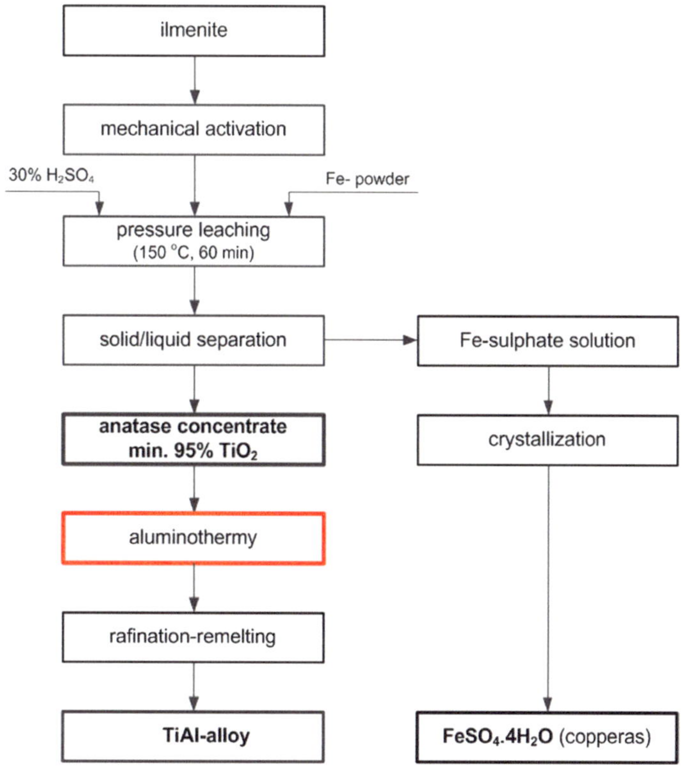

Figure 8. A process flowsheet of hydrometallurgical processing of $FeTiO_3$ connected with the aluminothermic production of titanium alloys.

concentrate decreases the temperature and subsequently the time of thermal reduction of $FeTiO_3$, which could be also used in the titanium alloys production [23]. For subsequent leaching in H_2SO_4 (40%) at 50°C, mechanically activated $FeTiO_3$ with Al in ratio 1:0.3 was tested. In this case, the necessary agglomerates of $FeTiO_3$/Al are created even after 15 min of mechanical activation. **Figure 9** shows the measured grain size distribution of the unmilled $FeTiO_3$ and the activated $FeTiO_3$/Al mixtures.

The SEM image in **Figure 10** shows the formed agglomerates. The contact pressure caused by the impact stress on the agglomerates during the second stage of ultrafine milling (agglomeration stage) leads to structures similar to briquettes with the highest bulk density [24, 25].

By leaching of agglomerated $FeTiO_3$/Al mixture with diluted H_2SO_4, the highly reactive atomic hydrogen (in nascent state) is created at the contact areas between $FeTiO_3$ and Al, which immediately reacts and causes a partial reduction of the quadrivalent titanium to trivalent titanium. A Ti^{3+}/Ti^{4+} dark violet to black solution is generated. The summation equation of the conversion of $FeTiO_3$ with diluted H_2SO_4 in the presence of hydrogen in nascent state is [24]:

$$2\mathrm{Fe(II)Ti(IV)O_3} + 5\,\mathrm{H_2\,SO_4} + 2\,\mathrm{H^{o}} \rightarrow 2\,\mathrm{Fe^{2+}} + 2\,\mathrm{SO_4^{2-}} + 2\,\mathrm{Ti^{3+}} + 3\,\mathrm{SO_4^{2-}} + 6\,\mathrm{H_2\,O} \quad (1)$$

Figure 11 shows that the maximum Ti recovery of activated $FeTiO_3$ reaches 23%. Activated $FeTiO_3$/Al mixture shows a 53% Ti dissolution after only 15 min of activation.

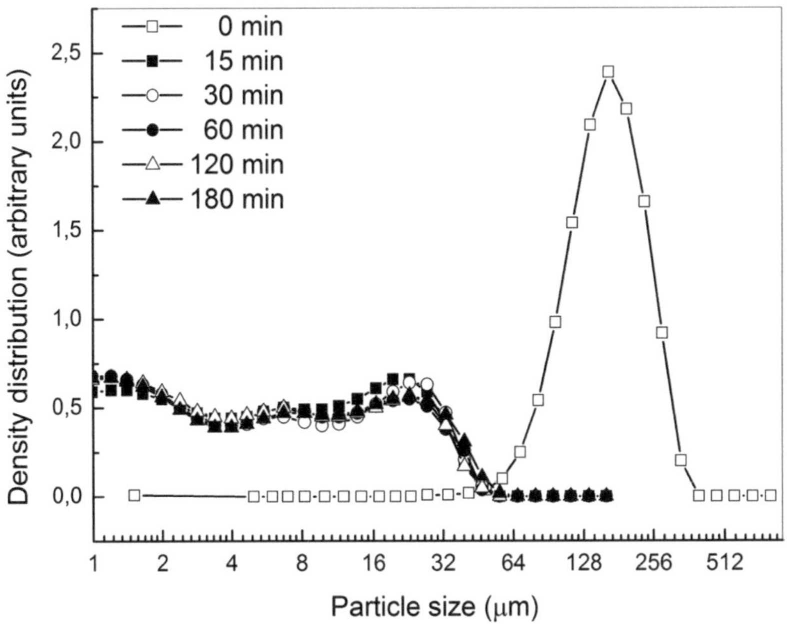

Figure 9. Grain size distributions of $FeTiO_3$/Al mixtures with various times of mechanical activation.

Figure 10. SEM image of formed $FeTiO_3$/Al agglomerates.

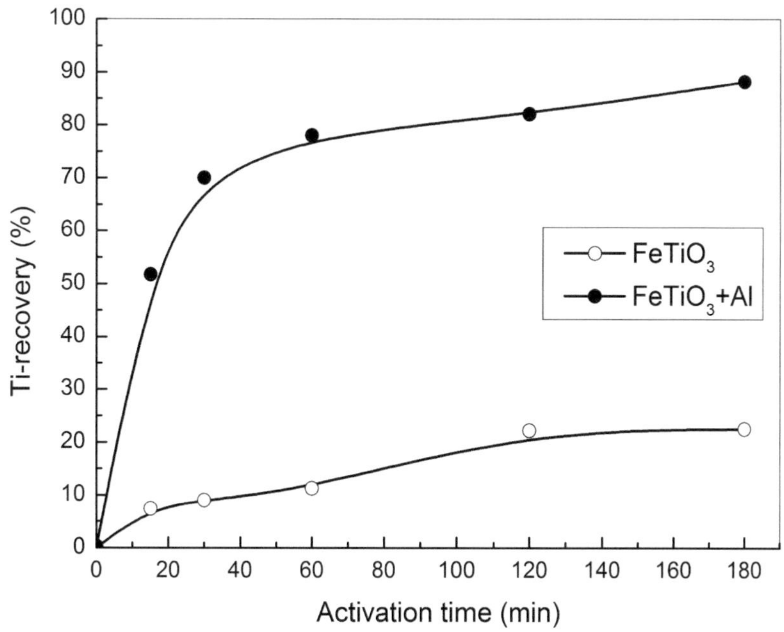

Figure 11. Influence of mechanical activation time of $FeTiO_3$ and $FeTiO_3$/Al mixtures on Ti extraction during 60 min of leaching.

Investigations on the influence of the initial H_2SO_4 concentration on the leaching of $FeTiO_3$/Al mixture (150 $g.l^{-1}$) activated for 60 min, with the use of 30, 40, 50, or 60% H_2SO_4, show a maximum Ti recovery of 79% after a leaching time of 60 min and the use of 40% acid, corresponding to a solid/acid ratio of 1:3.3. With the increase of the initial H_2SO_4 to 50% (solids/acid ratio 1:4.3), dissolved Ti^{3+} partially hydrolyses, which causes the decrease of Ti recovery to 42%. For technical scale, the solid/acid ratio should be <1:4 because an excess H_2SO_4 complicates the conditions of subsequent TiO_2 hydrolysis. Ti recovery after 60 min of leaching is ~78% when milling charge of 200 g was used. With a further increase to 300 g/charge, the Ti recovery decreases to 69%. For 600 g/charge, the Ti-recovery reaches to 56%, which is still a high value. Ti recovery for the unmilled $FeTiO_3$ was 0.3% only. A check of the s/l ratio per liter showed that the optimal ratio was 150 $g.l^{-1}$ of $FeTiO_3$/Al mixture. For technical processes, at least 300 $g.l^{-1}$ must be feasible. **Figure 12** evidences that the leaching rate of $FeTiO_3$/Al mixture at temperature 50°C is very fast. After 5 min of leaching, the Ti recovery was 69% and after 60 min of leaching, almost 80% of Ti was leached out.

In **Figure 13**, the flowsheet of the described process for production of synthetic TiO_2 was proposed [26]. In an open agitator vessel, $FeTiO_3$/Al mixture is dissolved at a temperature of 50°C since the strong bond in the briquetted mixture follows the shrinking core model [26]. Undesirable metals, Fe and Al, can be crystallized as a mixture of Fe, Al sulfates at temperature < 15°C. $Ti(OH)_3$ precipitates during hydrolysis and by calcination oxidizes to TiO_2 (anatase) with the following composition: 99% TiO_2, 0.59% Fe_2O_3, <0.1% Al_2O_3, <0.2% SiO_2, <0.03 ZrO_2, and 0.01% Cr_2O_3.

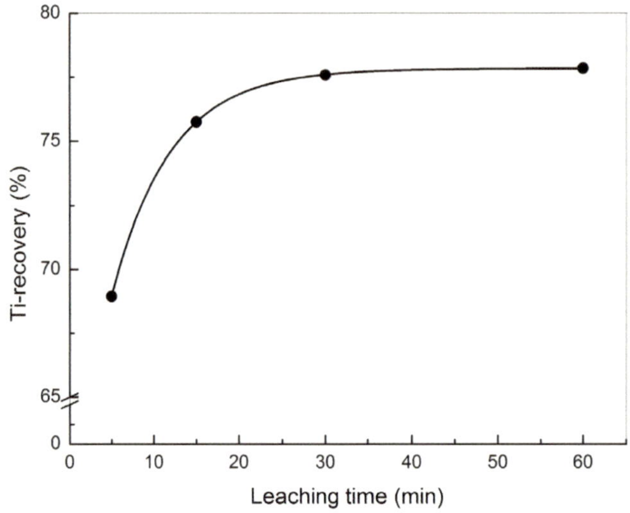

Figure 12. Influence of leaching time on Ti extraction of mechanically activated $FeTiO_3$/Al mixtures for 60 min with 40% H_2SO_4 and s/l ratio as 150 g.l^{-1}.

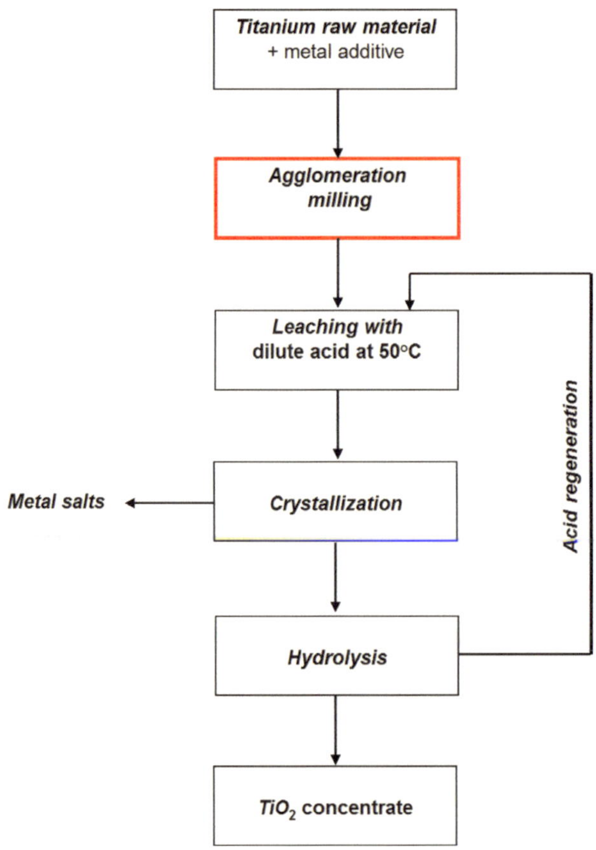

Figure 13. The flowsheet of a proposed process for production of synthetic TiO_2.

4. Conclusion

New technological processes of using titanium oxide concentrates (rutile, ilmenite) as starting materials for titanium alloys production were proposed. The first one is based on mechanical activation of natural rutile concentrate with milling, which caused the reducing of booster ($KClO_4$) amount around 30% and saving expenses for previous chlorination in subsequent aluminothermic reductions. Promising steps to replace the $KClO_4$ could be observed but are still under further investigations. The second process produces synthetic TiO_2 (anatase) from mechanically activated ilmenite concentrates treated by pressure leaching in dilute H_2SO_4 at 150°C. The third process presents the improvement of the previous one because ilmenite concentrate can be hydrometallurgically processed to synthetic TiO_2 by normal leaching in dilute H_2SO_4 at 50°C. Ilmenite concentrate has to be pre-treated by mechanical activation with aluminum as a reducing agent which leads to formation of $FeTiO_3$/Al agglomerates and a new effective reaction mechanism of leaching. After the separation of iron and aluminum by means of crystallization, it is possible to extract synthetic TiO_2 as a raw material for aluminothermic production of titanium alloys.

Acknowledgements

The authors would like to express their gratitude to the Deutsche Forschungsgemeinschaft DFG for the financial support of the project FR 1713/31-1.

Author details

Marcela Achimovičová[1,2]*, Christoph Vonderstein[3] and Bernd Friedrich[3]

*Address all correspondence to: marcela.achimovicova@tu-clausthal.de

1 Institute of Mineral and Waste Processing, Waste Disposal and Geomechanics, University of Technology-Clausthal, Germany

2 Institute of Geotechnics, Slovak Academy of Sciences, Košice, Slovakia

3 IME Process Metallurgy and Metal Recycling, RWTH Aachen University, Germany

References

[1] Butyagin PY. Structure disordering and mechanochemical reactions in solid states. Uspechi Khimiji. 1984;**53**(11):1769-1789

[2] Boldyrev VV, Tkáčová K. Mechanochemistry of solids: Past, present, and prospects. Journal of Materials Synthesis and Processing. 2000;**8**(3):121-132. DOI: 10.1023/A:1011347706721

[3] Gock E, Kähler J, Vogt V. Produktionsintegrierter Umweltschutz bei der Aufbereitung und Aufarbeitung von Rohstoffen. In: Brauer H, editor. Handbuch des Umweltschutzes und den Umweltschutztechnik. Vol.2. Berlin-Heidelberg-New York: Springer; 1996. pp. 79-237

[4] Gock E, Jacob KH. Direkter Aufschluß von Rutil mit Schwefelsäure. Erzmetall. 1980;**33**(6): 308-314

[5] Gock E, Szantho E. Laugung von Rutil nach mechanischer Aktivierung in einer Schwingmühle. Erzmetall. 1970;**23**:165-168

[6] Hassan-Pour S, Vonderstein C, Achimovičová M, Vogt V, Gock E, Friedrich B. Aluminothermic production of titanium alloys (Part 2): Impact of activated rutile on process sustainability. Metallurgical and Materials Engineering. 2015;**21**(2):101-114

[7] Stoephasius JC, Friedrich B. Modelling of metallothermic reactions—Impact of energy effects on the calculation of the reaction mixture. Erzmetall, World of Metallurgy. 2004;**57**(4):217-224

[8] Noda T. Titanium from slag in Japan.The Journal of the Minerals, Metals and Materials Society. 1965;**17**(1):25-32

[9] Lee HY, Poggi D. Mine, mill and smelting complex at Richards Bay, South Africa. The Metallurgical Society of CIM. 1978;**Annual**:93-96

[10] Berkovich SA. Recovery of titanium from ores. US patent 3903239. 1975

[11] Duyvesteyn W, Sabachy B, Verhulst D, Edmund V, Westsells PG. Processing titaniferous ore to titanium dioxide pigment. WO Patent 2001/00531 A1. 2001

[12] Thomas JB. Titanium dioxide pigment production from ilmenite. EU Patent 0186370. 1985

[13] Lakshmanan VI, Sridhar R, Rishea MM, Joseph DE, Laat R. Production of titanium metal from titanium-bearing ores involves leaching and selective removal of iron values by solvent extraction. US Patent 2001/007646-A1. 2001

[14] Gock E. Erzeugung von synthetischem Anatas aus Ilmeniten mit Dünnsäure. DE patent 3635010 A1. 1986

[15] Welham NJ, Llewellyn DJ. Mechanical enhancement of the dissolution of ilmenite. Minerals Engineering. 1998;**11**(9):827-841 DOI: 0892-6875(98)00070-3

[16] Li C, Liang B, Guo LH, Wu ZB. Effect of mechanical activation on the dissolution of Panzhihua ilmenite. Minerals Engineering. 2006;**19**(14):1430-1438. DOI: 10.1016/j.mineng.2006.02.005

[17] Sasikumar C, Rao DS, Srikanth S, Mukhopadhyay N, Mehrotra SP. Dissolution studies of mechanically activated Manavalakurichi ilmenite with HCl and H_2SO_4. Hydrometallurgy. 2007;**88**(1-4):154-169. DOI: 10.1016/j.hydromet.2007.03.013

[18] Baláž P. Extractive Metallurgy of Activated Minerals. 1st ed. Amsterdam: Elsevier; 2000. p. 292

[19] Achimovičová M, Hassan-Pour S, Gock E, Vogt V, Baláž P, Friedrich B. Aluminothermic production of titanium alloys (Part 1): Synthesis of TiO_2 as input material. Metallurgical and Materials Engineering. 2014;**20**(2):141-154

[20] Li C, Liang B, Song H, Xu J, Wang X. Preparation of porous rutile titania from ilmenite by mechanical activation and subsequent sulphuric acid leaching. Microporous and Mesoporous Materials. 2008;**115**:293-300. DOI: 10.1016/j.micromeso.2008.01.045

[21] Tao T, Glushenkov AM, Chen Q, Hu H, Zhou D, Zhang H, Boese M, Liu S, Amal R, Chen Y. Porous TiO_2 with a controllable bimodal pore size distribution from natural ilmenite. CrystEngComm. 2011;**13**:1322-1327. DOI: 10.1039/C0CE00533A

[22] Stoephasius JC, Friedrich, B. Modelling of metallothermic reactions local reaction rates during aluminothermic γ-TiAl-Nb production. Erzmetall-World of Metallurgy. 2005;**58**(2):63-69

[23] Achimovičová M, Gock E, Turianicová E, Kostova NG, Velinov N, Kaňuchová M, Baláž P. Study of the mechanochemical reduction of ilmenite concentrate by addition of aluminum. Acta Physica Polonica A. 2014;**126**(4):867-870

[24] Gock E, Vogt V, Achimovičová M. Verfahren zum behandeln eisenhaltiger Titan-Rohstoffe. DE Patent 102014206776 A1. 2015

[25] Gock E, Vogt V, Achimovičová M. Procedures for treating ferrous titanium raw materials. International Journal of Research in Chemical, Metallurgical and Civil Engineering. 2016;**3**(1):49-53. DOI: 10.15242/IJRCMCE.IAE0316403

[26] Braun RL, Lewis AE, Wadsworth M. In-place leaching of primary sulfide ores: Laboratory leaching data and kinetics model. In: Aplan FF, editor. Solution Mining Symposium. Dallas: 103 AIME Annual Meeting; 1974. pp. 295-323

Chapter 5

Hydrogen Reduced Rutile Titanium Dioxide Photocatalyst

Fumiaki Amano

Additional information is available at the end of the chapter

http://dx.doi.org/10.5772/intechopen.68603

Abstract

For TiO_2 photocatalysts, recombination of photoexcited electrons and holes would occur in crystalline defects such as oxygen vacancies, Ti^{3+} ions, and surface states. Therefore, it is believed that the density of crystalline defects should be decreased to improve the photocatalytic activity of TiO_2 particles. Contrary to this common knowledge, the introduction of crystalline defects by hydrogen reduction treatment is shown to increase the lifetime of photoexcited electrons in rutile TiO_2 photocatalysts with an increase of n-type electrical conductivity. The photocatalytic activities of H_2-reduced rutile TiO_2 were higher than those of anatase TiO_2 and mixed-phase TiO_2. This chapter explains the effect of donor doping on the photocatalytic activity of rutile TiO_2, the relationship between its physicochemical properties and photocatalytic performances, and the mechanism of the enhanced activity of H_2-reduced TiO_2. Particle size dependence on the enhanced activities suggests the formation of a space charge layer in large TiO_2 crystallites.

Keywords: rutile, n-type semiconductor, conduction electron, donor density, oxygen vacancy

1. Introduction

For utilization of light energy including solar light, the development of highly active photocatalytic materials is expected. Mainly oxide photocatalysts have been studied so far, but the design guideline did not become clear even in the case of representative TiO_2. In crystalline type of TiO_2 photocatalysts such as anatase, rutile, and brookite, it is believed that anatase TiO_2 is active based on the studies such as the oxidative degradation of organic compounds, H_2 evolution from water, and photoinduced super hydrophilicity. When anatase TiO_2 is annealed to decrease the density of crystalline defects, the crystal structure is transformed into a thermodynamically stable rutile phase. Generally speaking, the photocatalytic performance of rutile-type

TiO_2 is inferior except for the case in specific photocatalyst reactions. The reasons of lower activity are attributed to the smaller BET-specific surface area, the energetically lower conduction band bottom, and the shorter lifetime of the photoexcited carriers, compared to those of anatase TiO_2.

When oxide photocatalyst absorbs light with energy larger than the bandgap, an electron is excited to the conduction band, and a positive hole generates in the valence band. The lifetime of the photoexcited electron and the hole must be long enough to promote reductive and oxidative reactions on the surface efficiently. However, most of the photoexcited carriers are deactivated by recombination at crystalline defects such as impurities and disorder of the atomic arrangement in the bulk, on the surface, and at the interface. Therefore, highly crystalline particles are considered to show high photocatalytic activity if the band structure and the BET-specific surface area are the same.

The photocatalytic activity of rutile TiO_2 is frequently low compared to that of anatase TiO_2 photocatalyst. However, Maeda recently revealed that a rutile TiO_2 can induce overall water splitting to evolve H_2 and O_2 under UV irradiation [1]. We have also found that photocatalytic activity of rutile TiO_2 was improved by H_2 reduction treatment. The reaction can be written using Kröger-Vink notation (**Table 1**). Hydrogen reduction of TiO_2 creates both oxygen vacancy and electrons as shown in Eq. (1). Therefore, this treatment is recognized as a donor doping. The electron is trapped in a Ti^{4+} lattice site to form Ti^{3+} ions (Eq. (2)). The enhanced activity by introducing lattice defects is against the common knowledge in photocatalyst chemistry: crystalline defects should be decreased.

$$O_O^{\times} + H_2 \rightarrow V_O^{\bullet\bullet} + 2e' + H_2O \tag{1}$$

$$Ti_{Ti}^{\times} + e' = Ti'_{Ti} \tag{2}$$

Despite the reputed lower activity of rutile TiO_2 than anatase, it is important to determine the physical properties affecting the photocatalytic efficiency of rutile TiO_2. This chapter explains the donor doping effect on rutile TiO_2 photocatalysts and the effect of H_2 reduction treatment

Notation	Meaning
$Ti_{Ti}^{\times}$	Ti^{4+} ion in titanium lattice site
$\mathbf{M}'_{Ti}$	$\mathbf{M}^{3+}$ ion in titanium lattice site
$\mathbf{M}_{Ti}^{\bullet}$	$\mathbf{M}^{5+}$ ion in titanium lattice site
e'	Conduction electron
$O_O^{\times}$	O^{2-} ion in oxygen lattice site
$OH_O^{\bullet}$	OH^- ion in oxygen lattice site
$V_O^{\bullet\bullet}$	Oxygen vacancy, with double positive charge

Note: **M** corresponds to metal cation.

Table 1. Kröger-Vink notation of species in TiO_2 lattice.

on the photocatalytic activity of rutile TiO_2. The properties of the H_2-reduced rutile TiO_2 were characterized using diffuse reflectance UV-vis-NIR spectroscopy, electron spin resonance (ESR) spectroscopy, sheet resistance measurements, and electrochemical Mott-Schottky analysis. The role of oxygen vacancies, Ti^{3+} species, and conduction band electrons in the enhancement of photocatalytic and PEC activities of H_2-reduced TiO_2 is discussed based on the experimental results.

2. Doping in oxide photocatalysts

For semiconductor materials in electronic use, the density of electrons and holes is controlled by the doping of impurities into crystalline materials. Doping means incorporation of foreign impurities into the crystalline lattice of the parent semiconductor, and it is significantly different from surface modification. The donor doping implies an introduction of electrons to create an n-type semiconductor. The n-type conductivity is increased by donor doping and decreased by an acceptor doping-introducing hole. The electrons in the conduction band and the trapping site are one of crystalline defects in a wide meaning.

In semiconductor photocatalysts, doping of impurities is used to control the band structures. An impurity level and a subband can be formed in the bandgap by substituting a different ion for an ion constituting a crystal, and it is applied for the development of visible light-responsive photocatalysts. However, photocatalytic activity under visible-light irradiation has not yet been in practical use because of the low quantum yield. The doped impurities frequently resulted in deactivation of the doped photocatalysts, suggesting that the dopants and the created defects work as a recombination center decreasing the lifetime of photoexcited electrons and holes.

2.1. Perovskite oxide photocatalysts

There are many papers reporting the enhancement of photocatalytic activity by doping of cations and anions, although some of the reports lack reproducibility and experimental evidence. For perovskite oxide photocatalysts, some research groups reported that the photocatalytic activity was enhanced by acceptor doping, that is, doping of cations with valence lower than that of the parent cations. It is known that oxygen vacancies are easily formed in perovskite oxides, which resulted in the increase of electron density (Eq. (3)) and the origin of n-type semiconductivity.

$$O_O^{\times} \rightarrow V_O^{\bullet\bullet} + 2e' + 1/2\, O_2 \tag{3}$$

The electrons are trapped in shallow midgap states near conduction band bottom and can be easily excited from the donor levels to the conduction band. During photoexcitation, photogenerated holes would recombine with the electrons in addition to the photoexcited electrons. Therefore, a higher accumulation of electrons in n-type oxide would result in lower photocatalytic activity. Actually, this is true for perovskite oxide photocatalysts such as $SrTiO_3$ and $KTaO_3$ [2–4].

Ishihara et al. reported that the doping of Zr^{4+} to $KTaO_3$ particles was effective for improving the activity for overall water splitting under UV irradiation [2, 3]. The photocatalytic activity of nondoped $KTaO_3$ loaded with a NiO cocatalyst ($NiO/KTaO_3$) was negligible, but the doping of a small amount of tetravalent cations such as Zr^{4+} increases the rate of photocatalytic H_2 and O_2 evolution. $KTaO_3$ was originally an n-type semiconductor, since the electrical conductivity was monotonically increased with decreasing oxygen partial pressure. The doping of acceptors would decrease electron density in $KTaO_3$ according to Eqs. (4) and (5). The electrical conductivity of $KTaO_3$ was decreased with an increase of the amount of doped Zr^{4+}, which resulted in the enhanced photocatalytic activity.

$$ZrO_2 \xrightarrow{KTaO_3} Zr'_{Ta} + 2O_O^{\times} + 1/2V_O^{\bullet\bullet} \tag{4}$$

$$V_O^{\bullet\bullet} + 1/2\,O_2 + 2e' \rightarrow O_O^{\times} \tag{5}$$

Takata and Domen also reported that acceptor doping effectively enhanced the photocatalytic activity of cocatalyst-loaded $SrTiO_3$ particles for overall water splitting [4]. Ga-doped and Na-doped $SrTiO_3$ exhibited photocatalytic activity higher than that of nondoped $SrTiO_3$ by about 10 times. Ga^{3+} occupies the Ti^{4+} site and Na^+ occupies the Sr^{2+} site, which resulted in the decrease of electron concentration (Eqs. (6) and (7)). In contrast, the doping of a higher valence cations (Ta and La) led to the suppressed photocatalytic activity of $SrTiO_3$ (Eqs. (8) and (9)). Here, Ta^{5+} occupies Ti^{4+} site, and La^{3+} occupies the Sr^{2+} site increasing the electron density of $SrTiO_3$. Since the electrons are trapped in Ti^{4+} sites to create Ti^{3+} species (Eq. (2)), the defect species most responsible for recombination would be Ti^{3+} species in n-type perovskite oxide photocatalysts.

$$Ga_2O_3 + 1/2\,O_2 + 2e' \xrightarrow{SrTiO_3} 2Ga'_{Ti} + 4O_O^{\times} \tag{6}$$

$$Na_2O + 1/2\,O_2 + 2e' \xrightarrow{SrTiO_3} 2Na'_{Sr} + 2O_O^{\times} \tag{7}$$

$$Ta_2O_5 \xrightarrow{SrTiO_3} 2Ta_{Ti}^{\bullet} + 4O_O^{\times} + 1/2O_2 + 2e' \tag{8}$$

$$La_2O_3 \xrightarrow{SrTiO_3} 2La_{Sr}^{\bullet} + 2O_O^{\times} + 1/2O_2 + 2e' \tag{9}$$

2.2. TiO_2 photocatalysts

The roles of dopants in TiO_2 photocatalysts are complicated and controversial, since it might depend on the particle size, the crystallinity, the crystalline phase, and reaction conditions. Generally speaking, impurities and crystalline defects work as recombination centers. However, there are some reports pointing that donor doping enhanced the photocatalytic activity of Pt-loaded TiO_2 [5, 6]. This trend for Pt/TiO_2 photocatalysts is opposite to the case of perovskite oxides, which is activated by acceptor doping.

Karakitsou and Verykios reported the effect of aliovalent cation doping to the TiO_2 matrix on the photocatalytic activity of Pt/TiO_2 for H_2 evolution [5]. Since the doped TiO_2 was prepared at 900°C, the crystalline structure was rutile phase and the particle size was large (BET-specific surface area, ~1 m^2 g^{-1}). In contrast to the case of perovskite oxides mentioned above, the doping of cations with valence higher than Ti^{4+} (W^{6+}, Ta^{5+}, and Nb^{5+}) enhanced the photocatalytic activity, while the opposite was observed for acceptor doping (In^{3+}, Zn^{2+}, and Li^{+}). The measurement of electrical conductivity revealed that the cation doping changed the bulk electronic structure [7]. The authors concluded that n-type conductivity correlates with the enhanced photocatalytic activity of Pt/TiO_2 with rutile form.

Ying et al. investigated the role of particle size in cation-doped TiO_2 nanoparticles with anatase crystalline structure [6]. For TiO_2 nanocrystals with an average diameter less than 11 nm, the doping of Fe^{3+} enhanced the photocatalytic activity for $CHCl_3$ degradation. The optimal concentration of Fe^{3+} dopants decreased with increasing TiO_2 particle size, suggesting that the role of Fe^{3+} species is the inhibition of surface recombination. The Fe^{3+} doping might work less effectively for large TiO_2 particles, since the dominant recombination process is bulk recombination rather than surface recombination. The photocatalytic activity of TiO_2 with large particle size was increased by Nb^{5+} doping in a combination with Pt loading, while the activity was decreased by sole Nb^{5+} doping.

3. Deactivation of TiO_2 photocatalysts at high temperature

Improving crystallinity by annealing can enhance the photocatalytic activity of TiO_2 by decreasing defects that act as recombination centers. However, high temperature calcination frequently decreased photocatalytic activity. This deactivation is commonly attributed to a change of the crystal structure from anatase to rutile and a decrease in BET-specific surface area because of crystal growth. Both crystallinity and specific surface area, which are related to particle size, are changed by high temperature calcination. Therefore, it is difficult to characterize the effect of annealing except for the effects by crystal growth.

We investigated the effect of high temperature calcination on the photocatalytic activity of rutile TiO_2 with a small BET-specific surface area, ~2.3 m^2 g^{-1}, to diminish the change in the crystalline phase and surface area by crystal growth [8]. Photocatalytic activity toward H_2 evolution was examined using an aqueous ethanol solution, with *in-situ* photodeposited Pt nanoparticles, under UV irradiation. **Figure 1** shows that the rate of photocatalytic H_2 evolution was significantly decreased by calcination in air at temperatures higher than 500°C, although the BET-specific surface area showed a little change. These results indicated that the deactivation of rutile TiO_2 particles at high temperature calcination could not be attributed to the phase transition and the decreased surface area.

Then, the behavior of photoexcited electrons was examined using transient IR spectroscopy to probe the dynamics of photoexcited electrons [8]. Changes in IR absorption were recorded after the pump irradiation of 355-nm laser pulse in the presence of methanol. Owing to low signal levels under vacuum, methanol was added as an electron donor. **Figure 2** shows the millisecond-scale decay of the photoinduced IR absorption observed for calcined rutile samples. Since the photogenerated holes react with methanol within a millisecond, the decay of the signal attributed to photoexcited electrons is very slow. The signal magnitude decreased as

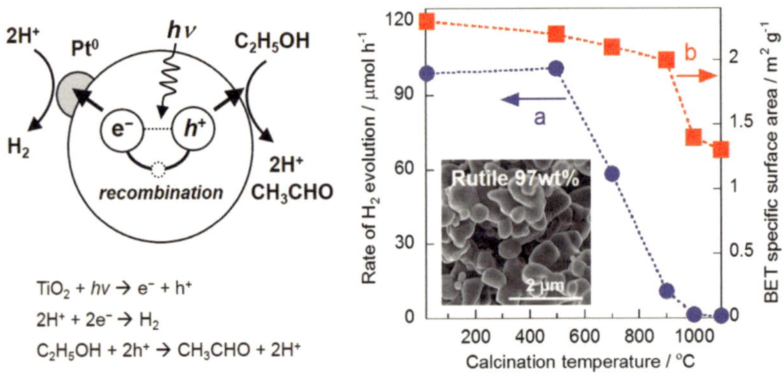

Figure 1. Effect of calcination temperature on (a) the rate of photocatalytic H_2 evolution from an aqueous solution of 50 vol% ethanol with H_2PtCl_6 (2.0 wt% as Pt) under UV irradiation from light emitting diode (peak wavelength 380 nm) and (b) BET specific surface area of rutile TiO_2 particles.

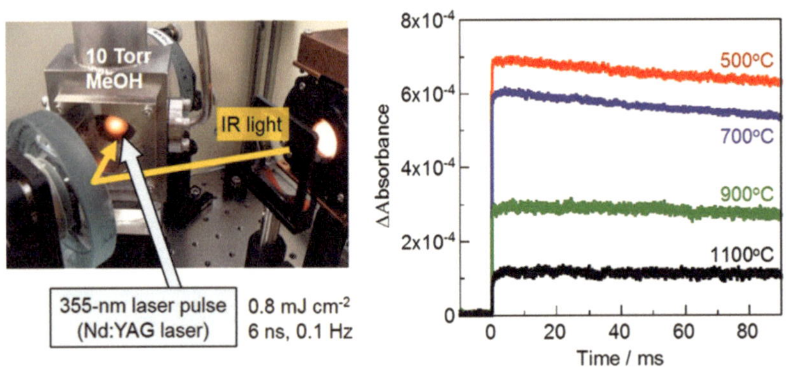

Figure 2. Transient IR spectroscopy setup and the transient absorption at 2000 cm^{-1} triggered by a 355-nm laser pulse with 6-ns duration in the presence of methanol of the rutile TiO_2 calcined in air at different temperatures: 500°C, 700°C, 900°C, and 1100°C.

calcination temperature of the sample increased, suggesting the more recombination of photoexcited carriers. The rutile TiO_2 particles calcined at high temperature showed the less density of long-lived photoexcited electrons, which resulted in the low photocatalytic activity. The reason for the deactivation at high temperature calcination is attributed to fast charge carrier recombination.

4. Hydrogen treatment of rutile TiO_2

4.1. Photocatalytic activity of reduced TiO_2

TiO_2 is often considered as a nonstoichiometric oxygen-deficient compound. According to Eq. (3), a small amount of electrons would be naturally doped. However, the n-type conductivity may be decreased by calcination in air at high temperature owing to the strong oxidation. To confirm the effect of electron density on the fast recombination observed in deactivated

TiO_2, we performed H_2 reduction treatment to the rutile TiO_2 particles calcined at 1100°C [8]. The TiO_2 samples were reduced by H_2 at 500 or 700°C, increasing both the density of long-lived charge carriers and photocatalytic activity (**Figure 3**). These results suggest that the density of oxygen vacancies and/or electrons is an important factor determining the photocatalytic activity.

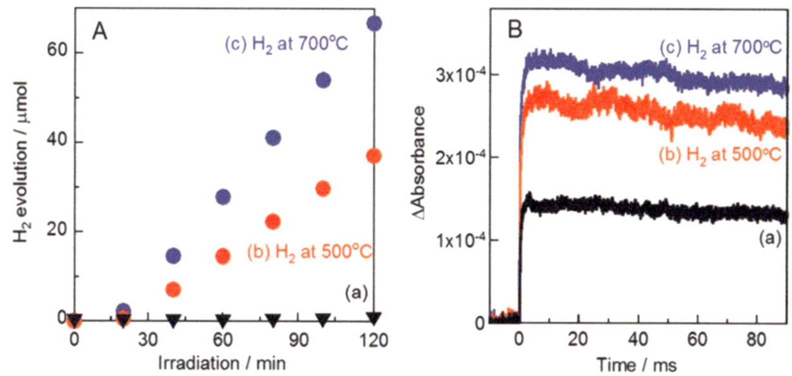

Figure 3. (A) Time course of photocatalytic H_2 evolution from an aqueous ethanol solution with Pt, and (B) transient IR absorption triggered by a 355-nm laser pulse in the presence of methanol: (a) TiO_2 calcined in air at 1100°C, (b) TiO_2 reduced with H_2 at 500°C after calcination at 1100°C, and (c) TiO_2 reduced with H_2 at 700°C after calcination at 1100°C.

H_2 reduction treatment has received extensive attention for improving the photocatalytic activity of anatase TiO_2 nanostructures since the recent report of the visible-light sensitivity of "hydrogenated black TiO_2" with defect disorders [9]. H_2 reduction treatment has also been reported to improve the PEC activity of TiO_2 nanowire arrays [10]. The H_2 treatment creates a high density of oxygen vacancies and electrons as shown in Eq. (1). We investigated the effects of H_2 treatment temperature on the photocatalytic activity of rutile TiO_2 particles to discuss the role of oxygen vacancies, Ti^{3+} ions, and conduction band electrons [11]. The photocatalytic activity was examined using an O_2 evolution from an aqueous solution of 50 mmol L^{-1} $AgNO_3$ as a sacrificial electron acceptor ($4Ag^+ + 2H_2O \rightarrow 4Ag^0 + O_2 + 4H^+$). Calcination above 900°C decreased the photocatalytic activity of rutile TiO_2 particles probably owing to strong oxidation, but its initial activity was restored by H_2 treatment at above 500°C. The photocatalytic activity of reduced TiO_2 was hardly changed after recalcination in air at 300°C, but significantly decreased after recalcination at 500°C.

4.2. ESR and UV-vis-NIR spectroscopy

Electron spin resonance (ESR) is active for paramagnetic species such as Ti^{3+} ions (electron trapped in titanium lattice site) and $O^{\bullet -}$ radicals (hole trapped in oxygen lattice site). **Figure 4A** shows ESR spectra of the H_2-reduced TiO_2 samples [11]. The TiO_2 calcined at 1100°C exhibited signals with $g = 2.061$ and 2.045. Kumar et al. have reported radical formation ($g_1 = 2.026$, $g_2 = 2.017$, and $g_3 = 2.008$) on rutile TiO_2 by calcination at 750°C and assigned the signals to trapped holes on the surface ($Ti^{4+}O^{2-}Ti^{4+}O^{\bullet -}$ radicals) [12]. It has been reported that strong oxidation of TiO_2 facilitates the transformation of n-type oxygen-deficient TiO_{2-x} to p-type metal-deficient $Ti_{1-y}O_2$ [13]. The high temperature calcination of TiO_2 may be considered as acceptor doping to create titanium vacancy (V''''_{Ti}) and hole ($h^{\bullet}$) according to Eq. (10). The hole would be trapped in the oxygen lattice site as an $O^{\bullet-}$ radical.

$$O_2 \rightarrow 2O_O^{\times} + V''''_{Ti} + 4h^{\bullet} \tag{10}$$

The signals attributable to $O^{\bullet-}$ radicals (trapped hole) disappeared after H_2 treatment at 400°C, indicating that strongly oxidized $Ti_{1-y}O_2$ was reduced to neutral TiO_2 by the mild H_2 treatment. The ESR spectrum of TiO_2 reduced at 500°C exhibited a sharp signal at $g = 2.002$ assigned to electrons trapped in oxygen vacancies and an intense signal at $g = 1.974$ assigned to Ti^{3+} ions (electron trapped in Ti lattice site) in rutile. This indicates that H_2 treatment at 500°C further reduced the TiO_2 to oxygen-deficient TiO_{2-x}. The signal at $g = 1.974$ was significantly broadened when the H_2 treatment temperature was increased to 700°C, indicating the high density of Ti^{3+} ions. The TiO_2 reduced at 700°C was assumed to be deeply doped n-type TiO_2.

The color of the TiO_2 samples changed from white to a pale ash color after H_2 reduction treatment [11]. **Figure 4B** shows diffuse reflectance UV-vis-NIR spectra of the TiO_2 samples. The onset of intense photoabsorption originating from interband transitions was located at ca. 415 nm corresponding to the bandgap of rutile, 3.0 eV. The spectrum of TiO_2 reduced at 500 and 700°C exhibited a broad absorption located in the visible and NIR region, which can be assigned to the transition of electrons in shallow traps and the conduction band. From the NIR absorption, the density of electrons in the TiO_2 reduced at 500°C was higher than that in the TiO_2 reduced at 700°C. Therefore, it is concluded that H_2 reduction at higher temperature increased the density of electrons in the conduction band.

4.3. X-ray photoelectron spectroscopy (XPS)

Figure 5 shows X-ray photoelectron spectra of the strongly oxidized TiO_2 and reduced TiO_2 [11]. There was no significant change in the Ti 2p spectra. The binding energy of 458.6 eV for Ti $2p_{3/2}$ was similar to the literature value for Ti^{4+} in TiO_2. This indicates that the amount of Ti^{3+} ions on the surface of reduced TiO_2 was too small to analyze by XPS. In contrast, there was a

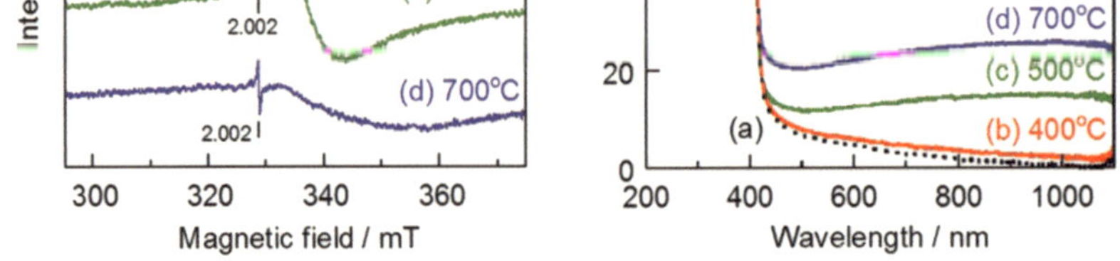

Figure 4. (A) ESR spectra and (B) diffuse reflectance UV-vis-NIR spectra of H_2-reduced TiO_2 samples: (a) TiO_2 calcined in air at 1100°C, and (b–d) TiO_2 treated with H_2 after calcination at 1100°C. The H_2 treatment temperatures are (b) 400°C, (c) 500°C, and (d) 700°C.

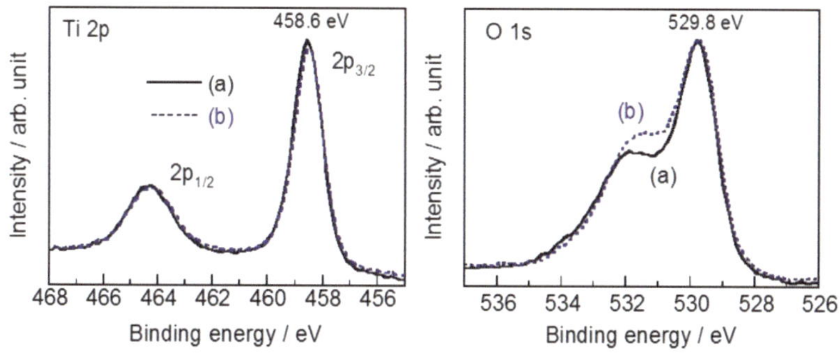

Figure 5. Ti 2p and O 1s X-ray photoelectron spectra of (a) TiO_2 calcined in air at 1100°C, and (b) TiO_2 treated with H_2 at 700°C after calcination at 1100°C.

significant difference observed in the O 1s spectra. The peak at 529.8 eV was assigned to lattice oxygen of TiO_2, and the shoulder peak at 531.5–532.0 eV was assigned to surface hydroxyl groups. Unexpectedly, the area assigned to the hydroxyl group was higher in intensity for reduced TiO_2 than that of oxidized TiO_2. This is because the oxygen vacancies at the top surface became filled by reaction with H_2O in the air. It is reported that oxygen vacancies induce dissociation of water molecules and form two hydroxyl groups via H^+ transfer to a neighboring lattice oxygen according to Eq. (11) [14]. Therefore, oxygen vacancies are not considered to be present on the surface of reduced TiO_2 under ambient conditions. This suggests that oxygen vacancies are not a catalytic site in reduced TiO_2 photocatalysts.

$$V_O^{\bullet\bullet} + H_2O + O_O^{\times} \rightarrow 2(OH)_O^{\bullet} \tag{11}$$

4.4. PEC and electrochemical properties

Thermally oxidized TiO_2 films were prepared by a simple calcination of a Ti sheet at 900°C [11]. **Figure 6A** shows PEC properties evaluated in dilute sulfuric acid (0.1 mol L^{-1} H_2SO_4, pH = 1) under UV irradiation (wavelength >330 nm). The thermally oxidized TiO_2 film was inactive for

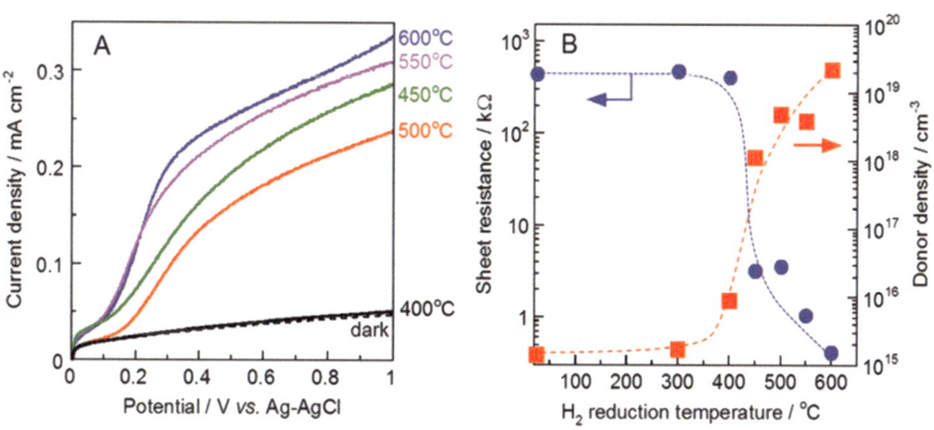

Figure 6. Effect of H_2 treatment temperature on (A) PEC voltammetry curves in 0.1 mol L^{-1} H_2SO_4 (pH = 1) and (B) sheet resistance and donor density of thermally oxidized TiO_2 films.

PEC water oxidation. The photocurrent was also negligible for the TiO_2 films treated with H_2 at 400°C. However, anodic photocurrent was observed at applied potentials larger than +0.1 V versus Ag-AgCl after H_2 treatment at higher temperature, suggesting the formation of long-lived holes in the reduced TiO_2 films.

Figure 6B shows the relationship between sheet resistance and donor density of the thermally oxidized TiO_2 films after H_2 treatment [11]. The sheet resistance was measured using a four-point probe. The donor densities were evaluated by Mott-Schottky analysis of the capacitance of the space charge layer. The Mott-Schottky plots of the TiO_2 films showed a positive slope of n-type conductivity. The resistance of the thermally oxidized TiO_2 films was high owing to their low donor density. H_2 treatment at above 450°C greatly reduced the sheet resistance and increased the donor density. H_2 treatment at 600°C increased the donor density by 2–3 orders of magnitude. The enhanced PEC and photocatalytic properties are because of the increase of n-type conductivity by H_2 treatment.

5. Highly efficient TiO_2 photocatalysts for H_2 evolution

5.1. Phase junction between anatase and rutile

A mixture of anatase and rutile phases has been reported to be more active than either pure phase alone. Incidentally, the highly efficient commercial photocatalyst Degussa (Evonik) P25 consists of mainly the anatase phase (~80%) with a reasonable amount of rutile (~15%). Because a synergetic effect between anatase and rutile phases is often not observed when separately synthesized powders are simply mixed together, close contact of the phases with each other is expected to be necessary. The high activity of the mixture of phases has been attributed to the separation of photoexcited charge carriers between the two phases. Anatase is considered to be an active component in mixed-phase TiO_2, while rutile is considered to act as an electron sink because of the lower conduction band energy than that of anatase. In contrast, an ESR study revealed that photoexcited electron transfer occurred from the conduction band of rutile to that of anatase in mixed-phase P25 [15]. This would be because the trapping sites of anatase lie below the energy level of the conduction band of the rutile phase. However, in attempts to explain electron transfer from the conduction band of rutile to that of anatase, it is reported that the conduction band of rutile lies ~0.4 eV above that of anatase based on calculation and X-ray photoelectron spectroscopy studies [16]. In this way, there is controversy over the alignment of the conduction band minima of rutile and anatase phases of TiO_2, experimental results suggest that the photoexcited electrons of rutile are less active than those of anatase. It is reputed that pure rutile is less active for photocatalytic H_2 evolution from water than pure anatase and mixed-phase TiO_2.

5.2. Reduced TiO_2 with pure rutile phase

As mentioned, TiO_2 P25 is a well-known commercial material with high photocatalytic activity. The mixture of anatase and rutile phases in P25 is reportedly more active than the individual polymorphs of TiO_2. Contrary to this viewpoint, we demonstrated that H_2-reduced rutile TiO_2 is much more active than mixed-phase P25 for photocatalytic H_2 evolution from aqueous ethanol solution [17, 18]. To confirm that the anatase phase does not work as an active component, we investigated H_2 reduction treatment of pure single-phase rutile particles [17]. P25 was first calcined in air at 900°C to induce complete phase transition from anatase to rutile, and then the pure rutile phase was reduced by H_2 at 700°C (**Figure 7A**). The photocatalytic H_2 evolution rate was enhanced to 344 μmol h^{-1} after H_2 reduction treatment at 700°C (**Figure 7B**).

Highly efficient rutile photocatalyst is easily fabricated from P25 by H_2 reduction treatment at 700°C for 2 h [18]. The H_2-reduced rutile TiO_2 outperforms anatase-rich TiO_2 because of the wider absorption range caused by its bandgap of rutile (3.0 eV) smaller than that of anatase (3.2 eV). The apparent quantum yield of H_2-reduced rutile TiO_2 was estimated to be 46% for photocatalytic H_2 evolution under 390-nm irradiation, which was 3.3 times higher than that of mixed-phase P25.

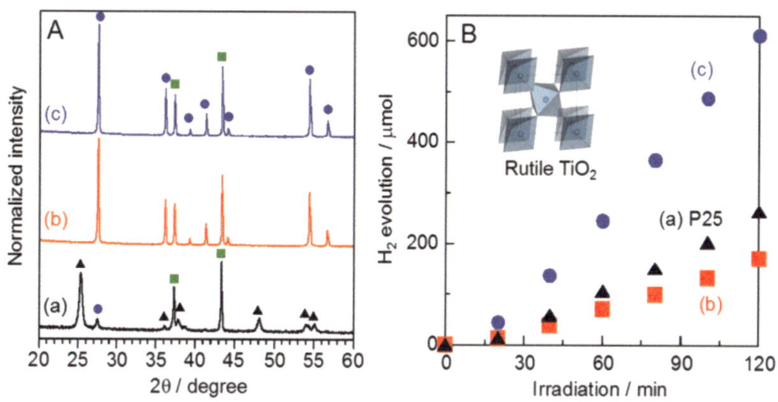

Figure 7. (A) XRD patterns of (a) P25, (b) P25 after calcination at 900°C, and (c) P25 treated with H_2 at 700°C after calcination at 900°C. Symbols ▲, •, and ▪ indicate peaks due to anatase TiO_2, rutile TiO_2, and NiO, respectively, which was added as an internal standard. (B) Photocatalytic H_2 evolution from aqueous ethanol solution over TiO_2 samples (50 mg) with 2.0 wt% Pt under UV irradiation from 380-nm LEDs.

6. Effect of particle size

6.1. Reduced TiO_2 with large particle size

Crystalline size is one of factors affecting the lifetime of photogenerated charge carriers in oxide photocatalysts. For photocatalytic O_2 evolution by water oxidation, anatase TiO_2 nanoparticles exhibit poor activity, while larger rutile particles are more efficient. Large WO_3 particles with low surface area-to-volume ratios are suited to providing long-lived photogenerated holes for water oxidation [19]. This is because slow bulk recombination is the dominant process in larger particles, rather than fast surface recombination.

Controlling the crystalline size of H_2-reduced TiO_2 is expected to improve the photocatalytic activity for O_2 evolution by water oxidation. The crystalline size and electron density of rutile

TiO_2 particles (BET-specific surface are 17 m^2 g^{-1}) was controlled by high-temperature calcination and subsequent H_2 reduction [20]. The photocatalytic activity of TiO_2 for water oxidation was significantly improved by H_2 reduction at 700°C, after calcination at 1100°C (**Figure 8**). The effect of H_2 reduction treatment was obtained only if the rutile particle was previously calcined at temperatures higher than 1000°C. The improved activity was probably due to a combination of the increased crystalline size and the increased electron concentration. The H_2-reduced TiO_2 exhibited high apparent quantum yield for O_2 evolution, 41% under irradiation at 365 nm.

Photocatalytic efficiencies per unit of BET-specific surface area were calculated to consider surface reactivities [20]. The surface reactivity for water oxidation, defined as the O_2 evolution rate per unit of surface area, was significantly improved by H_2 treatment when the samples were previously calcined at >1000°C. It was found that H_2 treatment also improved the surface reactivities for photocatalytic H_2 evolution and photocatalytic CO_2 evolution (oxidative decomposition of acetic acid). The H_2 treatment effectively improved the surface reactivity of TiO_2 calcined at high temperature, without dependence on a particular photocatalytic reaction. Thus, the improvement was not due to the surface modification such as a creation of catalytic active sites, but due to the change of electronic properties.

6.2. Mechanism of the enhanced activity

Figure 9 shows a schematic illustration of the Fermi level in reduced TiO_2. An increase in the electron concentration of n-type semiconductors results in an improvement of the electrical conductivity and an upward shift of the Fermi level toward the conduction band edge [21]. When n-type TiO_2 contacts with water, space charge layer forms at the interface by electron transfer from conduction band, with a simultaneous potential drop inside TiO_2 (**Figure 10**). This is so-called band bending. The higher Fermi level of n-type TiO_2 resulted in the increase of the surface barrier of Schottky type and the electric field in space charge layer. The intrinsic

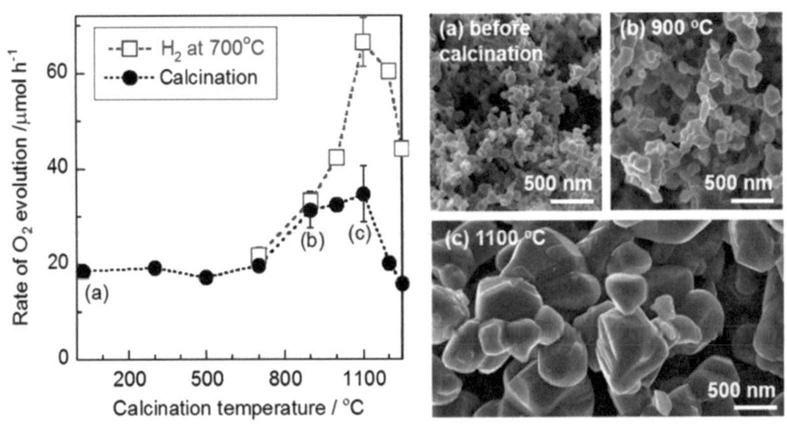

Figure 8. Effect of calcination temperature on the rate of photocatalytic O_2 evolution from water in the presence of sacrificial $AgNO_3$ over TiO_2 samples treated by (•) calcination in air and (□) reduction with H_2 at 700°C after calcination. SEM images of (a) TiO_2 (rutile 96 wt%, BET-specific surface area 17 m^2 g^{-1}), (b) TiO_2 calcined at 900°C, and (c) TiO_2 calcined at 1100°C.

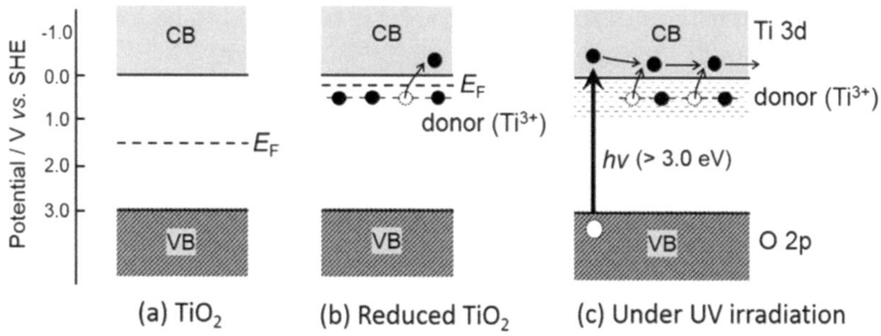

Figure 9. Fermi level (E_F) of (a) stoichiometric TiO_2 insulator and (b) reduced TiO_2 with n-type conductivity, in which Ti^{3+} ions are donor and (c) reduced TiO_2 under photoirradiation.

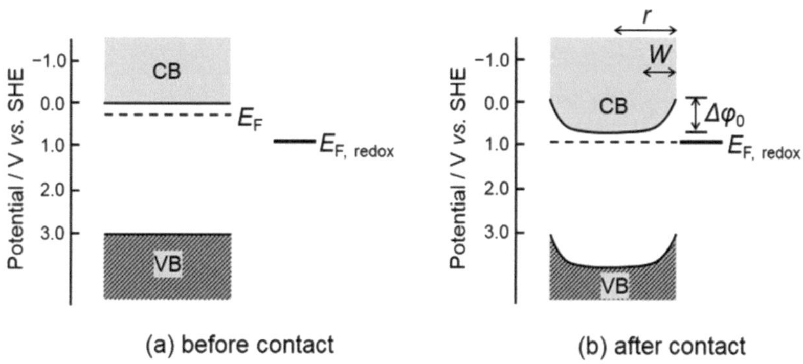

Figure 10. Interface of n-type semiconductor and solution (a) before and (c) after contact in thermal equilibrium. E_{redox} is the electrode potential of redox species in the solution, $\Delta\phi_0$ is potential drop in the semiconductor, W is the width of space charge layer, and r is the radius of the semiconductor particle ($r > W$).

electric field in the space charge layer separates photoexcited electrons and holes, preventing their recombination. This facilitates the transfer of holes from the valence band to the reactants. The width of space charge layer (W) is determined by the donor density (N_D) and the potential drop in the layer ($\Delta\varphi_0$) according to Eq. (12) [21].

$$W = \sqrt{2\varepsilon_0 \varepsilon \Delta\phi_0 / eN_D} \tag{12}$$

where ε_0 is the permittivity of vacuum, ε is the dielectric constant of semiconductor, and e is electronic charge. As the electron concentration increases, the space charge layer narrows. Thus, there exist an optimum electron concentration and an optimum particle size depending on this concentration. For example, W can be calculated to be 220 nm at $N_D = 10^{17}$ cm^{-3}, assuming $\Delta\varphi_0 = 500$ mV, and taking ε for rutile to be 86. The radius of photocatalyst particle should be larger than this thickness to obtain a band bending. The calcination temperature dependence observed in **Figure 8** can be explained in terms of the relation between the particle size and the thickness of space charge layer. High-temperature calcination would be necessary to increase the TiO_2 particle size to an optimum value, 500 nm–1 μm, to form space charge layer thickness. The formation of space charge layer is suggested to be involved in the activation mechanism of H_2-reduced rutile TiO_2 with large particle size.

7. Conclusion

H_2 reduction treatment enhanced the photocatalytic activity and PEC properties of rutile TiO_2 with large particle size. One of the most important factors deciding the photocatalytic activity of reduced TiO_2 is the density of electrons in shallow traps and conduction band, rather than the density of oxygen vacancies. H_2 treatment at 500°C created Ti^{3+} ions (trapped electrons), while treatment at 700°C increased the density of conduction electrons, resulting in an improvement of the electrical conductivity of the TiO_2 by 2–3 orders of magnitude. The enhanced activity of the reduced TiO_2 suggests that n-type conductivity governed by electron density plays an important role in suppressing fast recombination by facilitating charge transport. The suppression of recombination in the reduced TiO_2 was caused by not only the high electrical conductivity but also the band bending in space charge layer. The built-in electric field facilitates charge separation and charge transfer. Thus, the surface reactivities of reduced TiO_2 with large particle size were enhanced without dependence on reactions. It was demonstrated that H_2-reduced rutile TiO_2 exhibited photocatalytic activity higher than that of anatase TiO_2 and mixed-phase TiO_2.

Author details

Fumiaki Amano

Address all correspondence to: amano@kitakyu-u.ac.jp

Faculty of Environmental Engineering, The University of Kitakyushu, Wakamatsu-ku, Kitakyushu, Japan

References

[1] Maeda K. Direct splitting of pure water into hydrogen and oxygen using rutile titania powder as a photocatalyst. Chemical Communications. 2013;**49**:8404–8406

[2] Ishihara T, Nishiguchi H, Fukamachi K, Takita Y. Effects of acceptor doping to $KTaO_3$ on photocatalytic decomposition of pure H_2O. The Journal of Physical Chemistry B. 1999;**103**:1–3

[3] Mitsui C, Nishiguchi H, Fukamachi K, Ishihara T, Takita Y. Photocatalytic decomposition of pure water over NiO supported on $KTa(M)O_3$ ($M = Ti^{4+}, Hf^{4+}, Zr^{4+}$) perovskite oxide. Chemistry Letters. 1999;**28**:1327–1328

[4] Takata T, Domen K. Defect engineering of photocatalysts by doping of aliovalent metal cations for efficient water splitting. The Journal of Physical Chemistry C. 2009;**113**:19386–19388

[5] Karakitsou KE, Verykios XE. Effects of altervalent cation doping of TiO_2 on its performance as a photocatalyst for water cleavage. Journal of Physical Chemistry. 1993;**97**:1184–1189

[6] Zhang Z, Wang CC, Zakaria R, Ying JY. Role of particle size in nanocrystalline TiO_2-based photocatalysts. The Journal of Physical Chemistry B 1998;**102**:10871–10878

[7] Akubuiro EC, Verykios XE. Effects of altervalent cation doping on electrical conductivity of platinized titania. Journal of the Physics and Chemistry of Solids. 1989;**50**:17–26

[8] Amano F, Nakata M, Asami K, Yamakata A. Photocatalytic activity of titania particles calcined at high temperature: Investigating deactivation. Chemical Physics Letters. 2013;**579**:111–113

[9] Chen X, Liu L, Yu PY, Mao SS. Increasing solar absorption for photocatalysis with black hydrogenated titanium dioxide nanocrystals. Science. 2011;**331**:746–750

[10] Wang G, Wang H, Ling Y, Tang Y, Yang X, Fitzmorris RC, Wang C, Zhang JZ, Li Y. Hydrogen-treated TiO_2 nanowire arrays for photoelectrochemical water splitting. Nano Letters. 2011;**11**:3026–3033

[11] Amano F, Nakata M, Yamamoto A, Tanaka T. Effect of Ti^{3+} ions and conduction band electrons on photocatalytic and photoelectrochemical activity of rutile titania for water oxidation. The Journal of Physical Chemistry C. 2016;**120**:6467–6474

[12] Kumar CP, Gopal NO, Wang TC, Wong MS, Ke SC. EPR investigation of TiO_2 nanoparticles with temperature-dependent properties. The Journal of Physical Chemistry B. 2006;**110**:5223–5229

[13] Nowotny MK, Sheppard LR, Bak T, Nowotny J. Defect chemistry of titanium dioxide. Application of defect engineering in processing of TiO_2-based photocatalysts. The Journal of Physical Chemistry C. 2008;**112**:5275–5300

[14] Schaub R, Thostrup P, Lopez N, Lagsgaard E, Stensgaard I, Nurskov JK, Besenbacher F. Oxygen vacancies as active sites for water dissociation on rutile TiO_2(110). Physical Review Letters. 2001;**87**:2661041

[15] Hurum DC, Agrios AG, Gray KA, Rajh T, Thurnauer MC. Explaining the enhanced photocatalytic activity of Degussa P25 mixed-phase TiO_2 using EPR. The Journal of Physical Chemistry B 2003;**107**:4545–4549

[16] Scanlon DO, Dunnill CW, Buckeridge J, Shevlin SA, Logsdail AJ, Woodley SM, Catlow CRA, Powell MJ, Palgrave RG, Parkin IP, Watson GW, Keal TW, Sherwood P, Walsh A, Sokol AA. Band alignment of rutile and anatase TiO_2. Nature Materials. 2013;**12**:798–801

[17] Amano F, Nakata M, Ishinaga E. Photocatalytic activity of rutile titania for hydrogen evolution. Chemistry Letters. 2014;**43**:509–511

[18] Amano F, Nakata M, Yamamoto A, Tanaka T. Rutile titanium dioxide prepared by hydrogen reduction of Degussa P25 for highly efficient photocatalytic hydrogen evolution. Catalysis Science & Technology. 2016;**6**:5693–5699

[19] Amano F, Ishinaga E, Yamakata A. The effect of particle size on the photocatalytic activity of WO_3 particles for water oxidation. The Journal of Physical Chemistry C. 2013;**117**:22584–22590

[20] Amano F, Nakata M. High-temperature calcination and hydrogen reduction of rutile TiO_2: A method to improve the photocatalytic activity for water oxidation. Applied Catalysis B. 2014;**158-159**:202–208

[21] Rajeshwar K. Fundamentals of semiconductor electrochemistry and photoelectrochemistry. In: Licht S, editor. Encyclopedia of Electrochemistry. Semiconductor Electrodes and Photoelectrochemistry. Vol. 6. Weinheim: Wiley-VCH; 2002. pp. 1–53

Section 3

Different Types of TiO2 Materials

Chapter 6

Mesoporous Titania: Synthesis, Properties and Comparison with Non-Porous Titania

Barbara Bonelli, Serena Esposito and
Francesca S. Freyria

Additional information is available at the end of the chapter

http://dx.doi.org/10.5772/intechopen.68884

Abstract

Some relevant physico-chemical and photocatalytic properties of ordered mesoporous TiO_2 as obtained by template-assisted synthesis methods are reported. After a review of the crucial aspects related to different synthesis procedures reported by the literature, the focus is pointed on the (often) superior physico-chemical properties of ordered mesoporous TiO_2 with respect to (commercial) bulk TiO_2. Those are essentially higher specific surface area and ordered mesoporosity; possibility to control the formation of different crystalline phases by varying the synthesis conditions and possibility to obtain films, nanoparticles with different morphologies and/or materials with hierarchical porosity. Although mesoporous TiO_2 is extensively studied for many applications in the fields of photocatalysis, energy and biomedicine, this chapter focuses on the use of mesoporous TiO_2 in environmental photocatalysis, by putting in evidence how the physico-chemical properties of the material may affect its photocatalytic behaviour and how mesoporous TiO_2 behaves in comparison with commercial TiO_2 samples.

Keywords: template-assisted synthesis, ordered mesoporosity, TiO_2, anatase, photocatalysis, doping

1. Introduction

TiO_2 is one of the most studied metal oxides: this is mainly due to its intrinsic properties, i.e. occurrence of different polymorphs and amorphous phases, low toxicity to humans, relatively low cost, good chemical and thermal stability and good electronic and optical properties [1]. Such characteristics make TiO_2 the oxide of choice for application in (photo)catalysis, adsorption/separation processes, sensing, energy storage/conversion and biotechnology/biomedicine [1].

Although several commercial products are available on the market, the intrinsic properties of TiO_2 may be enhanced in the presence of porous samples that can be obtained by different synthesis procedures [2–9].

Besides increasing the specific surface, porosity facilitates some physico-chemical phenomena like adsorption and diffusion of chemical species, especially in the presence of mesopores ($\O$ = 2.0–50 nm). Conversely, micropores ($\O \leq 2.0$ nm) may have a detrimental effect on the diffusion of larger moieties.

For instance, mesoporous TiO_2 (M-TiO_2) was found to be highly active in several photocatalytic processes, since mesopores promote the diffusion of chemical species, either reactants or products [5, 6], and simultaneously enhance M-TiO_2 photocatalytic activity by facilitating access to the reactive sites at the photocatalyst surface. Moreover, the possibility to synthesise nanoparticles (NPs) allows obtaining a material that is highly dispersible in solution: consequently, M-TiO_2 is not only considered for environmental photocatalytic processes (usually carried out in aqueous phase) but is attracting increasing attention also for applications in biomedicine [7], due to the possibility of controlling NP morphology and size during the synthesis.

Moreover, highly organised films of M-TiO_2 may be obtained by evaporation-induced self-assembly (EISA) process [8, 10, 11] in the presence of diblock and triblock copolymers [12–14]. The availability of both EISA process and Pluronic amphiphilic triblock copolymers allowed overcoming the main issue related to the synthesis of M-TiO_2, i.e. the high reactivity of Ti precursors, which are usually unstable when exposed to water or even to moisture [15].

Discovered and developed in the 1990s, the first type of ordered mesoporous oxide ever obtained was SiO_2, likely due to the high pliability of the Si–O–Si bond, which is a peculiarity of silicon [16, 17]. The synthesis methods leading to the production of ordered mesoporous SiO_2-based materials have developed much faster [18] with respect to other mesoporous oxides [19]. Reasons for that have been (i) the lack of suitable metal precursors for sol-gel synthesis and (ii) their lower stability with respect to Si-precursors. Nonetheless, the low thermal stability of some mesoporous oxides may induce formation of undesired crystalline phases and/or grain growth. Such processes are likely at the high temperatures reached during calcination, the process often adopted to remove organic moieties deriving from the template and/or hydrolysis of the (metal alkoxide) precursor [20].

It is possible to obtain M-TiO_2 materials with either disordered or ordered mesoporosity. Disordered mesostructures are obtained by template-free methods, whereas the use of a structure-directing agent and/or a template is necessary to obtain ordered mesostructures. Historically, the first attempts to obtain M-TiO_2 led to materials characterised by a disordered porosity, which were mainly studied for photocatalytic applications [5]. Ordered mesoporous frameworks are supposed to improve the photocatalytic performance of TiO_2, since an ordered porous structure favours diffusion of both reactants and products. The latter class of materials is therefore more interesting for photocatalytic applications in environmental remediation and is addressed in this chapter. Other types of ordered TiO_2 structures, i.e. hollow fibres, nanotubes (not addressed here), are instead obtained by other methods like electrodeposition, electrospinning [21], etc.

Concerning the use of templates, both soft-template methods and hard-template methods are available [2]. The former imply the use of surfactants that form micelles (direct or inverse), whereas hard-template methods imply the use of preformed mesoporous solids like SiO_2 or carbon, as well as (natural and/or synthetic) polymers (**Figure 1**).

The physico-chemical properties of the final product depend on several factors, like the chosen Ti precursor, which affects the hydrolysis and condensation reactions occurring during formation of the mesoporous network, and the calcination temperature. Calcination, usually necessary to remove the organic moieties, deriving by both the surfactants used in soft-template methods and the hydrolysis of Ti alkoxides used as Ti precursors, markedly affects the type of TiO_2 polymorph formed as well as the size of crystalline grains.

The physico-chemical properties of the final material are very important since they ultimately affect the performance of M-TiO_2 in the target application and may be responsible of a better (or worst) performance with respect to commercial products. Among the latter ones, so far, Degussa P25 is the most used and studied commercial TiO_2: due to its remarkable photocatalytic performance, it is used as benchmark material in most of the literature works.

Far from having the purpose of reviewing in detail the procedures for the preparation of ordered M-TiO_2, only some crucial aspects of the most acknowledged synthesis methods will be addressed in this chapter. Afterwards, the chapter will show some correlations between the peculiar physico-chemical properties of M-TiO_2 and its performance in environmental photocatalysis, although the properties of M-TiO_2-based systems are attracting considerable attention also for other applications, including H_2 production, fabrication of electrodes in Li-ion batteries, dye-sensitised solar cells and biomedicine [22].

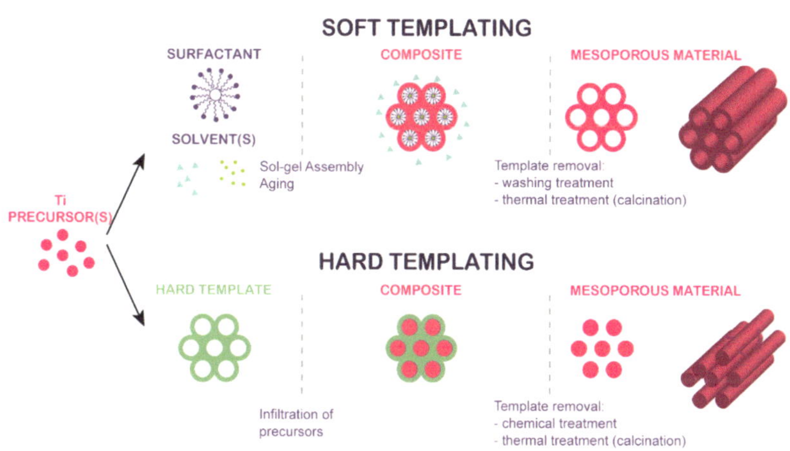

Figure 1. Simplified scheme of the steps leading to the production of ordered M-TiO_2 by means of template-assisted syntheses.

2. Template-based methods for the synthesis of ordered mesoporous titania: main features and critical issues

2.1. Hard-template methods: nanocasting

Nanocasting (**Figure 1**, lower part) consists of using preformed (natural or synthetic) mesoporous solids to synthesise porous materials of different compositions: such method has been extensively applied to produce porous carbons, metal oxides and metal sulphides as negative replicas of preformed hard templates [23, 24].

Usually, nanocasting envisages three fundamental steps: (i) the precursor is infiltrated within the mesoporous channels of a porous solid (i.e. preformed mesoporous SiO_2 or carbon); (ii) the precursor reacts forming a composite of the negative replica and the hard template and (iii) the hard template is removed (either chemically or thermally).

As a result, a negative replica is obtained of the hard template, which in the case of M-TiO_2 is mostly mesoporous SiO_2 (e.g. SBA-15, SBA-16, MCM-41, MCM-48, KIT-6, FDU-12), mesoporous carbon (e.g. CMK-1, CMK-3), porous Al_2O_3 or polystyrene spheres [5].

Nanocasting allows obtaining M-TiO_2 characterised by crystallinity and thermal stability, because it overcomes the problem of thermal collapse of the TiO_2 framework, which may occur during phase transition. Some problems, however, may arise during the synthesis: for instance, if the precursor infiltration is performed in aqueous phase, some undesired precipitation and crystallisation of TiO_2 in solution may lead to an incomplete filling of the hard-template pores, ultimately blocking the channels and avoiding any further precursor infiltration [15].

In order to avoid such undesired phenomenon, some parameters like precursor/template ratio, calcination temperature and immersion time have to be strictly controlled [25].

A careful evaluation of the weight ratio between the Ti alkoxide and the solid template [25] is crucial, as a too high amount of precursor would lead to formation of bulky material outside the hard-template pores, with consequent loss of surface area and poor precursor infiltration [25].

Conversely, some positive confinement effects due to the hard template in contact with TiO_2 allow obtaining anatase M-TiO_2 at temperatures that are unusual for bulk TiO_2, for instance, with mesoporous silica KIT-6 as hard template, anatase (instead of rutile) M-TiO_2 formed by calcination at 750°C [25]. Such result is particularly sound since anatase, though characterised by a larger band gap than rutile ($E_g \cong 3.2$ eV and 3.0 eV for anatase and rutile, respectively), has a large surface area that is very useful for (photo)catalytic applications.

As for the soft-template routes discussed in the following paragraph, the choice of Ti precursor is the main issue. While metal nitrates and chlorides are successfully employed [26] for producing replicas of other metal oxides, Ti nitrates and chlorides are mostly unstable in water (or even when exposed to moisture).

Such problem has been successfully overcome by Yue et al. [27]: they prepared M-TiO_2-negative replicas of both SBA-15 and KIT-6 mesoporous silica by pre-hydrolysing the Ti precursor (either Ti isopropoxide ($Ti[OCH(CH_3)_2]_4$) or tetrabutyl titanate ($Ti[OC_4H_9]_4$) and dissolving it in HNO_3. The so-obtained acidic solution of Ti nitrate complexes was used for the infiltration step. Another interesting result was the formation of rutile M-TiO_2 at calcination temperatures as low as 100°C, ascribed to either the confinement effect [27] of KIT-6 mesoporous channels or the ability of nitrate ions to stabilise the rutile phase within the SiO_2 walls. Indeed, when other Ti precursors are employed (like chloride, sulphate and isopropoxide), anatase M-TiO_2 preferentially forms and transition to rutile M-TiO_2 is not observed at temperatures as high as 600°C [28].

Nanocasting is feasible also by using naturally occurring hard templates: a fascinating work reports on a nanocrystalline rutile TiO_2 obtained by using the chitin scales present on the wings of a *Morpho* butterfly, which were coated with a uniform oxide film by adopting a computer-controlled surface sol-gel process [29]. In addition, the calcination process was a crucial step: when the composite was fired at 450°C, mostly anatase TiO_2 formed, whereas firing at 900°C mainly led to rutile, though some anatase was still present in the final product [29].

Hierarchical structures of TiO_2 films (**Figure 2**) were obtained by combining PMMA (poly(methyl methacrylate)) microspheres and sol-gel chemistry, in the presence of an amphiphilic diblock copolymer as a structure-directing agent [30]. Poly(dimethylsiloxane)-block-methyl methacrylate poly(ethylene oxide) (PDMS-bMA(PEO)) was used as a structure-directing agent for the preparation of the mesopore structure, whereas PMMA microspheres acted as a template for the micrometre-scale structure (**Figure 3**). Both the hard and the soft templates were then removed either by acetic acid or calcination, leading to a macro/mesoporous network, where the macropores generated by the hard template are supposed to favour mass transport phenomena and improve accessible surface area.

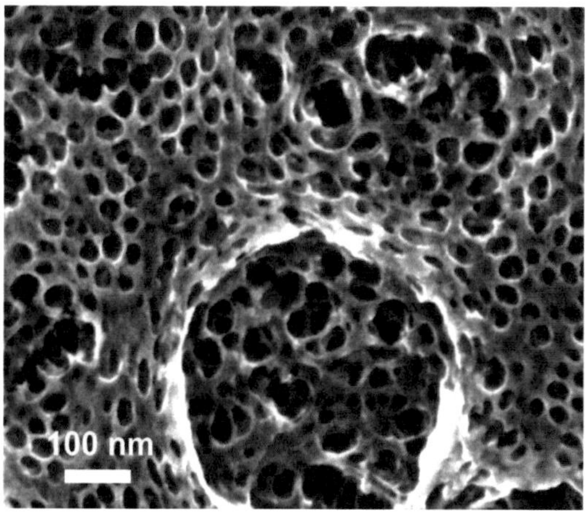

Figure 2. SEM image showing the occurrence of macro- and mesopores in a hierarchically structured TiO_2 film. (Reprinted with permission from Ref. [30]. Copyright 2009 American Chemical Society).

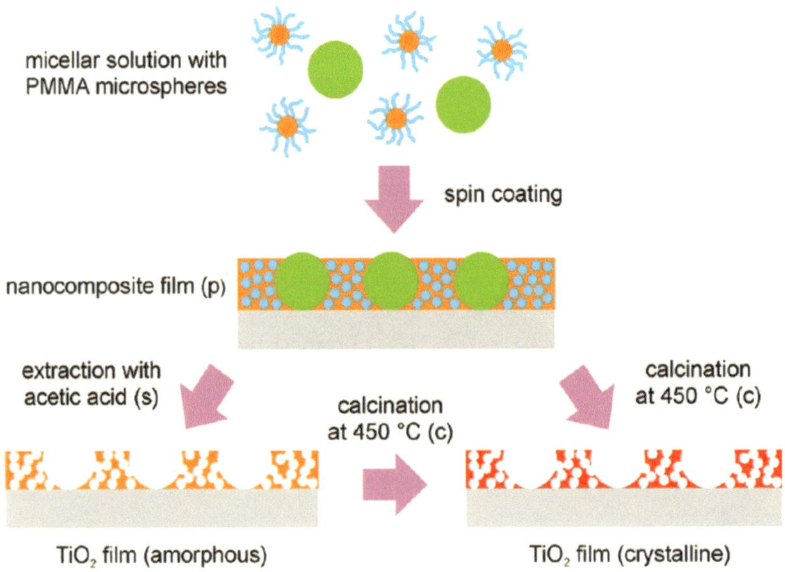

Figure 3. Scheme of the procedure adopted to produce hierarchically structured TiO_2 films. (Reprinted with permission from Ref. [30]. Copyright 2009 American Chemical Society).

Similarly, a macro/mesoporous TiO_2 was obtained by using poly(styrene-co-acrylic acid) colloidal spheres and triblock copolymer Pluronic P123 as macro- and mesoporous structure-directing agents [31], leading to a material with enhanced photoelectrocatalytic activity towards the removal of Rhodamine B.

2.2. Soft-template methods

The first report on an aqueous soft-templating route for the synthesis of M-TiO_2 dates back to 1995 [32]: soft-template methods employ different types of structure-directing agent, like charged surfactants (either anionic or cationic), neutral surfactants (e.g. alkylamines) and block copolymers [12, 14, 20, 32–38].

The crucial issue is the control of hydrolysis and condensation rates of the Ti precursor during the cooperative assembly between the structure-directing agent and the inorganic phase.

The second major problem is the thermal stability of the so-obtained M-TiO_2 framework, especially during the high-temperature calcination step (required to remove the template), since undesired structural collapse and/or crystallisation may occur, leading to the loss of mesoporosity and/or the grain growth.

The Ti precursor stabilisation is particularly difficult when charged surfactants are used: in a pioneering work, Antonelli et al. used acetylacetonate as a ligand to decrease the reactivity of Ti isopropoxide through the formation of Ti acetylacetonate tris(isopropoxide) in the presence of an alkyl phosphate surfactant [32]. The proposed method showed, however, some limits: on the one side, it was not possible to completely remove phosphorous, likely due to the strong interaction between surfactant and M-TiO_2 framework; on the other side, the method was unsuccessful when another charged surfactant (either anionic or cationic) was used.

In a subsequent work, Antonelli obtained a worm-like M-TiO_2 by using an amine as template [39]: the material was characterised by short-range order and relatively low thermal stability, as well as high specific surface area (≅700 $m^2 g^{-1}$) and the possibility to tune pore dimensions by changing the length of the amine carbon chain (12–18 C atoms). At variance with the alkyl phosphate surfactant used in Ref. [32], where it was not possible to remove phosphorous completely by calcination, when an amine molecules is used, weaker H-bonding interaction occurs between the template and Ti oligomers, finally stabilising the latter species. The main drawback of this method was the longer ageing time required in order to obtain a stable material.

Other attempts with an amine as soft template were made, for instance, with hexadecylamine [40]: the basic molecule seemed to positively affect both hydrolysis and condensation of the Ti precursor, allowing an effective cooperative assembly between the organic and inorganic phase.

However, an effective development of M-TiO_2 was reached only with the advent of block copolymers and EISA method [20, 41]. Block copolymers are non-ionic surfactant (**Figure 4a**) consisting of distinct homopolymer subunits (blocks) linked by covalent bonds like the triblock copolymer known under the commercial name of Pluronic P123 (nominal chemical formula $HO(CH_2CH_2O)_{20}(CH_2CH(CH_3)O)_{70}(CH_2CH_2O)_{20}H$).

Such molecules have a high self-assembly capability and may form different mesostructures, depending on their concentration in an alcohol-rich solution [8, 42, 43]: at concentrations below the CMC (critical micelle concentration), each copolymer behaves as a free molecule. At the CMC, copolymer molecules tend to form spherical micelles, with the hydrophobic part in contact with the alcoholic solution and the hydrophilic part shielded within the micelle, in order to minimise the free energy. At concentration above the CMC, coalescence of spherical micelles into cylindrical ones occurs. If the copolymer concentration increases, phase separation may occur, with micelles self-assembling in hexagonal, cubic or lamellar mesophases.

Other types of synthesis imply the use of hydrophobic solvents like cyclohexane and diblock copolymers, known under the commercial name of Brij-n (**Figure 4c**): in those cases, inverse micelles form, with a hydrophilic core that acts as a nanoreactor for the polymerisation of TiO_x. The surfactant is then removed by extractions/centrifugation/washing, and then a final calcination step brings about the total removal of the organic part as well as M-TiO_2 crystallisation [44]. After calcination, M-TiO_2 NPs form aggregates with interparticle porosity (**Figure 4c**).

Concerning the morphology of the final material, it is possible to obtain films in a rather simple way, through the so-called EISA method coupled to the sol-gel technique [20, 43] in ethanol/water mixtures (**Figure 2**). An acid (i.e. HCl) is added to control the sol-gel chemistry of the Ti precursor, which forms Ti hydroxides and/or oligomers. The sol is then deposited on a solid substrate, and solvent evaporation is induced by regulating the relative humidity (RH).

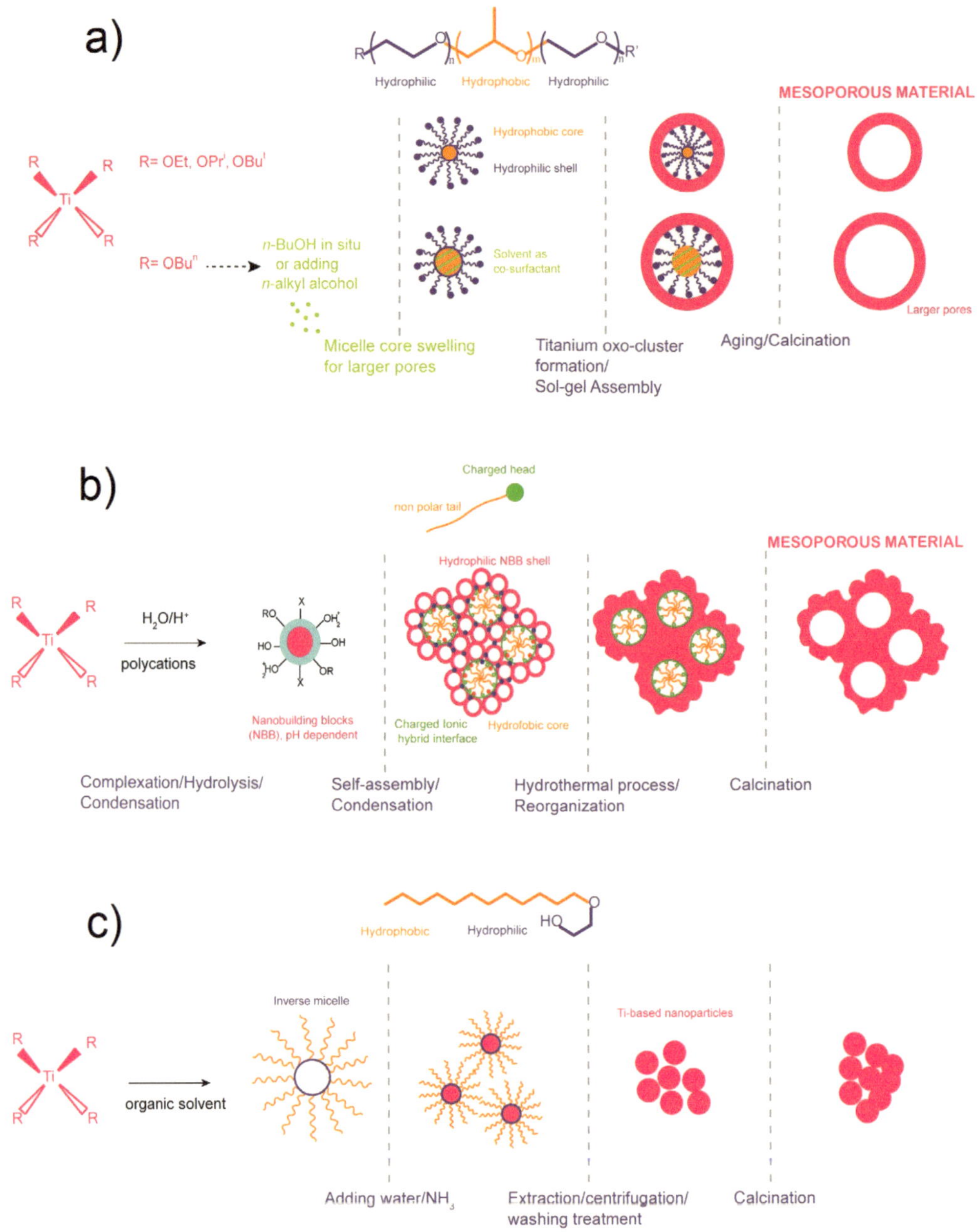

Figure 4. Schemes of three synthesis routes that are possible by using as soft template: (a) a triblock copolymer, e.g. Pluronic P123. The optional addition of a swelling agent (like *n*-BuOH) allows tuning mesopore dimension [52]; (b) an ionic surfactant, like cetyltrimethylammonium bromide (CTAB). NBB resulting from partial condensation of the precursor may have a charged surface due to pH-dependent protonation of ≡Ti–OH to $\equiv Ti\text{–}OH_2^+$ and are stabilised by the negative charge of the anion [50, 51]; and (c) diblock copolymer, such as Brij-n, which forms inverse micelles in cyclohexane [44].

http://dx.doi.org/10.5772/intechopen.68884

As the solvent evaporates, the concentration of both Ti species and the copolymer increases, and, simultaneously, the assembly of the inorganic and the organic phases is favoured. Induced evaporation also helps the removal of HCl (and other volatile species), improving the long-range order of the obtained film. Indeed, in a previous work where $TiCl_4$ was used as Ti precursor, the molecules of HCl likely induced some hydrolysis in the oxide framework, and rather disordered films were obtained [41].

The order of the final M-TiO_2 films obtained by the EISA method (**Figure 5**) depends on the amount of H_2O, HCl and RH and crystallisation process, as extensively discussed by Crepaldi et al. [20]. In particular, RH results to be a key parameter during the self-assembly step, as it both affects hydrolysis/condensation of Ti species and polarity of the PEO chain [20]. Moreover, RH was found to affect the order, thickness and transparency of the obtained film.

To further control the hydrolysis and condensation of the Ti precursor during the assembly process, addition of ligands like acetic acid [45] and acetylacetonate [46] usually leads to more ordered and/or more stable materials.

In the EISA method, the initial solution is prepared by dissolving anhydrous $TiCl_4$ into an alcohol-rich solution where the block copolymer has been pre-dissolved. The occurring reaction is

$$\mathrm{TiCl_4} + x\mathrm{EtOH} + y\,E_2\mathrm{O} \rightarrow \mathrm{TiCl}_{4-x-y}(\mathrm{OEt})_x(\mathrm{OH})_y + (x-y)\mathrm{HCl} \qquad (1)$$

Concerning the Ti precursor, as $TiCl_4$ leads to the formation of HCl through reaction (1) with consequent pH lowering and disordering of the obtained solid [30], stabilised alkoxides are used either as such or mixed with $TiCl_4$ itself [47, 48].

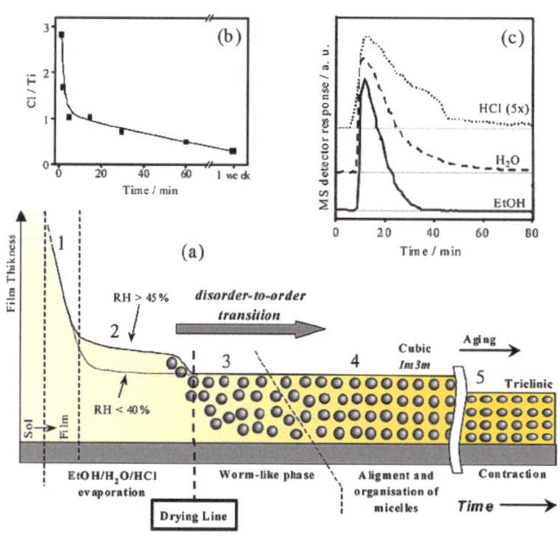

Figure 5. Scheme of the steps leading to the formation of M-TiO_2 by means of the EISA method. (Adapted from Ref. [20]. Copyright 2003 American Chemical Society).

Ti alkoxides are more easy to handle with respect to $TiCl_4$, but they need some acid as a stabiliser to control the hydrolysis: besides providing acidic conditions, HCl also forms complexes with the Ti alkoxide. For instance, when $Ti(OEt)_4$ is used, the following reaction occurs:

$$Ti\,(OEt)_4 + (4 - x - y)HCl + y\,H_2O \rightarrow TiCl_{4-x-y}\,(OEt)_x\,(OH)_y + (4 - x)EtOH \qquad (2)$$

Starting from different precursors, the same (partially hydrolysed) species $TiCl_{4-x-y}(OEt)_x(OH)_y$ form in the initial sol [49]. Such species are low-molecular-weight oligomers that, being resistant to hydrolysis, act as nanobuilding blocks (NBB) and cooperate with the hydrophilic portion of the copolymer micelles (for instance, by forming H-bonding with the PEO segment in the Pluronic surfactants).

Concerning NBB, their surface is pH-dependent, due to the protonation of ≡Ti—OH groups to $\equiv Ti{-}OH_2^+$, and in the presence of ionic surfactants, they can be stabilised by the charge of ions present in the solution [50, 51], as depicted in **Figure 4b**, where the ionic template may be hexadecyltrimethylammonium bromide (CTAB). Formation of the hybrid composite depends on the CTAB/Ti ratio and pH of the solution [50, 51]. After self-assembly and condensation, NBB are likely located between the micelle and the inorganic framework: the subsequent hydrothermal treatment can promote the condensation of such NBB, finally leading to a robust inorganic mesostructure.

In order to obtain larger mesopores, it is possible to use a swelling agent, like, for instance, an *n*-alkyl alcohol (*n*BuOH), which solubilises the hydrophobic/hydrophilic interface of the micelle, thereby causing it to swell (**Figure 4a**) [52]. The degree of swelling is proportional to the amount of alcohol, which not only acts as a swelling agent but, being likely located at the hydrophilic/hydrophobic interface, also stabilises the liquid crystal phase and determines its curvature at the interface with the inorganic mesostructure [52].

Calcination is the most effective process for the organic template removal, but it is also a crucial step, since the mesoporous network may collapse at high temperature, with consequent loss of specific surface area. Nonetheless, the thermal treatment also induces crystallisation of the initially amorphous material: the degree of crystallinity is of paramount importance for applications like photocatalysis; therefore, the temperature and time of the calcination have to be carefully controlled. A high degree of crystallinity is desirable, as it implies less surface defects, and therefore a better photocatalytic performance, as defects may act as recombination centres, lowering the photocatalyst performance. An opposite concomitant effect is the grain growth, favoured at high temperatures [5].

In order to obtain thermally stable materials, with suitable grain size and few surface defects, different post-synthesis thermal treatments have been proposed in the literature. For instance, it is possible to obtain cubic M-TiO_2 with anatase phase stable up to 400°C obtained after 4 h calcination at 400°C (heating rate 1°C/min) [53]. This material was considered as a promising one for photocatalytic (and optoelectronic) applications, whereas the hexagonal M-TiO_2 was not stable above 250°C.

M-TiO_2 thin films stable up to 600°C [54] due to the presence of rather thick walls (9.0–13 nm), where synthesised by using Pluronic F127 (a triblock polymer with chemical composition:

$PEO_{106}PPO_{70}PEO_{106}$, EO = ethylene oxide, PO = propylene oxide) as the structure-directing agent and tetrabutyl titanate as the precursor, in the presence of acetylacetonate [54]. According to the authors, the use of Pluronic F127 as the structure-directing agent favoured the formation of thick TiO_2 framework.

Another work reported [55] on the preparation of a M-TiO_2 stable up to 650°C obtained by a carbonisation step of the organic template with H_2SO_4 followed by thermal treatment at 350°C in inert atmosphere (N_2). Such process led to the formation of tubular C deposits that acted as stabilisers of TiO_2 during the calcination at a high temperature carried out through different steps (first at 550 and 650°C in N_2 and then at 450°C in air to burn out the carbon). The final material was a highly ordered 2D hexagonal mesostructure, with a crystalline framework arising by the connection of anatase nanocrystals. The ordered M-TiO_2 obtained had high surface area (~193 $cm^2\ g^{-1}$), large pore volume (~0.23 $cm^3\ g^{-1}$) high thermal stability (~650°C) and uniform mesopore size (4.6–5.1 nm).

Stabilisation of M-TiO_2 with larger mesopores (Ø > 5.0 nm) is still a challenge and requires some post-treatment methods. For instance, a modified EISA approach envisages the use of ethylene diamine molecules [56], which effectively protect the M-TiO_2 primary particles from collapsing finally delaying the phase transition of anatase to rutile. According to this method [56], a M-TiO_2 with large pore size (10 nm), high surface area (122 $m^2\ g^{-1}$) and thermal stability up to 700 °C was obtained.

Other factors controlling the phenomena occurring during the synthesis of M-TiO_2 materials (not be addressed here) have been extensively reviewed in the literature [2, 5, 8].

3. Photocatalytic applications of M-TiO_2

Photocatalytic processes involving TiO_2 are among the most common advanced oxidation processes (AOPs) proposed for the degradation of several inorganic and organic environmental pollutants [57, 58]. Ordered mesoporous structures are highly desirable for photocatalytic applications, since sufficiently large mesopores facilitate diffusion of reactants/products and high specific surface area improves adsorption, especially when bulky organic moieties are implied, like in wastewater and groundwater remediation.

3.1. Photocatalytic properties of pure M-TiO_2

Positive effects ascribable to the occurrence of an ordered mesoporous structure were observed in the photocatalytic degradation of methylene blue (MB), a dye commonly used in this type of studies as a model molecule of recalcitrant organic pollutants that are resistant to biodegradation and for which a photocatalytic treatment is necessary.

For instance, a M-TiO_2 occurring as pure anatase phase showed good photocatalytic activity towards MB removal [59]. The best photocatalyst was a hexagonal M-TiO_2 obtained by sol-gel synthesis using CTAB as template and Ti isopropoxide as precursor. The material obtained through a 6-days sol-gel synthesis showed mesopores with 6.86 nm diameter and a specific surface area of 284 m^2 g^{-1}. Longer synthesis time led to a decrease of both values and to the simultaneous disappearance of the hexagonal mesophase. The sample showed indeed a sizeable MB degradation with respect to Degussa P25, especially after M-TiO_2 was recycled and contacted with a fresh MB solution. Such superior photocatalytic behaviour was assigned to both higher surface area and higher anatase content of M-TiO_2 with respect to Degussa P25 [60].

Satisfactory results concerning the photobleaching of MB were obtained with a M-TiO_2 using SBA-15 silica as hard template, the activity of the sample being again ascribed to a compromise between its high surface area and the percentage of anatase. The former parameter was likely responsible for a very efficient dye adsorption at the surface of M-TiO_2, the latter likely reduced the electron/hole (e^-/h^+) recombination, which is slow in crystalline and defect-free materials [28].

A M-TiO_2 photocatalyst obtained through the EISA method by employing ethylene diamine as a stabiliser showed better photocatalytic activity than Degussa P25 towards the degradation of 2,4-dichlorophenol, a toxic chlorinated compound produced by environmental transformations of some chlorinated herbicides and/or antimicrobial agents [56]. Several M-TiO_2 were studied and compared to Degussa P25: the most efficient degradation was obtained in the presence of a M-TiO_2 calcined at 700°C, whereas the performance of M-TiO_2 calcined at higher temperatures decreased. Moreover, the performance of the M-TiO_2 photocatalyst resulted stable after recycling. The photocatalytic behaviour of the studied materials was explained not only on the basis of a higher surface area (122 m^2 g^{-1}) with respect to Degussa P25 (ca. 50 m^2 g^{-1}): according to the authors, a trade-off exists between the occurrence of pure anatase phase and a well-ordered mesoporous structure facilitating diffusion of reactants/products. At higher calcination temperatures, i.e. 900°C rutile M-TiO_2, formed, with a lower activity. Nonetheless, the sample calcined at 800°C, though occurring as pure anatase, showed both a partial collapse of mesoporous walls and the formation of larger particles, leading to a worst photocatalytic performance.

A performance comparable with Degussa P25 towards the degradation of dimethyl phthalate (a persistent antiparasite) was obtained with M-TiO_2 prepared by using Pluronic P123 in weak acidic solution of acetic acid [60]. Similarly to what mentioned before, both the high surface area and the crystallinity of the anatase phase contribute to the catalytic activity of the sample.

As a whole, although different experimental conditions are adopted during photocatalytic tests carried out in different laboratories, the occurrence of pure anatase M-TiO_2 usually favours better photocatalytic performances under UV radiation. However, this is just a general rule, but several exceptions are observed. For instance, another M-TiO_2 obtained by soft-template route under high-intensity ultrasound irradiation showed better photocatalytic performance than P25 in the UV-assisted degradation of *n*-pentane in air (**Figure 6**). The studied M-TiO_2 was not pure anatase, but a bi-crystalline material, containing ca. 20% brookite and 80% anatase, the formation of brookite being ascribed to the ultrasound treatment adopted during the synthesis [61]. The M-TiO_2 samples after calcination were characterised by high surface area (112–128 $m^2 g^{-1}$) and the occurrence of both anatase and brookite phase (**Figure 7**).

The band gap of the as-prepared M-TiO_2 materials (SM-1 and SM-2) was estimated through Tauc's plot since the band gap of brookite should be ca. 0.16 eV higher than that of anatase and could contribute to a more efficient UV light absorption, indicating that brookite could be a more powerful photocatalyst. According to the authors, composite materials of brookite and anatase can also suppress e^-/h^+ recombination, similarly to what widely accepted for Degussa P25 (i.e. that its high photocatalytic activity is partially due the coexistence of 80% anatase and 20% rutile, which can inhibit the recombination of excited e^-/h^+ [62]).

Other authors found a positive effect of the incorporation of Degussa P25 particles in M-TiO_2 thin films [63]: the films were tested in the photocatalytic degradation of Acid Black Dye, used as a model molecule of textile water pollutants. As expected, Degussa P25 incorporation led to a decrease of surface area in the composite with respect to M-TiO_2 films: however, a 5.0 wt.% Degussa P25 content led to an increase in the dye degradation efficiency. The positive role of Degussa P25 was ascribed both to an increase of the M-TiO_2 film thickness and to the presence of rutile in Degussa P25 leading to a band-gap change in the composite.

Another major use of TiO_2 is in disinfection processes: to this respect, wormhole like anatase M-TiO_2 was studied for the photocatalytic disinfection of *Escherichia coli*, showing better performances with respect to Degussa P25 [64].

3.2. Photocatalytic properties of metals doped M-TiO_2

Processes like doping and modification of TiO_2 are usually required in order to decrease the recombination rate of e^-/h^+ pairs and to extend the absorption range towards the Vis [65]. The latter effect is particularly important in the perspective of exploiting solar light, especially for photocatalytic processes of environmental remediation that have to be carried out under solar illumination in view of actual large-scale applications [66, 67].

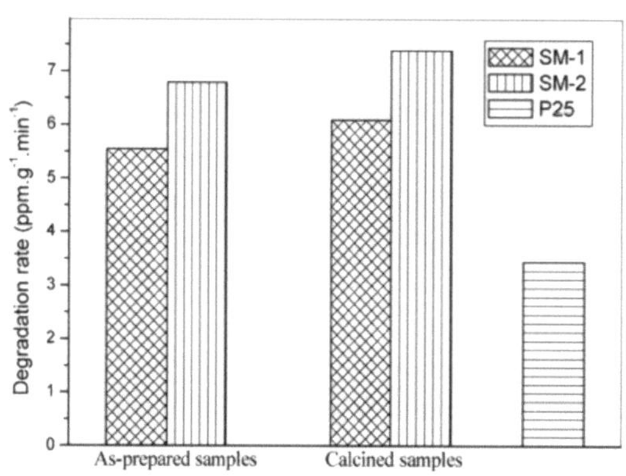

Figure 6. Rates of UV-assisted degradation of *n*-pentane as catalysed by two M-TiO_2 (SM-1 and SM-2) obtained under high-intensity ultrasound irradiation are compared to the rate obtained by using Degussa P25 (after Ref. [61]). (Copyright 2002 American Chemical Society).

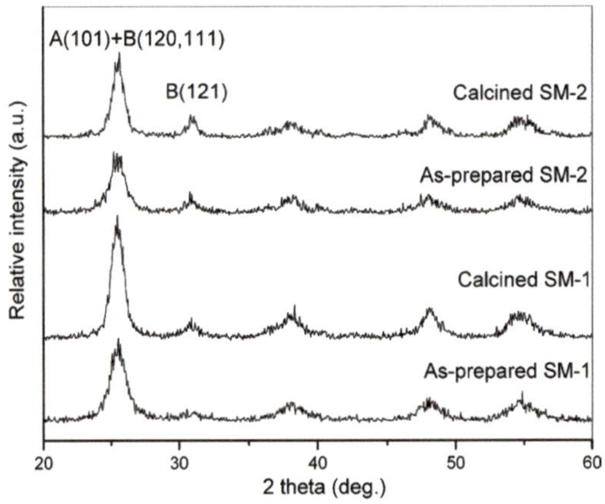

Figure 7. XRD patterns of the as-prepared and calcined M-TiO_2 (SM-1 and SM-2) obtained under high-intensity ultrasound irradiation: A denotes anatase; B denotes brookite [61]. (Copyright 2002 American Chemical Society).

As expected, the photocatalytic activity of M-TiO_2 is significantly improved after doping with metals and non-metals. Besides doping, other methods are used to modify the photocatalytic properties of M-TiO_2, like, for instance, the synthesis of solid solutions or of different TiO_2 composites with noble metals, other metal oxides and quantum dots (not addressed here) [5].

Fe is one of the most used metals for doping: on the one side, it extends the absorption of TiO_2 in the Vis range; on the other side, Fe species present at the surface may give rise to Fenton-like reactions [68], finally enhancing the reactivity merely due to the photocatalytic process (vide infra).

A Fe-doped M-TiO_2 was used to obtain composites with hollow glass microbeads: the latter were able to prevent aggregation of M-TiO_2 NPs [69]. As expected, Fe doping induced a red shift of the absorption band. Consequently, an effective photodegradation of methyl orange (a model molecule of azo dyes) in aqueous solution was achieved under visible light ($\lambda > 420$ nm) irradiation, revealing the potential applicability of such nanocomposites in some industry fields, like water purification.

A Fe-doped M-TiO_2 (with 2.5 wt.% Fe content) obtained by direct synthesis resulted very active towards the catalytic degradation of Acid Orange 7 (a model molecule of azo dyes) not only under UV irradiation but also in dark conditions in the presence of H_2O_2 [70, 71]. The dark process was studied in detail, showing that since not all Fe species entered the M-TiO_2 framework, surface Fe^{3+} species were very active in Fenton-like reaction. Though limited Fe leaching was observed, the Fe-doped M-TiO_2 was still active after reactivation in air. Preliminary results concerning the test reaction under UV-Vis illumination provided further support to this picture. The authors used the same synthesis protocol for V-doping, but in that case, it was not possible to obtain an actual doping, since all the V species resulted present at the surface of the catalyst. In any case, the studied M-TiO_2 materials showed remarkable high specific surface area (150–120 $m^2\ g^{-1}$) and pure anatase NPs.

The effect of other metals used for doping is more complex. For instance, when Ce was used to produce doped thin films of M-TiO_2 [72], Ce species were likely located at the surfaces/grain boundaries of M-TiO_2 crystallites, due to the larger size of Ce^{3+}/Ce^{4+} ions with respect to Ti^{4+} ions. High levels of Ce doping, instead, adversely affected the crystallisation of nanocrystalline anatase since Ce–O–Ti bonds at the grain boundaries inhibited crystallite growth.

The optimal amount of Ce doping corresponded to a ratio Ce/Ti = 0.3 mol%: the corresponding Ce-doped M-TiO_2 showed a remarkable photocatalytic activity towards the degradation of MB (**Figure 8**). The enhanced photocatalytic activity of the Ce-doped M-TiO_2 with Ce/Ti = 0.3 mol% was assigned to enhanced electron transport and oxygen storage capabilities in the presence of Ce, along with the highly nanocrystalline nature of the TiO_2 framework.

3.3. Photocatalytic properties of non-metal doped M-TiO_2

Doping with non-metals (mainly C, N, S, F and I_2 [5]) is also supposed to extend TiO_2 absorption towards the Vis range. The mechanisms responsible of such phenomenon are complex and usually related to either the narrowing of TiO_2 band gap or the creation of intermediate steps within the band gap [66], due to the non-metal atoms substituting oxygen atoms in the framework.

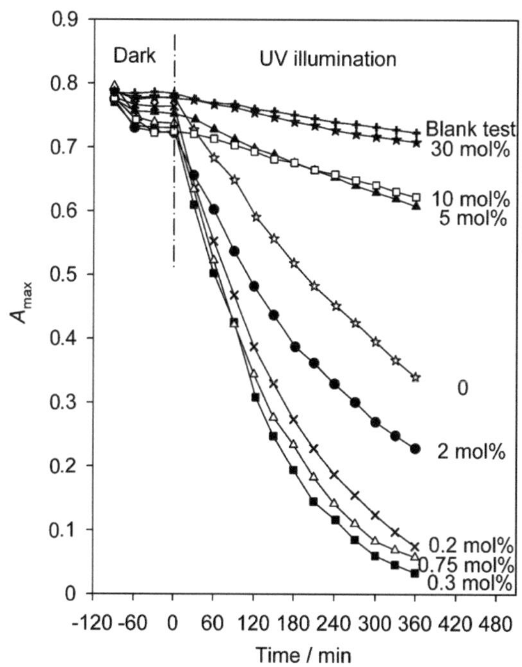

Figure 8. Changes in absorption maximum of MB as a function of exposure time during photocatalytic tests carried out with Ce-doped M-TiO_2 thin films. The best results are obtained with a Ce/Ti = 0.3 mol% composition (after Ref. [72]). (Copyright 2009 American Chemical Society).

N is one of the most used non-metals, as doping may be carried out rather simply by thermally treating M-TiO_2 under $NH_{3(g)}$ flow or by heating a M-TiO_2 produced in solutions containing a N source (NH_3, urea, etc.).

The ultimate effect of N of the light absorption capacity of the sample is however complex and may be due to different processes:

(i) Band-gap narrowing due to N_{2p} states close in energy to O_{2p} states

(ii) Formation of impurity energy levels above the valence band

(iii) Partial doping in oxygen vacancies

All this notwithstanding, literature reports on N-doped M-TiO_2 materials with improved photocatalytic properties. For instance, by a template-free combustion method, a wormhole M-TiO_2 was obtained, where N doping was due to the presence of urea during combustion [73]. The so-obtained N-doped M-TiO_2 occurred as nanocrystalline anatase phase, showing high surface area (234 m^2 g^{-1}) and type-IV H3-mesoporosity.

Two photocatalytic reactions were studied, namely, Rhodamine B degradation and *p*-anisyl alcohol oxidation to *p*-anisaldehyde in aqueous solution under direct sunlight. Notwithstanding the high band gap (3.24 eV) of the N-doped sample, the good activity was assigned to a better utilisation of holes due to the low-charge diffusion barrier associated with wormhole mesoporosity along with the occurrence of crystalline NPs, finally confirming the importance of having an ordered mesoporous photocatalyst.

N-doped anatase M-TiO_2 was prepared via soft-template route by using CTAB as template and by treating in NH_3 (70%)/N_2 atmosphere the calcined samples. The material was characterised by small crystallite size, large surface area (420–126 m^2 g^{-1} for calcination temperatures in the 400–800°C range), high crystallinity and Vis light response. The N-doped anatase M-TiO_2 photocatalysts showed much higher photocatalytic activity than N-doped Degussa P25 for the degradation of phenol under both UV and Vis light irradiation, owing to more oxidising hydroxyl radicals, which were the oxidative species mainly responsible for the degradation of phenol [74]. The authors concluded that the materials might be beneficial to solar-driven applications in the photodegradation of organic pollutants.

However, the effect of doping on the photocatalytic performance is not straightforward, as it has been observed that N-doped TiO_2 shows visible light-responsive photocatalytic activity but lower UV light-responsive photocatalytic activity. The visible light photocatalytic activity originates from new N 2*p* levels near the valence band. The oxygen vacancies and the associated Ti^{3+} species act as the recombination centres for the photoinduced e^-/h^+, finally reducing photocatalytic activity although contributing to Vis light absorbance [75].

4. Conclusions

The production of mesoporous titania by template-assisted methods allows obtaining materials characterised by ordered mesoporous structure, controllable crystallinity, high surface area and tuneable pore size. Moreover, powder nanoparticles or films may be obtained, as well as hierarchical porosity materials.

From the point of view of photocatalytic applications, the type of crystalline phase and ordered mesopores are crucial factors: the former may be responsible of more efficient UV light absorption and slower electron-hole recombination rate; the latter positively affect diffusion processes and mass transfer phenomena. Nonetheless, the possibility of obtaining high surface area materials, also after calcination, positively affects any kind of heterogeneous catalytic process, besides photocatalytic ones, as mesoporous titania may be an efficient support for other types of catalytically active phases.

Several mesoporous titania materials reported by the literature have remarkable photocatalytic properties and are competitive, at least on a lab scale, with commercial samples. It must be considered, however, that template-assisted syntheses require high-cost reagents and/or energy-intensive steps, like calcination.

However, for applications requiring tailored photocatalysts, the progresses made on the side of synthesis procedures will surely allow the development of promising (photo)catalysts based on mesoporous titania.

Author details

Barbara Bonelli[1]*, Serena Esposito[2] and Francesca S. Freyria[1,3]

*Address all correspondence to: barbara.bonelli@polito.it

1 Department of Applied Science and Technology, INSTM Turin-Polytechnic Unit, Polytechnic University of Turin, Turin, Italy

2 Department of Civil and Mechanical Engineering, University of Cassino and Southern Lazio, Cassino, Italy

3 Department of Chemistry, Massachusetts Institute of Technology, Cambridge, MA, USA

References

[1] Chen X, Mao S. Titanium dioxide nanomaterials: Synthesis, properties, modifications and applications. Chemical Reviews 2007;**107**(7):2891-2959. DOI: 10.1021/cr0500535

[2] Li W, Wu Z, Wang J, Elzatahry AA, Zhao D. A perspective on mesoporous TiO_2 materials. Chemistry of Materials. 2014;**26**:287-298. DOI: 10.1021/cm4014859

[3] Boettcher SW, Fan J, Tsung C-K, Shi Q, Stucky GD. Harnessing the sol–gel process for the assembly of non-silicate mesostructured oxide materials. Accounts of Chemical Research. 2007;**40**:784-792. DOI: 10.1021/ar6000389

[4] Zhang R, Elzatahry AA, Al-Deyab SS, Zhao D. Mesoporous titania: From synthesis to application. Nano Today. 2012;**7**(4):344-366. DOI: 10.1016/j.nantod.2012.06.012

[5] Zhou W, Fu H. Mesoporous TiO_2: Preparation, doping, and as a composite for photocatalysis. ChemCatChem. 2015;**5**:885-894. DOI: 10.1002/cctc.201200519

[6] Ismail AA, Bahnemann DW. Mesoporous titania photocatalysts: Preparation, characterization and reaction mechanisms. Journal of Materials Chemistry. 2011;**21**(32):11686-11707. DOI: 10.1039/C1JM10407A

[7] Vivero-Escoto JL, Chiang YD, Wu KC-W, Yamauchi Y. Recent progress in mesoporous titania materials: Adjusting morphology for innovative applications. Science and Techn ology of Advanced Materials. 2012;**13**:013003. DOI: 10.1088/1468-6996/13/1/013003

[8] Pan JH, Zhao XS, Lee WI . Block copolymer-templated synthesis of highly organized mesoporous TiO_2-based films and their photoelectrochemical applications. Chemical Engineering Journal. 2011; **170**(2-3):363-380. DOI: 10.1016/j.cej.2010.11.040

[9] Pan JH, Dou H, Xiong Z, Xu C, Ma J, Zhao XS. Porous photocatalysts for advanced water purifications. Journal of Materials Chemistry. 2010;**20**(22):4521-4528. DOI: 10.1039/B925523K

[10] Brinker CJ, Lu Y, Sellinger A, Fan H. Evaporation-induced self-assembly: Nanostructures made easy. Advanced Materials. 1999;**11**:579-585. DOI: 0935-9648/99/0705-0579

[11] Lu Y, Ganguli R, Drewien CA, Anderson MT, Brinker CJ, Gong W et al. Continuous formation of supported cubic and hexagonal mesoporous films by sol-gel dip-coating. Nature. 1997;**389**:364-368. DOI: 10.1038/38699

[12] Zhao D, Feng Q, Huo Q, Melosh N, Fredrickson GH, Chmelka BF, Stucky GD. Triblock copolymer synthesis of mesoporous silica with periodic 50-300 angstrom pores. Science. 1998;**279**:548-552. DOI: 10.1126/science.279.5350.548

[13] Zhao D, Huo Q, Fen J, Chmelka BF, Stucky GD. Nonionic triblock and star diblock copolymer and oligomeric surfactant syntheses of highly ordered, hydrothermally stable, mesoporous silica structure. Journal of the American Chemical Society. 1998;**120**(24):6024-6036 . DOI: 10.1021/ja974925i

[14] Zhao D, Yang P, Melosh N, Feng J, Chmelka BF, Stuky G. Continuous mesoporous silica films with highly ordered large pore structure. Advanced Materials. 1998;**10**(16):130-1385. DOI: 10.1002/(SICI)1521-4095(199811)10:16<1380::AID-ADMA1380>3.0.CO;2-8

[15] Livage J, Henry M, Snachez C. Sol-gel chemistry of transition metal oxides. Progress in Solid State Chemistry. 1988;**18**:259-341. DOI: 10.1016/0079-6786(88)90005-2

[16] Kresge CT, Leonowicz ME, Roth WJ, Vartuli J, Beck JS. Ordered mesoporous molecular sieves prepared with liquid crystal templates. Nature. 1992;**359**:710-712

[17] Beck JS, Vartuli JC, Roth WJ, Leonowicz ME, Kresge CT, Schmitt KD et al. A new family of mesoporous molecular sieves prepared with liquid crystal templates. Journal of the American Chemical Society. 1992;**114**:10834-10843. DOI: 10.1021/ja00053a020

[18] Wan Y, Zhao D. On the controllable soft-templating approach to mesoporous silicates. Chemical Reviews. 2007;**107**(7):2821-2860. DOI: 10.1021/cr068020s

[19] Kung HH, Ko EI. Preparation of oxide catalysts and catalyst supports—A review of recent advances. Chemical Engineering Journal. 1996;**64**(2):203-214. DOI: 10.1016/S0923-0467(96)03139-9

[20] Crepaldi EL, Soler-Illia GJdAA, Grosso D, Cagnol F, Ribot F, Sanchez C. Controlled formation of highly organized mesoporous titania thin films: From mesostructured hybrids to mesoporous nanoanatase TiO_2. Journal of the American Chemical Society. 2003;**125**(32):9770-9786. DOI: 10.1021/ja030070g

[21] Hou H, Shang M, Wang L, Li W, Tang B, Yang W. Efficient photocatalytic activities of TiO_2 hollow fibers with mixed phases and mesoporous walls. Scientific Reports. 2015;**5**:15228. DOI: 10.1038/srep15228

[22] Antonio E. H. Machado, Karen A. Borges, Tatiana A. Silva, Lidiaine M. Santos, Mariana F. Borges, Werick A. Machado, Bruno P. Caixeta, Marcela Dias França, Samuel M. Oliveira, Alam G. Trovó and Antonio O.T. Patrocínio (2015). Applications of Mesoporous Ordered Semiconductor Materials — Case Study of TiO_2, Solar Radiation Applications, Ms. Segun R. Bello (Ed.), InTech, DOI: 10.5772/59602. Available from: https://www.intechopen.com/books/solar-radiation-applications/applications-of-mesoporous-ordered-semiconductor-materials-case-study-of-tio2

[23] Yang HF, Zhao DY. Synthesis of replica mesostructures by the nanocasting strategy. Journal of Materials Chemistry. 2005;**15**:1217-1231. DOI: 10.1039/B414402C

[24] Lu AH, Schüth F. Nanocasting: A versatile strategy for creating nanostructured porous materials. Advanced Materials. 2006;**18**:1793-1805. DOI: 10.1002/adma.200600148

[25] Zhang Z, Zuo F, Feng P. Hard template synthesis of crystalline mesoporous anatase TiO_2 for photocatalytic hydrogen evolution. Journal of Materials Chemistry. 2010;**20**(11):2206-2212. DOI: 10.1039/B921157H

[26] Ren Y, Ma Z, Bruce PG. Ordered mesoporous metal oxides: Synthesis and applications. Chemical Society Reviews. 2012;**41**:4909-4927. DOI: 10.1039/C2CS35086F

[27] Yue W, Xu X, Irvine JT, Attidekou PS, Liu C, He H et al. Mesoporous monocrystalline TiO_2 and its solid-state electrochemical properties. Chemistry of Materials. 2009;**21**:2540-2546. DOI: 10.1021/cm900197p

[28] Yue W, Randorn C, Attidekou PS, Liu C, Irvine JTS, Zhou W. Syntheses, Li insertion, and photoactivity of mesoporous crystalline TiO_2. Advanced Functional Materials. 2009;**19**:2826-2833. DOI: 10.1002/adfm.200900658

[29] Weatherspoon MR, Cai Y, Srinivasarao M, Sandhage KH. 3D rutile titania-based structures with *Morpho* butterfly wing scale morphologies. Angewandte Chemie, International Edition. 2008;**47**:7921-7923. DOI: 10.1002/anie.20081311

[30] Kaune G, Memesa M, Meier R, Ruderer MA, Diethert A, Roth SV et al. Hierarchically structured titania films prepared by polymer/colloidal templating. ACS Applied Materials and Interfaces. 2009;**1**(12):2862-2869. DOI: 10.1021/am900592u

[31] Sun W, Zhou S, You B, Wu L. Facile fabrication and high photoelectric properties of hierarchically ordered porous TiO_2. Chemistry of Materials. 2012;**24**:3800-3810. DOI: 10.1021/cm302464g

[32] Antonelli DM, Ying JY. Synthesis of hexagonally packed mesoporous TiO_2 by a modified sol–gel method. Angewandte Chemie International Edition in English. 1995;**34**(18):2014-2017. DOI: 10.1002/anie.199520141

[33] Kluson P, Kacer P, Cajthlam T, Kalaji M. Preparation of titania mesoporous materials using a surfactant mediated sol-gel method. Journal of Materials Chemistry. 2001;**11**:644-651. DOI: 10.1039/b004760k

[34] Zimny K, Roques-Carmes T, Carteret C, Stébé MJ, Blin JL. Synthesis and photoactivity of ordered mesoporous titania with a semicrystalline framework. Journal of Physical Chemistry C. 2012;**116**(11):6585-6594. DOI: 10.1021/jp212428k

[35] Wang D, Choi D, Yang Z, Viswanathan VV, Nie Z, Wang C et al. Synthesis and Li-ion insertion properties of highly crystalline mesoporous rutile TiO_2. Chemistry of Materials. 2008;**20**(10):3435-3442. DOI: 10.1021/cm8002589

[36] Yoshitake H, Sugihara T, Tatsumi T. Preparation of wormhole-like mesoporous TiO_2 with an extremely large surface area and stabilization of its surface by chemical vapor deposition. Chemistry of Materials. 2002;**14**(3):1023-1029. DOI: 10.1021/cm010539b

[37] Choi H, Antoniou MG, Pelaez M, de La Cruz AA, Shoemaker JA, Dionysiou DD. Mesoporous nitrogen-doped TiO_2 for the photocatalytic destruction of the cyanobacterial toxin microcystin-LR under visible light irradiation. Environmental Science and Technology. 2007; **41**(21):7530-7535. DOI: 10.1021/es0709122

[38] Zhao D, Peng T, Lu L, Cai P, Jiang P, Bian Z. Effect of annealing temperature on the photoelectrochemical properties of dye-sensitized solar cells made with mesoporous TiO_2 nanoparticles. Journal of Physical Chemistry C. 2008;**112**(22):8486-8494. DOI: 10.1021/jp800127x

[39] Antonelli DM. Synthesis of phosphorus-free mesoporous titania via templating with amine surfactants. Microporous and Mesoporous Materials. 1999;**30**(2-3):315-319. DOI: 10.1016/S1387-1811(99)00042-6

[40] Chen D, Cao L, Huang F, Imperia P, Cheng Y-B, Caruso RA. Synthesis of monodisperse mesoporous titania beads with controllable diameter, high surface areas, and variable pore diameters (14–23 nm). Journal of the American Chemical Society. 2010;**132**(12):4438-4444. DOI: 10.1021/ja100040p

[41] Yang P, Zhao D, Margolese DL, Chmelka BF, Stucky GD. Generalized syntheses of large-pore mesoporous metal oxides with semicrystalline frameworks. Nature. 1998;**396**:152-155. DOI: 10.1038/24132

[42] Bucknall DG Anderson HL. Polymers get organized. Science. 2003;**302**(5652):1904-1905. DOI: 10.1126/science.1091064

[43] Grosso D, Cagnol F, Soler-Illia GJDAA, Crepaldi EL, Amenitisch H, Brunet-Bruneau A et al. Fundamentals of mesostructuring through evaporation-induced self-assembly. Advanced Functional Materials. 2004;**14**(4):309-322. DOI: 10.1002/adfm.200305036

[44] Chandra P, Doke DS, Umbarkar SB, Biradar AV. One-pot synthesis of ultrasmall MoO_3 nanoparticles supported on SiO_2, TiO_2, and ZrO_2 nanospheres: An efficient epoxidation catalyst. Journal of Materials Chemistry A. 2014;**2**:19060-19066. DOI: 10.1039/c4ta03754e

[45] Fan J, Boettcher SW, Study GD. Nanoparticle assembly of ordered multicomponent mesostructured metal oxides via a versatile sol-gel process. Chemistry of Materials. 2006;**18**(26):6391-6396. DOI: 10.1021/cm062359d

[46] Zhang J, Deng Y, Gu D, Wang S, She L, Che R et al. Ligand-assisted assembly approach to synthesize large-pore ordered mesoporous titania with thermally stable and crystalline framework. Advanced Energy Materials. 2011;**1**:241-248. DOI: 10.1002/aenm.201000004

[47] Tian B, Yang H, Liu X, Xie S, Yu C, Fan J et al. Fast preparation of highly ordered nonsiliceous mesoporous materials via mixed inorganic precursors. Chemical Communications. 2002;**17**:1824-1825. DOI: 10.1039/B205006D

[48] Tian B, Lu X, Tu B, Yu C, Fan J, Wang L et al. Self-adjusted synthesis of ordered stable mesoporous minerals by acid–base pairs. Nature Materials. 2003;**2**:159-163. DOI: 10.1038/nmat838

[49] Crepaldi EL, Soler-Illia GJDAA, Grosso D, Sanchez C. Nanocrystallized titania and zirconia mesoporous thin films exhibiting enhanced thermal stability. New Journal of Chemistry. 2003;**27**:9-13. DOI: 10.1039/b205497n

[50] Soler-Illia GJdAA, Louis A, Sanchez C. Synthesis and characterization of mesostructured titania-based materials through evaporation-induced self-assembly. Chemistry of Materials. 2002;**14**:750-759. DOI: 10.1021/cm011217a.

[51] Peng T, Zhao D, Dai K, Shi W, Hirao K. Synthesis of titanium dioxide nanoparticles with mesoporous anatase wall and high photocatalytic activity. Journal of Physical Chemistry B. 2005;**109**(11):4947-4952. DOI: 10.1021/jp044771r.

[52] Liu K, Fu H, Shi K, Xiao F, Jing L, Xin B. Preparation of large-pore mesoporous nanocrystalline TiO_2 thin films with tailored pore diameters. Journal of Physical Chemistry B. 2005;**109**(40):18719-18722. DOI: 10.1021/jp054546p

[53] Alberius PCA, Frindell KL, Hayward RC, Kramer EJ, Stucky GD, Chmelka, BF. General predictive syntheses of cubic, hexagonal, and lamellar silica and titania mesostructured thin films. Chemistry of Materials. 2002;**14**:3284-3294. DOI: 10.1021/cm011209u

[54] Li H, Wang J, Li H, Yin S, Sato T. High thermal stability thick wall mesoporous titania thin films. Materials Letters. 2009;**63**(18-19):1583-1585. DOI: 10.1016/j.matlet.2009.04.017

[55] Zhang R, To B, Zhao D. Synthesis of highly stable and crystalline mesoporous anatase by using a simple surfactant sulfuric acid carbonization method. Chemistry—A European Journal. 2010;**16**:9977-9981. DOI: 10.1002/chem.201001241

[56] Zhou W, Sun F, Pan K, Tian G, Jiang B, Ren Z et al. Well-ordered large-pore mesoporous anatase TiO_2 with Remarkably high thermal stability and improved crystallinity: Preparation, characterization, and photocatalytic performance. Advanced Functional Materials. 2011;**21**(10):1922-1930. DOI: 10.1002/adfm.201002535

[57] Compagnoni M, Ramis G, Freyria FS, Armandi M, Bonelli B, Rossetti I. Photocatalytic processes for the abatement of N-containing pollutants from waste water. Part 1: Inorganic pollutants. Journal of Nanoscience and Nanotechnology. 2017;**17**(6):3632-3653. DOI: 10.1166/jnn.2017.14006

[58] Freyria FS, Armandi M, Compagnoni M, Ramis G, Rossetti I, Bonelli B. Catalytic and photocatalytic processes for the abatement of N-containing pollutants from wastewater. Part 2: Organic pollutants. Journal of Nanoscience and Nanotechnology. 2017;**17**(6):3654-3672. DOI: 10.1166/jnn.2017.14014

[59] Kao L-H, Hsu T-C, Cheng K-K. Novel synthesis of high-surface-area ordered mesoporous TiO_2 with anatase framework for photocatalytic applications. Journal of Colloid and Interface Science. 2010;**341**:359-365. DOI: 10.1016/j.jcis.2009.09.058

[60] Liu J, An T, Li G, Bao N, Sheng G, Fu J. Preparation and characterization of highly active mesoporous TiO_2 photocatalysts by hydrothermal synthesis under weak acid conditions. Microporous and Mesoporous Materials. 2009;**124**:197-203. DOI: 10.1016/j.micromeso.2009.05.009

[61] Yu JC, Zhang L, Yu J. Direct sonochemical preparation and characterization of highly active mesoporous TiO_2 with a bicrystalline framework. Chemistry of Materials. 2002;**14**:4647-4653. DOI: 10.1021/cm0203924

[62] Bickley RI, Gonzalez-Carreno T,Lees JS, Palmisano L, Tilly RJD. A structural investigation of titanium dioxide photocatalysts. Journal of Solid State Chemistry. 1991;**92**: 178-190. DOI: 10.1016/0022-4596(91)90255-G

[63] Sreethawong T, Ngamsinlapasathian S, Yoshikawa S. Positive role of incorporating P25 TiO_2 to mesoporous-assembled TiO_2 thin films for improving photocatalytic dye degradation efficiency. Journal of Colloid and Interface Science. 2014;**430**:184-192. DOI: 10.1016/j.jcis.2014.05.032

[64] Kim E-Y, Kim,DS, Ahn B-T. Synthesis of mesoporous TiO_2 and its application to photocatalytic activation of Methylene Blue and *E. coli*. Bulletin of the Korean Chemical Society. 2009;**30**(1):193-196

[65] Chen X, Shen S, Guo L, Mao SS. Semiconductor-based photocatalytic hydrogen generation. Chemical Reviews. 2010;**110**:6503-6570. DOI: 10.1021/cr1001645

[66] Malato S, Fernández-Ibáñez P, Maldonado MI, Blanco J, Gernjak W. Decontamination and disinfection of water by solar photocatalysis: Recent overview and trends. Catalysis Today. 2009;**147**(1):1-59. DOI: 10.1016/j.cattod.2009.06.018

[67] Chong MN, Jin B, Chow CWK, Saint C. Recent developments in photocatalytic water treatment technology: A review. Water Research. 2010;**44**:2997-3027. DOI: 10.1016/j.watres.2010.02.039

[68] Pignatello JJ, Liu D, Houston P. Evidence for an additional oxidant in the photoassisted fenton reaction. Environmental Science & Technology. 1999;**33**:1832-1839. DOI: 10.1021/es980969b

[69] Cui L, Wang Y, Niu M, Chen G, Cheng Y. Synthesis and visible light photocatalysis of Fe-doped TiO_2 mesoporous layers deposited on hollow glass microbeads. Journal of Solid State Chemistry. 2009;**182**(10):2785-2790. DOI: 10.1016/j.jssc.2009.07.045

[70] Piumetti M, Freyria FS, Armandi M, Geobaldo F, Garrone E, Bonelli B. Fe- and V-doped mesoporous titania prepared by direct synthesis: Characterization and role in the oxidation of AO7 by H_2O_2 in the dark. Catalysis Today. 2014;**227**:71-79. DOI: 10.1016/j.cattod.2013.11.013

[71] Piumetti M, Freyria FS, Armandi M, Saracco G, Garrone E, Bonelli B. Anti-oxidant/pro-oxidant activity of ascorbic acid: Effect in the degradation of Acid Orange 7 with H_2O_2 catalyzed by transition metal ions. Chemistry Today. 2015;**33**(3):40-45.

[72] Zhang Y, Yuwono AH, Wang J, Li J. Enhanced photocatalysis by doping cerium into mesoporous titania thin films. Journal of Physical Chemistry C. 2009;**113**(51):21406-21412. DOI: 10.1021/jp907901k

[73] Sivanarjani K, Gopinath CS. Porosity driven photocatalytic activity of wormhole mesoporous TiO_2-xNx in direct sunlight. Journal of Materials Chemistry. 2011;**21**:2639-2647. DOI: 10.1039/C0JM03825C

[74] Tian G, Chen Y, Pan K, Wang D, Zhou W, Ren Z et al. Efficient visible light-induced degradation of phenol on N-doped anatase TiO_2 with large surface area and high crystallinity. Applied Surface Science. 2010;**256**(12):3740-3745. DOI: 10.1016/j.apsusc.2010.01.016

[75] Wang J, Tafen DN, Lewis JP, Hong Z, Manivavnnan A et al. Origin of photocatalytic activity of nitrogen-doped TiO_2 nanobelts. Journal of the American Chemical Society. 2009;**131**(34):12290-12297. DOI: 10.1021/ja903781h

Chapter 7

Advanced Hybrid Materials Based on Titanium Dioxide for Environmental and Electrochemical Applications

Katarzyna Siwińska-Stefańska and
Teofil Jesionowski

Additional information is available at the end of the chapter

http://dx.doi.org/10.5772/intechopen.69357

Abstract

Constant technological progress, as well as the pursuit of "friendly" technologies, leads to intensive work on the development of a new generation of advanced products with strictly defined, unique physicochemical properties dedicated to specific applications. This group of materials includes hybrids based on titanium dioxide and its derivatives, characterised with specific, well-defined physicochemical and structural properties, chiefly determined during their synthesis. Different properties of titania nanoparticles depend on their morphology, crystallite size, and crystalline structure. Nanocrystalline titanium dioxide can be synthesised via different methods, among which chemical precipitation, microemulsion method (inversed micelles), sol-gel process and hydrothermal crystallisation are the most important ones. That is why, a crucial part of the following chapter will be paid to characterisation of synthesis routes used for titanium dioxide and titania-based hybrid production. Furthermore, application of TiO_2-based materials, including mixed oxide systems as well as graphene oxide–based hybrids, in electrochemical (electrode material) and environmental (photocatalysis) aspects, will be described in detail.

Keywords: titanium dioxide, TiO_2-based hybrid materials, advanced material synthesis, photocatalyst, electrode material

1. Introduction

Titanium dioxide, commonly known as titanium white, is in normal conditions a colourless crystalline solid. It is non-toxic, non-hygroscopic, inflammable and non-volatile. It shows high chemical stability: it does not dissolve in water, organic solvents, acids (except for concentrated hydrofluoric acid and sulphuric acid) or alkalis. It is amphoteric, but more acidic than basic. Titanium dioxide is thermally stable: it loses oxygen only at a temperature of a few hundred degrees Celsius and under the influence of reducing agents (carbon, magnesium, hydrogen and halogens). Its melting point is 1825°C, while its boiling point is close to 2500°C. Above 400°C, a reversible change in colour to yellow takes place as a result of thermal expansion of the crystalline lattice. Above 1000°C, the oxide forms of titanium are formed, characterised by a lower content of oxygen than in TiO_2, an undesirable colour change takes place and the electrical conductivity changes. Titanium dioxide does not show activity towards living organisms [1–4].

Titanium dioxide occurs in nature in three polymorphous varieties: tetragonal rutile, anatase and rhombic brookite. Anatase and rutile are of practical importance and are commonly used in many applications, while brookite is not used because of the instability of its structure [4–6].

The rapidly developing technologies for obtaining new functional materials based on titania are an especially important topic for both theoretical study and practical application. The continually growing requirements of different technologies require new directions to be sought in order to obtain materials with precisely designed physicochemical and structural properties. Hybrid systems based on TiO_2 constitute a new group of compounds exhibiting strictly designed physicochemical properties resulting from the effects of combining the characteristic behaviours of the individual compounds from which they are made. The presence of a foreign element in the matrix of pure titania can greatly affect the structural, textural, acid/base and catalytic properties [7]. The textural properties of the hybrid materials, such as pore size distribution, surface area, etc., are strongly dependent upon the conditions of synthesis, including the nature and composition of the precursors, solvent, complexing/templating agent, hydrolysis and calcination conditions [8]. Research into the production and potential applications of new functional materials based on titanium dioxide is only possible when the final materials have a strictly defined dispersive character, crystalline structure, morphology and porous structure [9–13].

2. Synthesis of titanium dioxide and its derivatives

The synthesis of titanium dioxide is one of the major research areas in 'green chemistry'. Titania is a chemically inert, thermally stable, insoluble, biocompatible, non-toxic material, and an excellent absorber of destructive UV radiation. Because of these properties, titanium dioxide and its derivatives have for some time enjoyed a great and still growing popularity in many applications; see **Figure 1** [1–4].

On an industrial scale, titanium dioxide pigments are obtained by two methods (see **Figure 2**) whose names refer to the substrate salts used:

• the sulphate method, in which TiO_2 is precipitated from a solution of ilmenite ore by concentrated sulphuric acid, leading to both rutile and anatase;

• the chloride method, in which titanium dioxide is obtained by oxidation of titanium tetrachloride ($TiCl_4$) obtained by reduction and chlorination of ilmenite ore; this method leads only to rutile.

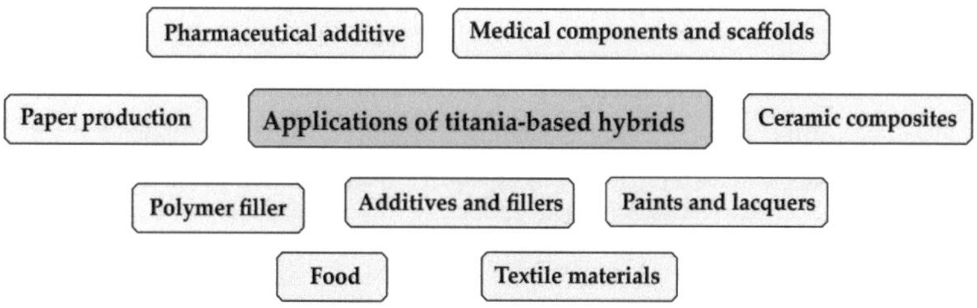

Figure 1. Applications of titania-based materials.

The use of the chloride process is on the increase. It is estimated that over half of the world's titanium dioxide is produced in this way. Although this process requires higher quality, previously enriched ore and more complex technology, it produces much less waste than the sulphate method, and the cost of production is also lower [1].

The properties of TiO_2 are determined by the morphology of its particles, the size of its crystals and its crystalline structure, which depend on the choice of method for its synthesis and final heat treatment [14]. Nanocrystalline TiO_2 particles are usually obtained by crystallisation (chemical precipitation) [15], the microemulsion method (reverse micelles) [16], the sol-gel method [17–21] or hydrothermal crystallisation [22–26]; see **Figure 2**. Each of these methods has its advantages and drawbacks, but a feature of all of them is the ability to obtain materials with strictly defined properties (**Table 1**).

Additionally, **Table 2** presents a comprehensive review of different methods of synthesis of titania-based materials.

2.1. Chemical precipitation

Co-precipitation is a wet chemical method and is one of the oldest methods for obtaining nanometric materials. The most common precursors used in this method are salts: nitrates,

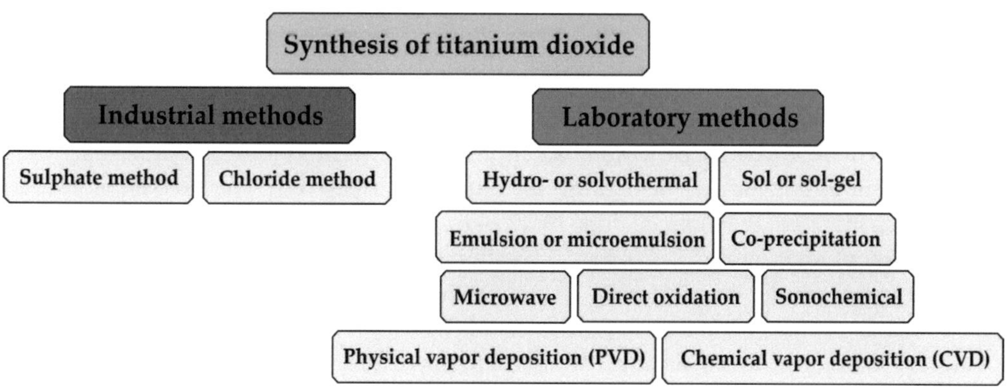

Figure 2. Synthesis of titanium dioxide.

Method	Advantages	Disadvantages
Sulphate method	Ability to control the crystalline structure; possibility of using low-quality ore	Susceptibility to agglomeration of particles; high energy costs; large quantities of waste produced and requiring disposal
Chloride method	High-quality products (including for medical applications); defined morphological structure and high degree of particle dispersion	Need to use high-quality substrates; production of only the rutile form, advanced technology; hazardous reaction environment, toxicity; danger of uncontrolled emission of chlorine gas
Co-precipitation method	Homogeneous mixing of reactant precipitates reduces the reaction temperature; a simple direct process for the synthesis of fine material powders which are highly reactive in low-temperature sintering	Not suitable for the preparation of a highly pure, stoichiometrically accurate phase; this method does not work well if the reactants have very different solubilities and precipitation rates; it does not meet universal experimental conditions for the synthesis of various types of metal oxides
Sol-gel route	Low temperature processing and consolidation is possible; smaller particle size and morphological control in powder synthesis; sintering at low temperature also possible; better homogeneity and phase purity than in traditional ceramic methods	Raw materials for this process are expensive (in the case of metal alkoxides) compared with mineral-based metal ion sources; products have high carbon content when organic reagents are used in the preparative steps and this inhibits densification during sintering; since several steps are involved, close monitoring of the process is needed
Hydrothermal route	Powders are formed directly from solution; it is possible to control particle size and shapes by using different starting materials and hydrothermal conditions; the resulting powders are highly reactive, which aid low-temperature sintering	Prior knowledge on the solubility of the starting materials is required; hydrothermal slurries are potentially corrosive; accidental explosion of the high-pressure vessel cannot be ruled out
Solvothermal method	An energy-efficient, environmentally friendly process; high-purity products can be synthesised; metastable and new phases can be accessed; simplified and precise control of the size, shape distribution, and crystallinity of the end product can be achieved via the adjustment of parameters such as reaction temperatures and time, the types of solvents, surfactants and precursors	The need for expensive autoclaves; safety issues during the reaction process; impossibility of observing the reaction process ("black box")

Table 1. Advantages and disadvantages of the most commonly used methods for the synthesis of titanium dioxide.

chlorates or chlorides, which dissolve in an appropriate solvent. Aqueous solutions are used most commonly, but the use of organic solvents is also possible. The precipitation reaction must be initiated; this may be done by changing the pH, concentration or temperature. Another way of starting the precipitation reaction is to carry out a reaction of hydrolysis, oxidation, reduction or complexation. Usually a base (potassium or sodium hydroxide) is added to the system containing the precursors of the oxide. The precipitation reaction itself involves reduction of the metal cation, and formation of a precipitate requires that the system reach saturation point. The precipitation method consists of three stages: nucleation, growth and agglomeration. To begin with, small crystallites are formed (nucleation), which over

http://dx.doi.org/10.5772/intechopen.69357

Product	Starting materials	Conditions of synthesis	Properties of obtained material	Potential application	Ref.
Co-precipitation method					
TiO_2	$TiCl_4$, HCl, H_2O_2, $NH_3{\cdot}H_2O$	Reaction: ice-water bath, then 10°C, next 50°C, Drying: 60°C, Calcination: 500°C for 2 h	Anatase and rutile structure	Photodegradation of helianthine	[29]
TiO_2	$Ti(OCH(CH_3)_2)_4$ (TTIP), C_3H_7OH, polyvinyl pyrrolidone (PVP)	Reflux: 2 h, Drying: 80°C for one day, Calcination: 800°C	Rutile structure, spherical shape of 10 nm in diameter	–	[30]
TiO_2/SnO_2	$TiCl_4$, $SnCl_4{\cdot}5H_2O$, $CO(NH_2)_2$	Reaction: room temperature, after adding $CO(NH_2)_2$—80°C for 8 h, ageing: room temperature for 24 h, drying: 60°C calcination: 550°C for 4 h	Rutile structure, BET surface area: 82 m^2/g, aggregated particles with a different size distribution, crystallite size: 17.4 nm	Degradation of methyl blue (MB)—90%	[31]
TiO_2/CeO_2 and TiO_2/SnO_2	$Ce(NO_3)_3{\cdot}6H_2O$, $SnCl_4{\cdot}5H_2O$, $Ti(SO_4)_2$, $NH_3{\cdot}H_2O$	Reaction: room temperature, 3 h, pH = 10 ageing: 48 h, drying: 105°C, 12 h, calcination: 500°C for 6 h	BET surface area: TiO_2/CeO_2 = 108 m^2/g, TiO_2/SnO_2 = 59 m^2/g, $TiO_2/CeO_2/SnO_2$ = 139 m^2/g, crystalline structure: TiO_2/CeO_2—anatase, TiO_2/SnO_2—rutile, $TiO_2/CeO_2/SnO_2$—amorphous	Reduction of NO with NH_3	[32]
TiO_2/ZrO_2	$TiCl_4$, $ZrOCl_2{\cdot}8H_2O$, $NH_3{\cdot}H_2O$, HCl	Reaction time: 3 h, ageing: 24 h, drying: 110°C (overnight) calcination: 500°C for 5 h	Amorphous structure, BET surface area: 234 m^2/g	Reduction of NO by NH_3	[33]
Sol-gel method					
TiO_2	$Ti(OC(CH_3)_3)_4$ (TBOT), HCl, CH_3OH, C_2H_5OH, CH_3COOH, F-127 (triblock copolymer)	Reaction time: 60 min, drying: 65°C, ageing: 24 h, calcination: 400, 500, 600, 700 and 800°C in air for 4 h	Anatase and rutile structure, BET surface area from 165 m^2/g (500°C) to 15 m^2/g (800°C) nanoparticles diameter: 10–15 nm (500°C), 50–75 nm (800°C), agglomerates: 1–3 μm (500°C), 1–5 μm (800°C)	Photodegradation of imazapyr (98%) and phenol (95%)	[54]

Product	Starting materials	Conditions of synthesis	Properties of obtained material	Potential application	Ref.
TiO_2	$Ti(OCH(CH_3)_2)_4$ (TTIP), C_3H_7OH, NaOH, HNO_3	Reaction: 80°C (5 h), then cooling to room temperature, ageing: 25°C for 24 h, drying: 100°C for 12 h, calcination: 200, 600 and 800°C for 2 h	Anatase-brookite (calcination at 200°C), anatase-brookite-rutile (calcination at 600 and 800°C), irregular clusters composed of spherical nanomeric primary particles (200 nm—pH = 2, 400 nm—pH = 9), BET surface area: pH = 2—calcination at: 200°C—186 m^2/g, 600°C—48 m^2/g, pH = 4—calcination at: 200°C—109 m^2/g, 600°C—42 m^2/g	Photodegradation of methylene blue (MB)—98%—samples calcined at 200 or 600°C by pH = 2	[55]
TiO_2/SiO_2	$Si(OC_2H_5)_4$ (TEOS), C_2H_5OH, HCl, $Ti(OCH(CH_3)_2)_4$ (TTIP)	Reaction: room temperature, ageing: room temperature for 72 h, drying: 80°C, calcination: 1000°C for 2 h	Anatase structure (400°C), rutile with a small amount of anatase (800°C), rutile (1000°C)	–	[57]
TiO_2/ZrO_2	$Ti(OC(CH_3)_3)_4$ (TBOT), HNO_3, C_2H_5OH, Pluronic P123 and Macrogol 20000 (triblock copolymers), $ZrOCl_2 \cdot 8H_2O$	Reaction: room temperature, after adding Zr precursor—80°C, ageing: room temperature for 24 h, drying: room temperature in air, calcination: 800°C for 5 h	Anatase structure, BET surface area: 149 m^2/g	Photodegradation of Rhodamine B (RhB)—90%	[59]
TiO_2, TiO_2/ZrO_2	$TiOCl_2$, $ZrO(NO_3)_2$, NaOH, cetyltrimethylammonium bromide (CTAB), C_2H_5OH	Reaction: room temperature, ageing: 80°C for 4 h, drying: 100°C for 6 h, calcination: 600-900°C for 2 h	Irregular spherical agglomerates, crystalline structure: 600°C—anatase TiO_2 and tetragonal ZrO_2 crystals, 800°C—rutile and anatase TiO_2 and tetragonal ZrO_2 crystals, 900°C—rutile TiO_2 and tetragonal and monoclinic ZrO_2 crystals, BET surface area: TT-600—88 m^2/g, ZT8-600—70 m^2/g, ZT14-600—62 m^2/g, ZT22-600—63 m^2/g, ZT32-600—61 m^2/g	Photodecolorization of the MB—sample ZT8-600—94.1%	[60]
ZrO_2/Al_2O_3 and TiO_2/Al_2O_3	2,4-Pentanedione, n-butanol, alkoxides of the respective metals	Reaction: room temperature, then 70°C drying, calcination: 500°C	Amorphous structure of TiO_2/Al_2O_3, crystalline size: TiO_2/Al_2O_3—4.9 nm, ZrO_2/Al_2O_3—4.1 nm, BET surface area: TiO_2/Al_2O_3—320 m^2/g, ZrO_2/Al_2O_3—200 m^2/g	Ethanol conversion: TiO_2/Al_2O_3—99%, ZrO_2/Al_2O_3—47%	[61]

http://dx.doi.org/10.5772/intechopen.69357

Product	Starting materials	Conditions of synthesis	Properties of obtained material	Potential application	Ref.
TiO_2/ZrO_2	$Zr(OCH_2CH_2CH_3)_4$, $Ti(OCH_2CH_2CH_3)_4$, C_3H_7OH, $NH_3{\cdot}H_2O$	Reaction: 18°C, pH = 9, then 27°C, refluxe: 1 h, ageing: 24 h, heating: 100°C for 24 h, calcination: 550 and 700°C for 5 h	Crystalline structure: anatase (66 wt.%), rutile (19 wt.%) and orthorhombic (15 wt.%) srilankite type—sample TiZr13 calcined at 700°C, amorphous—sample TiZr37 calcined at 550°C, BET surface area: sample TiZr37 calcined at 550°C—172 m^2/g	–	[62]
Hydrothermal method					
TiO_2	$Ti(OCH(CH_3)_2)_4$ (TTIP), C_2H_5OH, HNO_3	pH = 0.7, hydrothermal treatment—240°C for 4 h, drying: room temperature	Anatase and brookite structure, size of particles—21–23 nm BET surface area: 190 m^2/g (7 nm), 124 m^2/g (15 nm), 75 m^2/g (24 nm)	Decomposition of propan-2-ol to CO_2 (98%)	[65]
TiO_2	Anatase TiO_2, NaOH	Hydrothermal treatment—200°C for 24 h, drying: 70°C for 6 h	Anatase and brookite structure, TiO_2 nanowires	–	[66]
TiO_2/ZrO_2	$ZrOCl_2{\cdot}8H_2O$, $TiCl_4$, $TiCl_4$, $NH_3{\cdot}H_2O$	Hydrothermal treatment—220°C for 4 h, 25 bar drying: 120°C calcination: 500°C for 10 h	Amorphous structure, BET surface area: 209 m^2/g	–	[67]
TiO_2/ZrO_2	$TiOSO_4$, $ZrCl_4$	Hydrothermal treatment—200–240°C for 48 h, drying: 60°C calcination: 400–1000°C for 1 h	Anatase structure	Photodecolorization of the MB	[68]
Solvothermal method					
TiO_2	Polyethyleneimine (PEI), C_2H_5OH, $Ti(OC(CH_3)_3)_4$ (TBOT)	Mixing: 5 h, thermal treatment—180°C for 24 h, drying: 60°C, 6 h, calcination: 400°C for 2 h	Anatase structure, crystalline size: 12.1 nm, sample after calcination: anatase (82.3%) and rutile (17.7%) structure, quasi-spherical nanostructures with diameters of ca. 100–200 nm, BET surface area: 118 m^2/g	Photodegradation of methyl orange (MO)—74% and phenol—82%	[71]

Product	Starting materials	Conditions of synthesis	Properties of obtained material	Potential application	Ref.
TiO_2	HCl, TiF_4, C_3H_7OH, HF	Thermal treatment—180°C for 5.5–44 h, drying: in vacuum overnight, calcination: 600°C for 90 min	Anatase structure	–	[72]
TiO_2/SiO_2 and TiO_2/ZrO_2	Titanium(IV) n-butoxide, tetraethylorthosilicate, zirconium(IV) n-butoxide, toluene	Thermal treatment—300°C	Anatase structure, crystallite size from 11.0 to 9.0 nm, BET surface area: TiO_2/SiO_2—from 133 to 156 m^2/g, TiO_2/ZrO_2—from 95 to 106 m^2/g	Conversion of ethylene TiO_2/SiO_2—from 22.1 to 32.4%, TiO_2/ZrO_2—from 22.2 to 39.5%	[74]

Table 2. Synthesis of titania-based materials via different method.

time become more thermodynamically stable and larger (the growth stage); then as more time elapses, the permanent crystallites combine into persistent agglomerate forms (agglomeration). The properties of the final material are strongly influenced by the initial nucleation stage. An important part is also played by the particle agglomeration stage, which determines the morphological properties of the system [27, 28]. To obtain the final product, the dried precipitate is subjected to thermal treatment at the required temperature in an appropriate atmosphere.

Liu et al. [29] prepared titanium dioxide by five different methods: co-precipitation, the sol-gel route, hydrolysis, the hydrothermal method and sluggish precipitation. They investigated how the method used affected the physicochemical properties of the resulting TiO_2. Synthesis of TiO_2 by the co-precipitation method was carried out using titanium tetrachloride, hydrochloric acid, hydrogen peroxide and ammonia. Titanium tetrachloride as a precursor of TiO_2 was added to hydrochloric acid and deionised water. The resulting mixture was maintained at a temperature below 10°C, and hydrogen peroxide was added. Finally, ammonia was added to the solution (pH = 10). The resulting sample was calcined at 500°C for 2 h. XRD analysis showed that this procedure of TiO_2 synthesis leads to a mixture of anatase and rutile phases with anatase predominant. The material synthesised by the co-precipitation method also demonstrated with good photocatalytic activity in the decomposition of helianthine (with absorbency equal to 0.25 and transmission equal to 60.0).

Muhamed Shajudheen et al. [30] synthesised titanium dioxide using titanium tetraisopropoxide as a Ti precursor and poly(vinyl pyrrolidone) (PVP) as a capping agent. To a mixture consisting of titanium tetraisopropoxide and propan-2-ol, PVP and then water were added. The resulting white precipitate was refluxed for 2 h and then stirred continuously for 1 day, followed by calcination at 800°C. The proposed co-precipitation method allows synthesis of the rutile phase of titania in a single-step process without impurities and other phases, as was confirmed by XRD and Raman spectroscopy.

To alter the physicochemical properties of titanium dioxide, Huang et al. [31] prepared SnO_2/TiO_2 catalysts using five different preparation methods: the sol-gel method (SGM), the sol-hydrothermal method (SHM), the co-precipitation method (CM), a co-precipitation-hydrothermal method (CHM) and the hydrothermal method (HM). They determined the impact of the methodology for obtaining SnO_2/TiO_2 systems on the structure, chemical composition, particle sizes, specific areas, pore size distribution and energy band structure. The synthesis of SnO_2/TiO_2 by the co-precipitation method was carried out using $TiCl_4$ and $SnCl_4 \cdot 5H_2O$, which were dissolved in deionised water. In the next step, the obtained solution was added to an aqueous solution of urea. At this stage of the synthesis, the reaction was carried out at a temperature of 80°C for 8 h. At the final stage, the resulting material was calcined at 550°C for 4 h. XRD analysis of the sample obtained by the co-precipitation method revealed the presence of diffraction peaks indicating a pure rutile structure. Moreover, peaks corresponding to SnO_2 were not present in the pattern. This may indicate that the Sn^{4+} ion was successfully incorporated into the crystal lattice sites of the titania to form uniform SnO_2/TiO_2 solid solutions, or that the reflection bands attributed to SnO_2 overlapped with the crystalline plane of TiO_2. Moreover, the synthesised sample demonstrated good photocatalytic activity in the degradation of methyl blue. It was shown that the structure, crystallinity, dispersity, light adsorption properties and photocatalytic performance of SnO_2/TiO_2 photocatalysts are critically dependent on the preparation method.

Another example of the use of a precipitation method to obtain TiO_2/CeO_2 and TiO_2/SnO_2 systems is reported by Yu et al. [32]. As precursors of titanium dioxide, cerium oxide and tin oxide, they used titanium(IV) sulphate(VI), cerium(III) nitrate(V) hexahydrate and tin chloride pentahydrate. The precipitating agent was an aqueous solution of ammonia. Aqueous solutions of the oxide precursors were stirred for 1 h until the components dissolved completely, and then, ammonia solution was added to the reaction mixture. The process was carried out at room temperature, and the pH of the reaction mixture was maintained at 10. After all solutions had been added in the appropriate quantities, the system was stirred for a further 3 h. The material was then subjected to an ageing process for 48 h. The resulting precipitate was dried at 105°C for 12 h and then calcined at 500°C for 6 h. Physicochemical analysis revealed specific surface areas of 108 and 59 m^2/g respectively for the TiO_2/CeO_2 and TiO_2/SnO_2 systems. X-ray analysis showed the TiO_2/CeO_2 system to have an anatase crystalline structure, while for TiO_2/SnO_2, the diffractogram contained peaks corresponding to rutile. In both cases, the crystalline structure of cerium or tin oxide was not observed. The catalytic properties of the systems were also investigated; they demonstrated excellent performance in the reduction of nitrogen oxides.

Zhang et al. [33] synthesised TiO_2/ZrO_2 mixed oxide (with molar ratio = 1:1) by a co-precipitation method from $TiCl_4$ and $ZrOCl_2$ aqueous solutions, which were hydrolysed with ammonium hydroxide. The precursors of Ti and Zr were dissolved in deionised water, and then, HCl was added. Ammonia was added to the solution until pH = 10. Finally, the samples were dried at 110°C and then calcined at 500°C for 5 h. The Ti/Zr mixed oxide synthesised by a co-precipitation method was found to have a high specific surface area of 234 m^2/g, which is linked to the amorphous structure of the material. No diffraction peaks characteristic of TiO_2 or ZrO_2 were detected in the obtained sample. Moreover, the TiO_2/ZrO_2 mixed oxide exhibited good catalytic activity for the selective catalytic reduction of NO by NH_3.

2.2. Sol-gel method

The sol-gel route is a wet chemical method and is a multi-step procedure involving both chemical and physical processes such as hydrolysis, polymerisation, gelation, drying, dehydration and densification. In a typical sol-gel process, a colloidal suspension or sol is obtained as a result of hydrolysis and polycondensation of precursors, which are usually inorganic metal salts or organometallic compounds such as metal alkoxides. Polycondensation and the loss of solvent lead to a transformation from the fluid sol to the solid gel phase [34–38]. The sol-gel method is based on hydrolysis and condensation of metal alkoxides or metal salts [39]. The process involves the reaction of a metal chloride with metal alkoxide or an organic ether, which is an oxygen donor, according to Eqs. (1) and (2):

$$MCl_n + M(OR)_n \rightarrow 2MO_{n/2} + nRCl \quad (1)$$

$$MCl_n + (n/2)ROR \rightarrow MO_{n/2} + nRCl \tag{2}$$

In these reactions, the formation of ≡M–O–M≡ type bonds is favoured by the condensation between ≡M–Cl and ≡M–OR, according to the reaction (3):

$$\equiv M{-}Cl + \equiv M{-}O{-}R \rightarrow \equiv M{-}O{-}M\equiv + R{-}Cl \tag{3}$$

In the reaction with ether (4), an alkoxide is formed as a result of a reaction of alcoholysis with ≡M–Cl:

$$\equiv M{-}Cl + R{-}OR \rightarrow \equiv M{-}OR + R{-}Cl \tag{4}$$

These reactions run slowly, and usually, the formation of inorganic oxide is favoured by elevated temperature in the range 80–150°C. The main reaction (5) between metal chloride and metal alkoxide takes place at room temperature and leads to a solution of metal chloroisopropoxide:

$$\equiv M{-}Cl + \equiv M{-}OR \rightarrow \equiv M{-}OR\equiv + \equiv M{-}Cl \tag{5}$$

The properties of materials obtained by the sol-gel method depend on such factors as pH, the presence of admixtures of other substances, the volume ratio of water to precursor, and the rate of stirring of the system. The pH also influences the size and shape of pores [38].

Advanced materials in a wide variety of forms, such as spherical or ultrafine shaped powders, fibres, thin film coatings, dense or porous materials including high-purity inorganic oxides, and hybrid (inorganic-organic) materials can be synthesised using the sol-gel method [40–43]. The sol-gel process is a useful synthetic method for the preparation of amorphous as well as structurally ordered products [44–49]. It is of particular interest because it gives very good compositional and morphological control over the product's properties, such as specific surface area, nanoparticle size and degree of aggregation. The precipitates or gels obtained by sol-gel processing are typically amorphous, exhibiting a fairly high specific surface area and are in some cases even (meso)porous. A transition from the amorphous to the crystalline phase is typically induced by thermal treatment at temperatures higher than 300°C, leading in most cases to a deterioration of the pore system and an increase in the particle size, associated with a decrease in the specific surface area. In sol-gel processing, for better control of the hydrolysis and condensation process, many different modifiers of alkoxide precursors can be used, including acetylacetone [50–52], acetic acid [50, 53] and other complex ligands.

Mesoporous TiO_2 nanocrystals were synthesised by Faycal Atitar et al. [54] using the sol-gel method in the presence of triblock copolymer as the structure directing agent. In the typical synthesis, triblock copolymer was first dissolved in ethanol, and the resulting mixture was added to CH_3COOH and HCl. Next, $Ti(OC(CH_3)_3)_4$ (TBOT) was added to the mixture. The resulting TiO_2 nanocrystals were calcined at different temperatures (400, 500, 600, 700 and 800°C for 4 h) to demonstrate how their structural properties such as crystallite phases, morphology and mesoporosity affect the photocatalytic performance. X-ray analysis showed the samples to have a crystalline structure. The samples calcined at 400 and 500°C revealed the presence of anatase, but as the calcination temperature increased, the contribution of the anatase crystalline structure decreased in favour of rutile. All the synthesised materials were shown to be photocatalytically active, and the photocatalytic activity of mesoporous TiO_2 was strongly dependent on the final thermal treatment. The sample calcined at 500°C demonstrated higher activity in the decomposition of imazapyr (98%) and phenol (95%) compared with the commercially available Aeroxide TiO_2 (P-25). Based on the results obtained, the authors concluded that mesoporous titanium dioxide T-500 (calcined at 500°C) is an efficient material for the removal of organic pollutants from water.

The sol-gel route was also used by Mutuma et al. to obtain titanium dioxide [55]. The precursor used was titanium tetraisopropoxide (TTIP), which was dissolved in propan-2-ol and deionised water. The reaction mixture was heated to 80°C and maintained at that temperature for 5 h, after which it was cooled to room temperature. It was determined how the pH of the reaction system affected the physicochemical properties of the products. The pH was controlled by adding a precipitating agent in the form of a solution of sodium hydroxide or nitric(V) acid. The process was carried out at pH values of 2, 4, 7 and 9. The resultant systems were left to gel at room temperature for 24 h. The products were dried at 100°C for 12 h and then calcined at temperatures of 200, 600 and 800°C for 2 h. It was also determined how the process conditions influenced the crystalline and porous structures of the synthesised materials, which have a significant effect on their photocatalytic properties. X-ray analysis showed that titanium dioxide that had not undergone calcination had an anatase structure. Calcination at 200°C led to the appearance of brookite, although the intensity of the band corresponding to that crystallographic form decreased as the calcination temperature increased. When the calcination temperature increases to 800oC, bands corresponding to rutile appeared, indicating a transformation of anatase to rutile. In photocatalytic tests, the materials containing anatase-brookite (calcined at 200°C) or anatase-brookite-rutile (600 and 800°C) exhibited better photocatalytic properties than an anatase-rutile system (800°C). It was also found that the specific surface area of the products depends strongly on the pH of the reaction system. As the pH increased, the surface area of the synthesised materials decreased, irrespective of the calcination temperature. The experiments showed the systems with mixed crystalline structure to be an excellent photocatalytic material in the decomposition of non-biodegradable organic pollutants, for example from the textile industry.

Titanium dioxide powders were also prepared via the sol-gel method by Siwińska-Stefańska [56]. It was investigated how the conditions of preparation (addition of catalyst and chelating agent, temperature of calcination) affect the microstructural evolution, porous structure parameters and photocatalytic capability of the resulting TiO_2 powders. The results of dispersive analysis showed that an increase in the amount of catalyst used in the process of obtaining titanium dioxide results in an increase in particle diameter. Moreover, the diameter of particles tended to decrease with a decreasing quantity of chelating agent. The addition of chelating agent also caused significant changes in the crystalline structure and porous structure parameters of the resulting samples. The TiO_2 systems prepared by the sol-gel method with or without the addition of chelating agent exhibited relatively high photocatalytic activity in the decomposition of C.I. Basic Blue 9.

In a report by Italian researchers [57], the sol-gel method was used to obtain an SiO_2/TiO_2 system. The precursors of the dioxides were respectively TEOS and TTIP. First, TEOS was mixed with an organic solvent (ethanol) in the molar ratio TEOS:ethanol:water = 1:2:1. Hydrochloric acid was added to the mixture to maintain a pH of 1. After stirring for 6 h, TTIP was added to the system. The resulting sol was matured for 3 days to obtain a gel, which was then dried at room temperature for 7 days. The resulting materials then underwent calcination at temperatures of 600 and 800°C for 2 h. Detailed physicochemical analysis confirmed that the product consisted of titanium dioxide in rutile form and silicon dioxide. X-ray analysis showed that the system that had not been calcined had an amorphous structure, while calcination caused the formation of a crystalline structure. Calcination at 600°C leads to anatase, but when a temperature of 800°C is used, rutile appears. TEM microscopic images revealed a tendency for the agglomeration of particles in the samples. Variation in the molar ratio of the oxides was found not to have a significant effect on the morphology of the oxide system.

Siwińska-Stefańska et al. [58] reported the preparation of nano- and microstructured TiO_2 doped with Fe and Co by the sol-gel method and determined the effect of doping on the physicochemical properties of TiO_2. The doped materials were found to contain particles of smaller diameter and lower homogeneity than pure TiO_2. XRD analysis revealed that the addition of iron or cobalt to the titania preparation process has a significant effect on crystalline structure formation.

Fan et al. [59] prepared a mesoporous TiO_2/ZrO_2 nanocomposite from titanium tetrabutoxide, $ZrOCl_2 \cdot 8H_2O$, Pluronic P123 and Macrogol 20000 as double templates utilising the sol-gel method. In typical synthesis, to a solution of titanium tetrabutoxide and nitric acid, ethanol, Pluronic P123 and Macrogol 20000 were added. The resulting material was calcined for 5 h at 800°C. The structures and physicochemical properties of the products were determined by X-ray diffraction (XRD), Raman scattering studies and N_2 adsorption/desorption. The results proved that the use of double templates retarded the crystal phase transformation from anatase to rutile, and the obtained materials showed high thermal stability. Moreover, photocatalytic tests confirmed that the sample prepared with double templates exhibited higher photocatalytic activity in the decomposition of Rhodamine B (92%) than samples prepared with a single template (90 and 91%).

In another study, Shao et al. [60] obtained pure TiO_2 and TiO_2/ZrO_2 system using the sol-gel method. The crystalline structure and particle shape and size were found to be strongly dependent on the calcination temperature and on the ratio of Zr to Ti. XRD analysis showed the crystalline structure of the synthesised materials to be significantly affected by the conditions of calcination. As the content of ZrO_2 increased, the intensity of the bands corresponding to anatase (TiO_2) decreased in favour of those corresponding to tetragonal ZrO_2. When samples were treated at 800°C, the transformation of anatase to rutile was favoured, although a high content of zirconium dioxide retarded that effect. Further increase in the calcination temperature led to the transformation of tetragonal ZrO_2 to monoclinic.

Kraleva and Ehrich [61] obtained the oxide systems ZrO_2/Al_2O_3 and TiO_2/Al_2O_3 by the sol-gel route. They used 2,4-pentanedione as complexing agent and n-butoxide as solvent. The precursors of the component oxides were the alkoxides of the respective metals. The molar ratio of the precursors was 1:1. An appropriate quantity of the precursors was dissolved in the solvent, and the reaction system was stirred continuously for approximately 30 min. A complexing agent with a pH of 8 was then added, and the mixture was stirred for another 5 min. It was then heated to 70°C with the pH maintained at 8 for 10 min. Next, hydrolysis was performed by adding deionised water dropwise to the reaction mixture and stirring for 1 h. The sample was then left to cool at room temperature for 12 h, and then, the solvent was removed by pressure evaporation at 110°C. After drying, the system was calcined at 500°C. The resulting systems had high specific surface area (320 and 200 m^2/g respectively for TiO_2/Al_2O_3 and ZrO_2/Al_2O_3). The materials differed significantly in terms of crystallinity: the system containing titanium dioxide with aluminium oxide was completely amorphous, while that of zirconium dioxide with aluminium oxide had a crystalline structure. The synthesised materials were also shown to offer excellent performance as catalysts of the conversion of ethanol at 600°C, the products being H_2 and CO (syngas).

Kraleva et al. [62] used the sol-gel method to obtain TiO_2/ZrO_2 systems with different contents of ZrO_2 (3, 6, 13 and 37% mol.). The synthesised materials were subjected to detailed physicochemical analysis. Analysis of their porous structure parameters showed that as the content of ZrO_2 increased, there was an increase in the BET specific surface area. X-ray spectroscopy revealed that the addition of zirconium dioxide also has a significant effect on the crystalline structure and the phase composition of the resulting oxide systems. The TiO_2/ZrO_2 system obtained with a ZrO_2 content of 37% mol., calcined at 550°C, exhibited an amorphous structure. It was also observed that as the calcination temperature increased, diffraction bands appeared corresponding to srilankite—a mineral containing oxygen, titanium and zirconium.

2.3. Hydrothermal route

The hydrothermal method is one of the most advanced techniques for obtaining metals and their oxides. Hydrothermal synthesis is a non-conventional method defined as crystal synthesis or crystal growth under high temperature and high pressure water conditions from substances which are insoluble at ordinary temperature and pressure (<100°C, <1 atm). Water may act both as a catalyst and as a component of the continuous phase during synthesis. Among the wet chemical preparation methods, the hydrothermal route has been recognised as an energy and time saver, with faster kinetics of crystallisation than classic co-precipitation or sol-gel methods. The hydrothermal method has proven to be an excellent method for the synthesis of powders, fibres, single crystals, monolithic ceramic bodies, and coatings on metals, polymers, and ceramics [63]. By adjusting simple parameters such as temperature, pressure or precursor concentration, it is possible to alter the characteristics of the product particles, e.g. crystalline phase and particle size. In the hydrothermal method, the temperature of crystallisation is usually lower than in a typical thermal process. The agglomeration of particles can be prevented by carrying out crystallisation under high pressure. The products obtained without calcination and grinding are of high quality. Using this method, it is possible to control the shape and size of particles; nonetheless, the process is slow and is unsuitable for use on an industrial scale [64].

Chae et al. [65] report the synthesis of titanium dioxide using titanium tetraisopropoxide (TTIP) as the precursor of TiO_2 in an ethanol–water mixture as solvent. TTIP was added dropwise to a mixture of ethanol, water and nitric acid with pH = 0.7. After being well stirred, the solution underwent a reaction in a hydrothermal reactor at 240–300°C for 4 h. X-ray diffraction (XRD) analysis showed that hydrothermal processing at 240–260°C leads to a product with the greatest crystallinity, containing an anatase crystalline phase and a small quantity of brookite. Increasing the temperature of the hydrothermal reaction above 260°C caused the formation of agglomerates of primary particles. The size of the particles was strongly influenced by the concentration of the titanium dioxide precursor and by the molar ratio of ethanol to water and less so by the temperature and time of the reaction. An increase in the ethanol:water molar ratio led to smaller particles; also, when that ratio exceeded 8, a less crystalline product was obtained, with a tendency for the formation of aggregates. An increase in the concentration of TTIP in the reaction mixture retarded the increase in particle size. Porous structure analysis confirmed that smaller particle sizes in the resultant materials corresponded to higher specific surface areas.

Zhang et al. [66] used the hydrothermal method to obtain TiO_2 nanowires with anatase crystalline structure. An appropriate quantity of white TiO_2 powder with anatase structure was placed in a teflon-lined autoclave, and 10 M NaOH was added up to 80% of the capacity of the reactor. The mixture was heated for 24 h at 200°C, and the product was then dried for 6 h at 70°C. The resulting material underwent detailed analysis using X-ray diffraction (XRD) and scanning, transmission and high-resolution electron microscopy (SEM, TEM, HRTEM). XRD analysis confirmed the very high purity of the product. SEM images showed the titanium dioxide to have the form of numerous nanowires with uncontaminated surfaces. It was also found that the product had an anatase crystalline structure. The obtaining of titanium dioxide in nanowire form was conditional on the use of NaOH, which acted as a "soft" matrix and on the high process temperature. A lower reaction temperature would favour the formation of titanium nanorods. Advantages of the reported process include its low cost, the high purity of the products, and the large number of TiO_2 nanowires produced.

Caillot et al. [67] carried out hydrothermal synthesis of TiO_2/ZrO_2 oxide systems. It was determined how the process conditions affected the morphology, crystalline structure and specific surface area of the products. The precursors used were zirconium oxychloride ($ZrOCl_2{\cdot}8H_2O$) and titanium tetrachloride ($TiCl_4$), which were added to a solution of ammonia water ($NH_3{\cdot}H_2O$). Hydrothermal processing took place at a temperature of 220°C under a pressure of 25 bar for 4 h. The resulting sample was dried at 120°C and finally calcined at 500°C for 10 h. Thermogravimetric analysis of the TiO_2/ZrO_2 system following the hydrothermal process showed it to have high thermal stability. The diffractogram obtained for TiO_2/ZrO_2 following calcination at 500°C indicated an amorphous structure. Porous structure analysis showed the oxide system to have a specific surface area of 209 m^2/g.

Hirano et al. [68] investigated the catalytic properties and thermal stability of materials consisting of titanium dioxide and zirconium dioxide, obtained by the hydrothermal route from

$TiOSO_4$ (titanium(IV) sulphate(VI)) and $ZrCl_4$ (zirconium tetrachloride). Solutions were placed in hydrothermal reactors and heated at a temperature of 200 or 240°C for 48 h. The precipitate was dried at 60°C. Samples were additionally heated for 1 h at temperatures ranging from 400 to 1000°C. Diffractograms obtained for samples following hydrothermal treatment at 240°C for 48 h indicate the increasing presence of the monolithic structure of ZrO_2 as the concentration of Zr in the initial solution increases. Diffraction bands corresponding to anatase are also visible. Transmission electron spectroscopy showed that the addition of ZrO_2 causes a decrease in the sizes of crystallites. The photocatalytic activity of the products was tested in the decomposition of methylene blue (MB) under ultraviolet radiation. The TiO_2/ZrO_2 systems exhibited higher photocatalytic activity than a material consisting of pure TiO_2.

2.4. Solvothermal method

The solvothermal method is similar to the hydrothermal method, the difference lying in the type of solvent used: in the hydrothermal method it is water, while in the solvothermal method, it is a non-aqueous solvent. The range of temperatures used in the solvothermal method can be much greater than in the hydrothermal method and depends on the boiling point of the organic solvent used. In the solvothermal method, the control of the shape, size and crystallinity of TiO_2 particles is easier than in the hydrothermal method. The solvothermal method is considered a universal method for obtaining nanoparticles with a narrow range of size distribution. Using the solvothermal method, TiO_2 nanoparticles or nanotubes can be produced with or without a surfactant [69, 70].

Zhu et al. [71] described a method for obtaining mesoporous TiO_2 microspheres by a solvothermal route. The precursor used was titanium tetrabutoxide (TBOT), which was added to a solution of polyetherimide (PEI) and anhydrous alcohol. The resulting white suspension was transferred to an autoclave, where a reaction was carried out at 180°C for 24 h. The white precipitate was then washed with water and ethanol, dried for 6 h at 60°C, and calcined for 2 h at 400 or 500°C. The product was analysed using the XRD, SEM, TEM, HRTEM, XPS and BET techniques and UV-Vis absorption spectra. Photocatalytic activity was investigated based on the reaction of degradation of phenol and methyl orange (MO) under sunlight. Mesoporous anatase TiO_2 microspheres with high crystallinity were successfully obtained by the solvothermal method and exhibited high photocatalytic activity for both phenol and methyl orange.

The solvothermal method was used by Yang et al. [72] to synthesise titanium dioxide from titanium(IV) fluoride, which was dissolved in a mixture of deionised water and hydrochloric acid (used to stabilise the pH). The mixture was added, together with propan-2-ol and hydrofluoric acid, to a teflon-lined stainless steel autoclave. The reactor was placed in an electric oven at 180°C for between 5.5 and 44 h. X-ray analysis of the product showed the synthesised TiO_2 to have an anatase structure. The average particle size in the system was measured by scanning electron microscopy at 1.09 μm. Porous structure analysis showed the specific surface area of the titanium dioxide to be 1.6 m^2/g.

Oshima et al. [73] used the solvothermal method to obtain TiO_2 nanoparticles. Here, a polymer gel was used, which enabled strongly dispersed and homogeneous particles to be obtained. First, polyvinyl alcohol (PVA) was dissolved in water at 70°C. The reaction mixture was then cooled to room temperature; next, the precursor of titanium dioxide $[(NH_4)_8(Ti_4(C_6H_4O7)_4(O_2)_4{\cdot}8H_2O]$ was added, and water was evaporated off using microwaves. The resulting polymer gel was mixed with ethanol, which served as a solvent, and the mixture was placed in an autoclave and heated for 18 h at 230°C. Finally, the product was dispersed in water at 50–70°C. X-ray analysis showed that prior to the solvothermal process, the material had an amorphous structure, but the diffractograms obtained for the titanium dioxide following solvothermal treatment contained peaks indicating formation of the anatase crystalline structure. The particle size distribution was found by transmission electron microscopy to lie within the range 4.4–6.8 nm. Physicochemical analysis confirmed the soundness of the method used to obtain titanium dioxide, and that it leads to homogeneous particles without a tendency to form agglomerates.

Supphasrirongjaroen et al. [74] used the solvothermal method to synthesise TiO_2/SiO_2 and TiO_2/ZrO_2 systems. It was investigated how the addition of Si or Zr affected the photocatalytic activity of the oxide system. Titanium tetraisobutanol (TNB), tetraethoxysilane (TEOS) and zirconium tetraisobutanol were used as sources of titanium, silicon and zirconium. Titanium dioxide with admixed SiO_2 and ZrO_2 was obtained by dissolving TEOS and zirconium tetraisobutanol in toluene. The resulting materials were placed in an autoclave (300°C, 2 h). The synthesis products were subjected to physicochemical analysis to determine how the process temperature affects the photocatalytic activity of the product. It was found that the samples treated at room temperature have higher photocatalytic activity. The process temperature was also found to have a significant effect on the specific surface area: in almost every case, the surface area was larger for the samples that had undergone calcination at 350°C. It was also found that the materials containing zirconium exhibited higher photocatalytic activity than those with silicon. The researchers concluded that the addition of an appropriate metal can improve the physicochemical properties of inorganic materials.

A study of the literature shows that research on the synthesis of advanced materials based on titanium dioxide is chiefly focused on the skilful control of processes (through appropriate choices of methods and conditions) serving to generate changes in the properties of those materials. Key factors include the selection of appropriate raw materials, optimization of the pH of the reaction system, modification of the relative quantities of reagents and selection of an optimum temperature for thermal processing. These process parameters make it possible to synthesise materials with controlled physicochemical and structural properties, including grain size and shape, degree of crystallinity, crystallite size and phase or surface composition, as well as chemical and thermal stability. Temperature has a particularly significant effect on the crystalline structure of materials based on TiO_2, which in turn determines their potential applications. This applies both to the calcination temperature and to the conditions of synthesis. It is of particular interest to carry out reactions in hydrothermal or solvothermal conditions, leading to products not only having a precisely designed crystalline structure—with the use of a much lower temperature than in other conventional methods—but also exhibiting a unique and diverse morphology. These methods also enable greater control of the size and shape of particles. A fundamental weakness of these processes, however, is the difficulty of increasing their scale. When selecting an appropriate method for the synthesis of titanium dioxide or hybrid materials incorporating it, attention should be given to the possibility of obtaining those products on semi-industrial or full industrial scale. The transfer of optimum conditions of synthesis from the laboratory to larger-scale processes is often problematic and should continue to be the subject of intensive research.

3. Titanium dioxide–based hybrid materials as active photocalatysts

Photocatalysis is a phenomenon in which chemical reactions are accelerated by the action of light. The most important stage of the process is the light-induced excitation of electrons from the valence band to the conduction band. This takes place provided that the energy of the incident radiation is equal to or greater than the band gap of the photocatalyst and leads to the creation of electrons (e^-) and holes (h^+). The electrons combine with atmospheric oxygen to produce active O_2, while the holes combine with water or atmospheric water vapour to form OH• radicals. These hydroxide radicals are strong oxidising agents and can thus easily oxidise and decompose various organic pollutants (such as oils and fats). The active oxygen, on the other hand, triggers reduction reactions. In the photocatalysis process, oxidation and reduction reactions occur simultaneously. During photocatalysis, the created electrons and holes are subject to surface or voluminous recombination as well as taking part in redox reactions. The process of photocatalysis is affected by a number of factors: rate of reaction, mass of catalyst, wavelength, initial reagent concentration and luminous flux [75–80].

A key factor in the process is the photocatalyst used. Many semiconductor materials are available on the market for use in photocatalysis processes, but efforts are constantly being made to develop new materials that are highly active in the visible and near ultraviolet ranges, while also being biologically inert and photostable [81]. Among the wide range of photocatalysts in use, the most promising material is TiO_2, in view of its high photochemical activity. It is also regarded as a cheap, nontoxic material that is photostable and chemically and biologically inert [82].

Research has been carried out to investigate photocatalytic activity using two forms of TiO_2: anatase and rutile. The amorphous form of TiO_2 is considered to exhibit practically no such activity [83]. Photocatalytic activity is affected not only by the type of photocatalyst used, but above all by its physicochemical properties: specific surface area and pore type, degree of hydroxylation of the surface, particle size and degree of agglomeration and the degree of crystallinity and number of defects in the crystalline structure [84]. Differences in the performances of photocatalysts are attributed largely to physicochemical properties such as the width of the band gap, the rate of recombination of e^-–h^+ pairs and the number of hydroxyl groups on the TiO_2 surface.

The phase composition of the studied material is a very important factor in determining the photocatalytic activity of TiO_2. It has been frequently reported that anatase exhibits much greater photoactivity than rutile [85–87]. Tanaka et al. [88] and Kumar et al. [89] have suggested that the higher activity of anatase results from its lower capacity to adsorb oxygen, the higher degree of hydroxylation of the TiO_2 surface and the high specific surface area, which provides more active sites for the adsorption of pollutants. Too high a surface area entails the presence of a large number of structural defects, which means that the recombination of charge carrier pairs proceeds much more rapidly. The recombination rate is slowed by a larger number of OH^- groups on the photocatalyst surface, which trap the holes generated [83, 90–92].

Titanium dioxide is the most widely used catalyst for photocatalytic degradation of organic compounds, but there are some limitations in using TiO_2 for practical applications, which include:

- its large band gap;
- its low quantum yield;
- the low photon utilisation efficiency,
- the narrow available light spectrum (UV is the most responsive range).

Various strategies have been adopted for improving or enhancing the photocatalytic efficiency of TiO_2 (see **Figure 3**). These methods can be summarised as either morphological modifications, such as increasing surface area and porosity, or chemical modifications, by the incorporation of additional components into the TiO_2 structure. Modifications include:

- the addition of transition metal ions (such as Cr, V, Zr, Mn, Fe, Mo) [93–96];
- preparation of the reduced form TiO_{2-x};
- sensitisation using dyes [97–100];
- doping with non-metals (such as N, S, C) [78, 101–103];
- use of hybrid semiconductors such as TiO_2/SiO_2, TiO_2/Al_2O_3, etc. [10, 104, 105].

Table 3 contains information on selected methods of modifying titanium dioxide.

The absorption spectrum for titanium dioxide can be shifted towards the visible range by the incorporation of additional particles which cause significant changes in the material's semiconductor properties. Doping was the first technique used by researchers to modify the

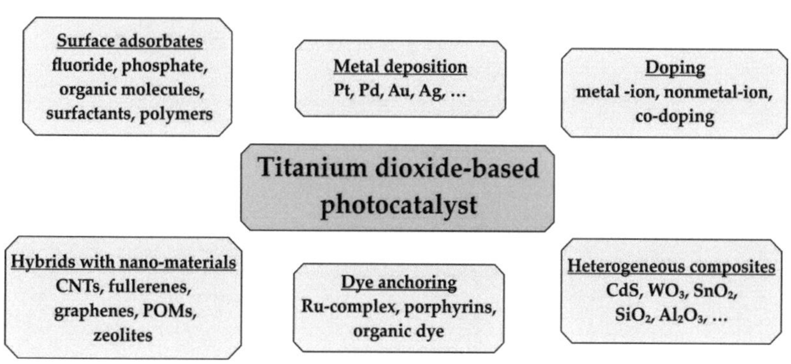

Figure 3. Various modification methods for titania-based photocatalysts.

Dopant	Precursor	Type of matrix	Conditions of preparation	Properties of obtained material	Potential application	Ref.
Doping with non-metal						
Iron	$FeCl_3 \cdot 6H_2O$, $TiCl_3$, $NH_3 \cdot H_2O$, HCl	–	Solution containing $TiCl_3$, ammonia and water was mixed at room temperature for 16 h, final pH = 8.5; hydrothermal treatment: 24 h at 110°C; drying: 80°C for ≥10 h, calcination: 500°C for 4 h	Anatase structure, the particle diameter decreases (from 25 to 15 nm) while the specific surface area increases (from 50 to 76 m^2/g) with the increasing iron content	Photodegradation of phenol	[106]
Tungsten	$Ti(OC_2H_5)_4$, toluene, $W(OC_2H_5)_5$	–	The resulting solution was stirred for 30 min and then atomised; the substrate temperature was kept at 500°C; deposition time was 45 min	Anatase structure	Destruction of resazurin redox dye	[107]
Doping with non-metal						
Nitrogen	NH_3 gas	TiO_2 nanorods prepared by hydrothermal route on a fluorine-doped tin oxide (FTO)-coated glass substrate	Thermal treatment of TiO_2 nanorods at 500°C for 1 h in the presence of NH_3 gas	Morphology: sample NTR-60—TiO_2 rods ~50 nm in length, sample NTR-150—TiO_2 rods ~900 nm in length and ~95 nm in width, rutile structure	Photooxidation of thiol molecules to the sulfonic acid species	[108]
	Trimethylamine	TiO_2 nanorods synthesised by the hydrothermal treatment	Hydrothermal treatment of TiO_2 at 200°C for 120 min with different amount (from 1 to 5% w/w compared to TiO_2) of trimethylamine	Anatase structure, crystallite size: is decreasing from 22 to 10 nm with increasing of N content, band gap: is decreasing from 3.22 to 2.85 eV with increasing of N content, BET surface area: from 81 to 101 m^2/g	Photodegradation of methyl orange (MO)—90% for sample NTO-4 (doping with 3%w/w N)	[109]

Dopant	Precursor	Type of matrix	Conditions of preparation	Properties of obtained material	Potential application	Ref.
Sulphur	Titanium butoxide, thiourea, methanol	–	Solution containing precursor of Ti with methanol was mixed with mixture of thiourea and methanol. Then obtained sample was calcined at 400, 500, 600 and 700°C for 3 h	Crystalline structure: sample calcined at 400–600°C—anatase structure, sample heated at 700°C—anatase and rutile structure, crystallite size: from 9 nm (sample TiS-400) to 50 nm (sample TiS-700), band gap: is decreasing from 3.21 to 3.07 eV with calcination treatment	Photodegradation of 4-chlorophenol—under UV light—98%—sample TiS-700, under visible light—40%—sample TiS-500	[110]
	Thiourea, ethanol, titanium tetraisopropoxide	–	Sample S-TiO_2: to solution containing thiourea and ethanol was added titanium tetraisopropoxide stirred at room temperature under aerated conditions for 48 h, calcination at 450°C for 4 h, sample S_8-TiO_2: pure, amorphous TiO_2 and S_8 were mixed and ground together thoroughly with a mortar, followed by annealing at 350°C for 4 h	Anatase structure, crystallite size: 13 nm for S-TiO_2 and S_8-TiO_2, and 14 nm, for TiO_2, particle size of 5–15 nm, BET surface area: 93 m^2/g for S-TiO_2 and 69 m^2/g for pure TiO_2		[111]
Heterogeneous composites						
ZnO	$Zn(NO_3)_2 \cdot 6H_2O$, hexamethylenetetramine (HMT, $C_6H_{12}N_4$)	The 3 durchin-like TiO_2 nanospheres	Hydrothermal synthesis: 100°C for 3 h, drying: 60°C	Anatase TiO_2 and hexagonal wurtzite-type ZnO, BET surface area—187 m^2/g	Photodegradation of nitrophenol and methyl orange	[112]
ZrO_2	Pluronic P123, $ZrOCl_2 \cdot 8H_2O$, titanium(IV) n-butoxide, HCl, C_2H_5OH	–	Synthesis: HCl + titanium(IV) n-butoxide + P123 in ethanol + $ZrOCl_2 \cdot 8H_2O$; deposition of thin film on indium tin oxide; calcination: 350, 400, 450, 500 and 800°C for 4 h	Anatase, rutile and tetragonal ZrO_2, BET surface area from 187 to 219 m^2/g,	Photodegradation of rhodamine B	[113]
	Ethanol, Pluronic P123, Macrogol 20000, titanium butyrate (TBT), HNO_3, $ZrOCl_2 \cdot 8H_2O$	–	Synthesis: ethanol, Pluronic P123, Macrogol 20000, titanium butyrate (TBT), HNO_3—room temparature, after adding $ZrOCl_2 \cdot 8H_2O$—80°C; calcination: 800°C for 5 h	Anatase structure, BET surface area—149 m^2/g	Photodegradation of rhodamine B	[59]

Dopant	Precursor	Type of matrix	Conditions of preparation	Properties of obtained material	Potential application	Ref.
Hybrids with nano-materials						
Graphene	Graphene oxide, poly(L-lysine) (PLL), ethylene glycol (EG)	TiO_2 hollow microspheres	Solvothermal method	–	Decomposition of MB dye	[114]
	Graphite powder, titanium dioxide, HCl, methanol, H_2SO_4, $KMnO_4$, H_2O_2, $NaNO_3$, methanol	–	Simple mixing at room temperature for 24 h and sonication, drying: dried in oven at 60°C	Anatase and rutile structure	Photodegradation of rhodamine B	[115]

Table 3. Modification of titanium dioxide.

electron structure of titanium dioxide [116]. Foreign ions or atoms are introduced into the titanium dioxide crystalline lattice with the aim of modifying and improving its physicochemical properties. The success of the process is largely dependent on the type and the quantity of dopant, which usually does not exceed a few percent. Both metals and non-metals may be used as dopants in titanium dioxide [83, 117–120].

The incorporation of metals (iron, chromium, tungsten, platinum, etc.) into titanium dioxide leads to a reduction in the band gap and thus an increase in absorption of radiation in the UV-Vis range. The activity of TiO_2 doped with metal ions depends largely on the valence of the dopant. If those ions have the same charge as the Ti ion in the crystal, the effect will be a change in the interactions between the metal atoms. The incorporation of metal ions of lower valence than Ti^{4+} (Y^{3+}, La^{3+}, Nd^{3+}, Pd^{2+}) favours a change in the size of the band gap and a reduction in the density of point defects [121–124].

Reports on the doping of titanium dioxide with iron confirm that the quantity of iron used, which may range from 0.05 to 50% Fe, has a significant effect on the increase in photocatalytic activity. In this process, a titanium atom in the anatase phase may be subject to substitution or else some of the iron atoms are incorporated into the titanium dioxide crystal lattice in anatase form to produce a composite, while some aggregates form the oxides Fe_2O_3 and Fe_3O_4. The doping process leads to the generation of shallow charge traps in the crystal structure, which decreases the recombination rate of electron-hole pairs. Introducing iron ions into the TiO_2 lattice not only leads to a lower electron-hole recombination rate but also increases excitability by visible light. Titanium dioxide doped with iron exhibits better photocatalytic activity under UV and also under visible light irradiation [106, 125–129].

The doping of TiO_2 with chromium not only increases its photocatalytic activity but also causes the photocatalyst to acquire ferromagnetic properties without losing its conductive properties. The enhancement of photocatalytic activity results from the formation of vast oxygen vacancies. The oxygen vacancies in TiO_2 act as electron traps which can bind the photoinduced electrons and play a significant role in inhibiting the recombination rate of photoinduced electron-hole pairs [130]. In turn, tungsten can be incorporated into the TiO_2 structure in oxide form (WO_x). This enhances the material's photocatalytic activity by reducing charge carrier recombination and by increasing light absorption in the visible portion of the spectrum [131].

Doping with non-metals is usually carried out to extend the photocatalytic activity of TiO_2 in the UV-Vis range. The introduction of non-metals into the oxygen sub-lattice may cause a change in the position of the valence band and thus reduce the band gap. Promising results have been obtained by doping titanium dioxide with non-metals such as nitrogen, carbon, iodine, sulphur and fluoride. Such doping narrows the band gap or leads to the appearance of new internal levels between the valence band and the conduction band [108–111]. When TiO_2 is doped with nitrogen, the dopant may replace an oxygen ion (in the case of a material with anatase structure) or a titanium ion (in the case of a rutile structure) [108, 109]. In case of doping with carbon, the dopant may replace oxygen or titanium or else occupy an inter-nodal position, depending on the energy of formation of the product and the presence of oxygen in the reaction environment. If doping is carried out in oxygen-rich conditions, substitution of carbon for titanium takes place or else it is incorporated in the inter-nodal position due to its small atomic size. If the environment contains little oxygen, the carbon atom takes the oxygen position, forming the structure C-Ti-O-C [117, 132, 133]. The doping of titanium dioxide with iodine, on the other hand, leads to increased visible light absorption and increased photocatalytic activity below the visible range. This phenomenon results from retardation of the recombination of electron-hole pairs due to the capture of electrons by the iodine. Substitution of iodine atoms for oxygen or titanium results in a narrowing of the band gap [134]. Doping titanium dioxide with sulphur is more difficult, due to the fact that the dopant replaces oxygen in the oxide crystal lattice, and the differences in the radii of the two atoms are significant [110, 111].

Reports on the doping of titanium dioxide indicate that a small quantity of dopant will not lead to major changes in the porous structure but may cause significant changes in the phase composition of TiO_2 and in the size of crystallites. In the case of doping with nitrogen, an increase in the quantity of the dopant has been found to increase the thermal stability of anatase and to alter the temperature of transformation of anatase to rutile [135–137].

Another method for modifying the electron structure of titanium dioxide is the formation of hybrid oxide systems. Among synthetic hybrid oxide systems, TiO_2/ZrO_2 materials, which thanks to the addition of zirconium dioxide have much greater surface area and mechanical strength than pure TiO_2, deserve special attention. An admixture of ZrO_2 prevents the phase change of anatase to rutile and causes a reduction in the particle diameter of the resulting hybrid. These factors contribute to the improved photocatalytic activity of TiO_2/ZrO_2 oxide systems. There are many publications with information on the application of TiO_2/ZrO_2 hybrids in photocatalysis. The use of TiO_2/ZrO_2 hybrid materials in the photo-oxidation of organic compounds or degradation of dyes originating from various industrial processes is well known. It also has applications in the photo-reduction of atmospherically harmful oxides, like CO_2 and NO_x, resulting for example from the combustion of fossil fuels. The advantages of the TiO_2/ZrO_2 hybrid are its mechanical strength, nontoxicity and corrosion resistance and the ability to conduct photocatalytic processes using sunlight. These advantages may stimulate increasing demand for this material in the near future [59, 113, 138].

Zhou et al. [113] determined the photocatalytic properties of a TiO_2/ZrO_2 system obtained by the sol-gel method. Physicochemical analysis showed the products to have an anatase crystalline structure. An increase in the molar content of zirconium dioxide leads to a decrease in the crystallinity of the resulting materials, while an increase in the temperature of calcination increases their crystallinity. The specific surface areas of the materials (for all variant methods of synthesis) lay in the range 187.0–219.2 m^2/g. Photocatalytic analysis indicated a fall in the effectiveness of photocatalysis as the temperature of calcination of the materials was increased.

The sol-gel method was also used by Fan et al. [59] to obtain a TiO_2/ZrO_2 system. It was found that the proposed method led to mesoporous materials with a well-crystallised anatase structure. The systems were found to have high specific surface areas, in the range 136.9–148.9 m^2/g.

Combining titania with zinc oxide can also lead to a hybrid oxide system with good photocatalytic properties. The resulting material can be used, for instance, in the degradation of organic impurities such as detergents, dyes and pesticides present in various types of wastewater. The TiO_2/ZnO hybrid material can be synthesised by both physical and chemical processes, which enables enhancement of its properties, for example by widening its light absorption spectrum. Additionally, the photocatalytic activity of oxides may help reduce the susceptibility of pollutants to form aggregate structures [139].

Cheng et al. [140] determined the photocatalytic properties of a hybrid material (UTZ) consisting of 3D nanospherical TiO_2 with a "hedgehog" shape and one-dimensional ZnO in the form of "nanospindles". The resulting system was highly homogeneous and contained the crystalline structure of anatase (TiO_2) and the hexagonal wurtzite structure (ZnO). The TiO_2/ZnO system was found to offer significantly better photocatalytic performance than pure ZnO or TiO_2 in the decomposition of methyl orange (MO) and nitrophenol. This high photocatalytic activity was probably due to the existence of a closely bound heterostructural surface between the ZnO and TiO_2, enabling charge separation and reducing the rate of recombination of electron-hole pairs.

The marked improvement in photocatalytic activity in the case of titania/graphene hybrids is linked to the fact that the graphene component enables the transfer and/or trapping of electrons photogenerated in the oxide semiconductor structure, allowing the holes to form reactive sites. Therefore, charge recombination is suppressed, leading to improvement of the photocatalytic performance [141].

Yan et al. [114] obtained a novel three-dimensional (3D) reduced graphene oxide/TiO_2 (rGO/TiO_2) hybrid composite by wrapping TiO_2 hollow microspheres with rGO sheets via a facile solvothermal route using poly(L-lysine) (PLL) and ethylene glycol (EG) as coupling agents. It was confirmed that the hybrid materials, containing mixed phases of TiO_2 (with content of rutile - 20.8%), demonstrate higher photocatalytic activity in the decomposition of MB dye. Ni et al. [142] synthesised high-photoactive GP strongly wrapped three-dimensional anatase TiO_2. The prepared material demonstrated excellent photocatalytic activity under UV irradiation for the degradation of MB, much higher than that of commercial P25 titania. Similar results were presented by Thomas et al. [115], who synthesised high-performance functionalised FLG (FFLG) decorated with TiO_2 photocatalyst, by simple mixing without any calcination or high-pressure treatment. The FFLG/TiO_2 system produced a higher rate of degradation of Rhodamine B (Rhd B) as compared with pure TiO_2 nanoparticles and FLG-TiO_2 (non-functionalised FLG).

Although titanium dioxide is an excellent candidate for photocatalytic applications, due to its band gap size, nontoxicity, chemical stability, inert nature and relatively low cost, it is subject to certain limitations, chiefly resulting from its relatively low activity in the visible light range and its high exciton recombination rate. For this reason, much research is carried out with the aim of improving and reinforcing the photocatalytic activity of TiO_2 and increasing its spectral sensitivity. This may be achieved by modifying TiO_2 during or after its synthesis, with the choice of a suitable method of activation (doping with metals or nonmetals, coupling with other semiconductor materials, increasing its crystallinity by calcination, or synthesis of hybrid materials). These solutions can lead to a material with enhanced photocatalytic properties, including increased sensitivity in the UV and visible light ranges, and with reduced recombination rate due to the provision of charge traps. New research trends also relate to the combination of TiO_2 with polymers or various forms of carbon nanotubes, fullerenes, graphene oxide (GO) or reduced GO (R-GO), with the aim of obtaining multifunctional materials with a wide range of applications. With this in mind, it should be emphasised how many opportunities and technological solutions are available for implementation with the goal of obtaining unique titania-based materials.

4. Titanium dioxide derivatives as effective electrode materials

Recent years have seen intense development in research aimed at seeking new materials and design solutions to enable further progress in the technology of lithium-ion batteries, which are seen as one of the leading technologies for energy storage. Currently, the greatest challenge in the design of these batteries is to find an optimum combination of cathode and anode materials, as these largely determine the cell's parameters, including capacity, voltage, reversibility of the charge/discharge reaction, and chemical stability. The electrode materials must not only be compatible with each other but also should form a synergic system together with the electrolyte and separator [143–145].

Among a range of available materials, titanium dioxide and its derivatives have recently gained popularity as anodes for Li-ion batteries because they allow the design of operational devices with only minor safety concerns. This class of materials offers improved chemical and thermal stability, low cost, biocompatibility, relatively high surface area and porosity, a broad electrochemical window, rate capability and enhanced cyclic performance by virtue of their superior electrical conductivity. These features make titania-based derivatives a good candidate to replace the commonly used carbon (graphene) as an anode material in LIBs. However, limitations include the low capacity, low electrical conductivity, poor rate capability and poor cycling performance of titanium dioxide. Much research has been carried out to overcome the difficulties related to the use of TiO_2 as an electrode material. Numerous scientific centres worldwide are working on ways of improving the electrochemical behaviour of titania and its derivatives. The chief aim is to enhance the electronic conductivity by producing different titania nanostructures to increase its capacity through the incorporation of selected metal oxides into its structure. Another approach is to combine or coat TiO_2 with carbonaceous materials or to introduce anionic or cationic dopants to form more open channels and active sites for Li ion transport [143–149].

Moreover, the electrochemical performance and the lithium intercalation/de-intercalation processes of titania-based materials typically depend on their crystallinity, structure, morphology, particle size and surface area. In particular, it has been found that nanostructured titanium oxide leads to better capacity, longer cycling life and higher rate capability than the bulk materials. Titania has several allotropic forms, the best-known being tetragonal rutile and anatase, and orthorhombic brookite. Even though anatase has been considered the most electroactive form, other allotropes such as rutile and brookites are also widely studied for potential use as anodes. Moreover, synthesis of this type of system can effectively improve their capacitive performance by creating products with excellent high-rate cycling ability and stability. The application of such hybrid materials as anodes in lithium-ion batteries should lead to charge redistribution in the lattice, facilitate the diffusion of Li^+, and finally increase lattice defects and conductivity [145–152].

Kubiak et al. [153] investigated the electrochemical performance of a mesoporous TiO_2 synthesised via a sol-gel method utilising an ethylene glycol-based titanium precursor in the presence of an amphiphilic molecule as the templating agent. The obtained material presents pure anatase TiO_2 without the presence of other phases, with a monomodal pore diameter close to 5 nm and BET surface area of 92 m^2/g. The mesoporous anatase titania shows excellent rate capability (184 mAh/g at C/5, 158 mAh/g at 2C, 127 mAh/g at 6C, and 95 mAh/g at 30C) and good cycling stability. The authors concluded that the electrochemical performance of anatase titania was determined not only by surface area and crystallite size but also by mesopore size. The presence of mesopores was important for high-rate performance and favourable to electrolyte ion transport.

Mancini et al. [154] found that new Cu or Sn/mesoporous anatase electrodes offer excellent electrochemical performance, especially in terms of fast insertion/extraction capacity. The capacity after 200 cycles is 123, 147 and 142 mAh/g for uncoated, Cu-coated and Sn-coated anatase electrodes, respectively, with capacity retention of about 80% for all electrodes. The good electrochemical behaviour of metal/mesoporous anatase TiO_2 is ascribed to the combined effects of mesopores and the electronically conductive metal layer. Moreover, the metal coating provides a lower polarisation of the electrodes, which indicates faster kinetics of the electrochemical processes. The researchers suggested that a thin metal coating may be a very promising method in the development of high-rate electrode materials for Li-ion batteries.

Kubiak et al. [155] produced nanosized rutile TiO_2 via a hydrolytic sol-gel route, applying a glycerol-modified precursor in the presence of an anionic surfactant. The proposed methodology led to rutile whiskers, which agglomerated to cauliflower-like aggregates of several micrometers, with a BET surface area of 181 m^2/g. This interesting morphology of rutile titania favours contact between the active material and the electrolyte. The obtained material shows excellent electrochemical performance in terms of capacity, cyclability, stability and reversibility, especially at high charge/discharge rates. The authors demonstrated that this high rate capability can be ascribed to shorter transport lengths for both electronic and Li^+ transport, as well as a larger electrode/electrolyte contact area due to the high surface area.

Mesoporous anatase TiO_2 was synthesised via a urea-assisted hydrothermal method by Jung et al. [156]. The authors investigated the influence of thermal treatment of mesoporous TiO_2 at 300, 400 and 500°C on its electrochemical performance. The prepared material was found to consist of monophasic TiO_2 sub-microspheres with uniform particle size (ca. 400 nm), a crystallite size of 14 nm and a BET surface area of 116 m^2/g. The capacity for the mesoporous TiO_2 calcined at 400°C after 80 cycles is 154 mAh/g, with capacity retention of about 94.5%. It was concluded that the large surface area introduced by the highly porous nano-structured building blocks of each TiO_2 sub-microsphere assisted in creating an easy and shorter diffusion pathway for ionic and electronic diffusion. These results indicate the good power performance of the synthesised material.

Zhang et al. [157] showed that hierarchical nanostructures and composition play key roles in the electrochemical performance of TiO_2 hollow microspheres used as anode materials. Mesoporous hollow TiO_2 microspheres with controlled size and hierarchical nanostructures were synthesised by hydrothermal methods. The results show that the hollow microspheres composed of mesoporous nanospheres exhibit a very stable reversible capacity of 184 mAh/g at 0.25C and an extremely high power of 122 mAh/g at the high rate of 10C. It was also shown that the hollow structure and large mesoporous channels of the material facilitate electrolyte transportation and lithium ion diffusion, and the small mesopores and small-sized nanoparticles increase the lithium storage capacity.

Metal oxides are one of the promising classes of materials to replace graphite anodes for LIBs, because these materials have diverse chemical and physical properties and can deliver high reversible capacities between 500 and 1000 mAh/g. The electrodes of metal oxides, which have high specific capacity, are prone to fail during their reaction with Li ions during the charge and discharge processes. To prevent these failures, TiO_2 is introduced into these electrodes to form TiO_2/metal-oxide composites. The TiO_2/metal-oxide composites for LIBs most commonly combine TiO_2, which has good cycling performance and capacity for LIBs, with other metal oxides with high capacity for LIBs such as SiO_2, ZnO, ZrO_2, etc [154, 155, 158].

Opra et al. [159] obtained nanostructured Zr-doped (1 at.%) TiO_2 (anatase) via a template sol-gel method on carbon fibre. The obtained material consisted of microtubes (length 10–300 μm, outer diameter 3–5 μm) composed of nanoparticles with a size of 15–25 nm. Moreover, Zr-doped TiO_2 shows significantly higher reversible capacity (140 mAh/g) after 20-fold cycling at a rate of 0.1C in the range 3–1 V in comparison with undoped titania (65 mAh/g). It was reported that the transport of Li^+ ions depends significantly on the structural characteristics of titania [17]. When Zr^{4+} ions are incorporated into the anatase structure, the difference in the ionic radii of the metal ions increases the lattice parameters and creates defects. The creation of defects leads to charge redistribution in the titania lattice and increases the conductivity (according to EIS results).

Gao et al. [160] reported the successful production of TiO_2/ZnO nanocomposite arrays for lithium-ion battery application. The sandwich-like TiO_2/ZnO framework with 3D interconnected construction shows stable cycling performance with a specific capacity of 340 mAh/g at a current density of 200 mA/g after 100 cycles. The authors noted that the uniform decoration of ZnO nanoparticles onto the TiO_2 nanosheet arrays plays a significant role in advancing their electrochemical performance.

Siwińska-Stefańska and Kurc [161] used a novel TiO_2-SiO_2-ZrO_2 (TSZ) ternary oxide system (with a TiO_2:SiO_2:ZrO_2 molar ratio of 8:1:1) synthesised via a sol-gel route as an anode material in a Li-ion battery. They combined titania with silica, which can react with a low discharge potential and can store a large quantity of lithium ions, as well as with zirconia, which is capable of suppressing SEI formation and enhancing electron transport to improve electrochemical performance. The specific discharge/charge capacity of the TSZ electrode is about 175 mAh/g.

Unique TiO_2 nanotube arrays (TNAs) grafted with MnO_2 nanosheets were synthesised as a Li-ion battery anode by Zhu et al. [162]. The obtained composite combines the advantages of both MnO_2, with its high capacity (1230 mAh/g), and TNAs, with excellent cycle stability and superior electrical conductivity. Sample TM-10 demonstrated a capacity of 610 mAh/g at a current rate of 350 mA/g and a capacity of 385 mAh/g at a rate of 700 mA/g even after 700 cycles. It was proved that the layer thickness of MnO_2 has a major impact on electrochemical performance.

Research on anode materials for lithium-ion batteries has also been focused on carbonaceous materials and transmission semiconductors such as TiO_2. Carbonaceous materials have high stability, but low volumetric capacity, mainly due to their large initial irreversible capacity. Metal oxide semiconductors have many advantages as electrode materials, including robustness, chemical and thermal stability, low cost, biocompatibility, and relatively high electronic conductivity. The synthesis of mesoporous oxide semiconductors like titania has become an important issue in the construction of smart nanosensors with electrodes decorated by metal oxides [163–166].

Qiu et al. [167] synthesised a mesoporous TiO_2/graphene composite using graphene oxide (GO) and cheap $TiOSO_4$ as precursors, via a facile one-step hydrothermal route. The obtained material exhibited a high discharge capacity (141.7 mAh/g) at the current density 5000 mA/g, an impressive value which is among the highest measured for any TiO_2/graphene composite. The authors suggested that the conductive graphene in the composite may facilitate electron transfer and contribute to the higher rate capability of the TiO_2/graphene composite electrode compared with the blank TiO_2 electrode.

Carbon-coated TiO_2/SiO_2 nanocomposites (CTSO) were produced using a simple hydrothermal approach by Zhang et al. [168] as anode materials for lithium-ion batteries. The CTSO anode exhibits superior high-rate capability and excellent cycling performance. The specific capacity of the obtained material is much higher than that of pure TiO_2 and silica-modified TiO_2 without carbon nanocoating (TSO), which indicates that the material and structural hybridisation has a positive synergistic effect on the electrochemical properties. CTSO (0.05) presented the best cyclability, with 264 mAh/g retained after 270 cycles at 30 mA/g, and superior high-rate performance (233 mAh/g at 150 mA/g after 600 cycles, and even 167 mAh/g at 300 mA/g after 1000 cycles).

Siwińska-Stefańska and Kurc [169] reported on the synthesis, electrochemical properties and performance of a new type of microsized titania/graphene oxide (TA/GO) composite applied as a new anode material in lithium-ion batteries. The material was characterised by the presence of microsized particles with anatase and rutile structure, and a BET surface area of 6.2 m^2/g. The specific discharge and charge capacities of TA/GO electrodes are approximately 1850–2010 and 2050–2100 mAh/g, respectively. Strong Ti–O–C chemical bonds give the composite resilient strength to facilitate the ordered assembly of TiO_2 nanoparticles and formation of a mesoporous structure with a high tap density, enable the rapid transport of Li ions and electrons within the composite structure and maintain a stable mesoporous structure during the discharge/charge process of the resultant LIBs.

Mesoporous TiO_2/CNTs 3D conductive network hybrid nanostructures were synthesised using a facile PEO-aided self-assembled process by Wang et al. [170]. The material demonstrated high Li storage capacity, superior rate performance and excellent long-term cycling stability. Mesoporous TiO_2/CNTs exhibits a reversible specific capacity of 203 mAh/g at 100 mA/g and a stable capacity retention of 91 mAh/g at 8000 mA/g (47.6C) over 100 cycles. The obtained material also retained approximately 90% (71 mAh/g) of its initial discharge capacity after 900 cycles at an extremely high rate of 15,000 mA/g (89C).

As this review of the subject literature shows, the search for new electrode materials is the subject of ongoing worldwide research efforts. The work is largely oriented towards the development of new types of batteries with high energy density and cyclability, and with as fast a charging rate as possible. The search for new electrode materials, and modification of existing ones, to achieve increased electrical capacity and the possibility of operation over a wider range of potentials relative to the lithium electrode is part of a current trend in scientific research.

Based on the current state of knowledge and our own research, we foresee continued growth in research work oriented towards the synthesis of functional materials containing titanium dioxide. Studies to date show that the obtaining of such materials is particularly important from both a theoretical and a practical standpoint. The good availability of methods for obtaining titania-based systems, their interesting physicochemical properties and their broad range of possible applications mean that these materials are coming to be used more and more widely in various branches of industry. Also of key importance is the synthesis of hybrid materials aimed at improving the physicochemical properties of TiO_2, including through the careful control of the quantities of particular components. This creates a wide area for potential research and represents an alternative to the popularly used methods of synthesis. Research in this area is directed towards obtaining functional materials with not only photocatalytic but also antibacterial properties, offering defined electrochemical or barrier effects. Also being intensively developed are combined methods, such as the microwave-assisted sol-gel process, which also introduce interesting theoretical considerations regarding the synthesis of titania-based materials. This creates better possibilities for control of the synthesis and of the physicochemical parameters of the products.

Analysis of the literature review presented confirms the justification for the continuation of research into the synthesis of functional materials based on TiO_2.

Acknowledgements

This work was supported by Poznan University of Technology research grant no. 03/32/DSPB/0706/2017.

Author details

Katarzyna Siwińska-Stefańska* and Teofil Jesionowski

*Address all correspondence to: katarzyna.siwinska-stefanska@put.poznan.pl

Faculty of Chemical Technology, Institute of Chemical Technology and Engineering, Poznan University of Technology, Poznan, Poland

References

[1] Wypych G. Handbook of Fillers. 2nd ed. Toronto: ChemTec Publishing; 1999. 912 p.

[2] Braun JH. Titanium dioxide: A review. Journal of Coating Technology. 1997;**69**(868):59-72

[3] Siwińska-Stefańska K, Krysztafkiewicz A, Ciesielczyk F, Paukszta D, Sójka-Ledakowicz J, Jesionowski T. Physicochemical and structural properties of TiO_2 precipitated in an emulsion system. Physicochemical Problems of Mineral Processing. 2010;**44**:231-244

[4] Winkler J. Titanium dioxide. Production, properties and effective usage. 2nd ed. Hanover: Vincentz Network; 2013. 150 p

[5] Chen X, Mao SS. Titanium dioxide nanomaterials: Synthesis, properties, modifications, and applications. Chemical Reviews. 2007;**107**(7):2891-2959. DOI: 10.1021/cr0500535

[6] Ragupathy S, Raghu K, Prabu P. Synthesis and characterization of TiO_2 loaded cashew nut shell activated carbon and photocatalytic activity on BG and MB dyes under sunlight radiation. Spectrochimica Acta Part A. 2015;**138**(5):314-320. DOI: 10.1016/j.saa.2014.11.087

[7] Azambre B, Zenboury L, Weber JV, Burg P. Surface characterization of acidic ceria-zirconia prepared by direct sulfation. Applied Surface Science. 2010;**256**(14):4570-4581. DOI: 10.1016/j.apsusc.2010.02.049

[8] Miller JM, Lakshmi LJ. Spectroscopic characterization of sol-gel-derived mixed oxides. The Journal of Physical Chemistry B. 1998;**102**(34):6465-6470. DOI: 10.1021/jp9810771

[9] Mohamed MM, Salama TM, Yamaguchi T. Synthesis, characterization and catalytic properties of titania-silica catalysts. Colloids and Surfaces A. 2002;**207**(1-3):25-32. DOI: 10.1016/S0927-7757(02)00002-X

[10] Tobaldi DM, Tucci A, Ŝkapin AS, Esposito L. Effects of SiO_2 addition on TiO_2 crystal structure and photocatalytic activity. Journal of the European Ceramic Society. 2010;**30**(12):2481-2490. DOI: 10.1016/j.jeurceramsoc.2010.05.014

[11] Wu ZY, Tao YF, Lin Z, Liu L, Fan XX, Wang Y. Hydrothermal synthesis and morphological evolution of mesoporous titania-silica. Journal of Physical Chemistry C. 2009;**113**(47):20335-20348. DOI: 10.1021/jp9037842

[12] Messina PV, Morini MA, Sierra MB, Schulz PB. Mesoporous silica-titania composed materials. Journal of Colloid and Interface Science. 2006;**300**(1):270-278. DOI: 10.1016/j.jcis.2006.03.039

[13] Nilchi A, Janitabar-Darzi S, Mahjoub AR, Rasouli-Garmarodi S. New TiO_2/SiO_2 nanocomposites—Phase transformations and photocatalytic studies. Colloids and Surfaces A. 2010;**361**(1-3):25-30. DOI: 10.1016/j.colsurfa.2010.03.006

[14] Hadjiivanov KI, Klissurski DG. Surface chemistry of titania (anatase) and titania-supported catalysts. Chemical Society Reviews. 1996;**25**(1):61-69. DOI: 10.1039/CS9962500061

[15] Scolan A, Sanchez C. Synthesis and characterization of surface-protected nanocrystalline titania particles. Chemistry of Materials. 1998;**10**(10):3217-3223. DOI: 10.1021/cm980322q

[16] Lal M, Chhabra V, Ayyub P, Maitra A. Preparation and characterization of ultrafine TiO_2 particles in reverse micelles by hydrolysis of titanium di-ethylhexyl sulfosuccinate. Journal of Materials Research. 1998;**13**(5):1249-1254. DOI: 10.1557/JMR.1998.0178

[17] Su C, Hong B-Y, Tseng C-M. Sol-gel preparation and photocatalysis of titanium dioxide. Catalysis Today. 2004;**96**(3):119-126. DOI: 10.1016/j.cattod.2004.06.132

[18] Yang P, Lu C, Hua N, Du Y. Titanium dioxide nanoparticles co-doped with Fe^{3+} and Eu^{3+} ions for photocatalysis. Materials Letters. 2002;**57**(4):794-801. DOI: 10.1016/S0167-577X(02)00875-3

[19] Bessekhouad Y, Robert D, Weber JV. Synthesis of photocatalytic TiO_2 nanoparticles: Optimization of the preparation conditions. Journal of Photochemistry and Photobiology A. 2003;**157**(1):47-53. DOI: 10.1016/S1010-6030(03)00077-7

[20] Kormann C, Bahnemann DW, Hoffmann R. Preparation and characterization of quantum-size titanium dioxide. Journal of Physical Chemistry. 1988;**92**(18):5196-5201. DOI: 10.1021/j100329a027

[21] Li B, Wang X, Yan M, Li L. Preparation and characterization of nano-TiO_2 powder. Materials Chemistry and Physics. 2002;**78**(1):184-188. DOI: 10.1016/S0254-0584(02)00226-2

[22] Kim C-S, Moon BK, Park J-H, Chung ST, Son S-M. Synthesis of nanocrystalline TiO_2 in toluene by a solvothermal route. Journal of Crystal Growth. 2003;**254**(3-4):405-410. DOI: 10.1016/S0022-0248(03)01185-0

[23] Wang C, Deng Z-X, Zhang G, Fan S, Li Y. Synthesis of nanocrystalline TiO_2 in alcohols. Powder Technology. 2002;**125**(1):39-44. DOI: 10.1016/S0032-5910(01)00523-X

[24] Kang M, Kim B-J, Cho SM, Chung C-H, Kim B-W, Han GY, Yoon KJ. Decomposition of toluene using an atmospheric pressure plasma/TiO_2 catalytic system. Journal of Molecular Catalysis A. 2002;**180**(1-2):125-132. DOI: 10.1016/S1381-1169(01)00417-4

[25] Kominami H, Kohno M, Takada Y, Inoue M, Inui T, Kera Y. Hydrolysis of titanium alkoxide in organic solvent at high temperatures: A new synthetic method for nanosized, thermally stable titanium(IV) oxide. Industrial & Engineering Chemistry Research. 1999;**38**(10):3925-3931. DOI: 10.1021/ie9901170

[26] Kolen'ko YV, Burukhin AA, Churagulov BR, Oleynikov NN. Synthesis of nanocrystalline TiO_2 powders from aqueous $TiOSO_4$ solutions under hydrothermal conditions. Materials Letters. 2003;**57**(5-6):1124-1129. DOI: 10.1016/S0167-577X(02)00943-6

[27] Zhang TC, Surampalli RY, Lai KCK, Hu Z, Tyagi RD, Lo IMC. Nanotechnologies for Mater Environment Application. Virginia: American Society of Civil Engineers; 2009. 630 p. DOI: 10.1061/9780784410301

[28] Michael D, Mingos P. Comprehensive Organometallic Chemistry III. 1st ed. Oxford: Elsevier Science; 2007. 958 p.

[29] Liu Z, Wang R, Kan F, Jiang F. Synthesis and characterization of TiO_2 nanoparticles. Asian Journal of Chemistry. 2014;**26**(3):655-659. DOI: 10.14233/ajchem.2014.15462

[30] Muhamed Shajudheen VP, Viswanathan K, Anitha Rani K, Uma Maheswari A, Savavana Kumar S. A simple chemical precipitation method of titanium dioxide nanoparticles using polyvinyl pyrrolidone as a capping agent and their characterization. International Journal of Chemical Molecular Nuclear, Materials and Metallurgical Engneering. 2006;**10**(5):552-555

[31] Huang M, Yu S, Lin B, Dongn L, Zhang F, Fan M, Wang L, Yu J, Deng Ch. Influence of preparation methods on the structure and catalytic performance of SnO_2-doped TiO_2 photocatalysts. Ceramics International. 2014;**40**(8):13305-13312. DOI: 10.1016/j.ceramint.2014.05.043

[32] Yu M, Li C, Zeng G, Zhou Y, Zhang X, Xie Y. The selective catalytic reduction of NO with NH_3 over a novel Ce-Sn-Ti mixed oxides catalyst: Promotional effect of SnO_2. Applied Surface Science. 2015;**342**:174-182. DOI: 10.1016/j.apsusc.2015.03.052

[33] Zhang Y, Zhu X, Shen K, Xu H, Sun K, Zhou Ch. Influence of ceria modification on the properties of TiO_2-ZrO_2 supported V_2O_5 catalysts for selective catalytic reduction of NO by NH_3. Journal of Colloid and Interface Science. 2012;**376**(1):233-238. DOI: 10.1016/j.jcis.2012.03.001

[34] Pierre AC, Pajonk GM. Chemistry of aerogels and their applications. Chemical Reviews. 2002;**102**(11):4243-4266. DOI: 10.1021/cr0101306

[35] Lu ZL, Lindner E, Mayer HA. Applications of sol-gel processed interphase catalysts. Chemical Reviews. 2002;**102**(10):3543-3578. DOI: 10.1021/cr010358t

[36] Wight AP, Davis ME. Design and preparation of organic-inorganic hybrid catalysts. Chemical Reviews. 2002;**102**(10):3589-3614. DOI: 10.1021/cr010334m

[37] Schwarz JA, Contescu C, Contescu A. Methods for preparation of catalytic materials. Chemical Reviews. 1995;**95**(3):477-510. DOI: 10.1021/cr00035a002

[38] Hench LL, West JK. The sol-gel process. Chemical Reviews. 1990;**90**(1):33-72. DOI: 10.1021/cr00099a003

[39] Arnal P, Corriu RJP, Leclercq D, Mutin PH, Vioux A. A solution chemistry study of nonhydrolytic sol-gel routes to titania. Chemistry of Materials. 1997;**9**(3):694-698. DOI: 10.1021/cm960337t

[40] Milea CA, Bogatu C, Duţă A. The influence of parameters in silica sol-gel process. Bulletin of the Transilvania University of Braşov Series I: Engineering Sciences. 2011;**4**(53):59-66

[41] Chan Y-L, Pung S-Y, Sreekantan S. Degradation of organic dye using ZnO nanorods based continuous flow water purifier. Journal of Sol-Gel Science and Technology. 2013;**66**(3):399-405. DOI: 10.1007/s10971-013-3022-9

[42] Wu L-A, Jiang X, Wu S, Yao R, Qiao X, Fan X. Synthesis of monolithic zirconia with macroporous bicontinuous structure via epoxide-driven sol-gel process accompanied

by phase separation. Journal of Sol-Gel Science and Technology. 2014;**69**(1):1-8. DOI: 10.1007/s10971-013-3157-8

[43] Ciesielczyk F, Przybysz M, Zdarta J, Piasecki A, Paukszta D, Jesionowski T. The sol-gel approach as a method of synthesis of $xMgO{\cdot}ySiO_2$ powder with defined physicochemical properties including crystalline structure. Journal of Sol-Gel Science and Technology. 2014;**71**(3):501-513. DOI: 10.1007/s10971-014-3398-1

[44] Vrancken KC, Possemiers K, Voort PVD, Vansant EF. Surface modification of silica gels with aminoorganosilanes. Colloids and Surfaces A. 1995;**98**(3):235-241. DOI: 10.1016/0927-7757(95)03119-X

[45] Cerveau G, Corriu RJP, Lepeytre C, Mutin PH. Influence of the nature of the organic precursor on the textural and chemical properties of silsesquioxane materials. Journal of Materials Chemistry. 1998;**8**(12):2707-2714. DOI: 10.1039/A805794J

[46] Corriu R. A new trend in metal-alkoxide chemistry: The elaboration of monophasic organic-inorganic hybrid materials. Polyhedron. 1998;**17**(5-6):925-934. DOI: 10.1016/S0277-5387(97)00261-1

[47] Corriu RJP, Leclerq D. Recent developments of molecular chemistry for sol-gel processes. Angewandte Chemie. 1996;**35**(13-14):1420-1436

[48] Shea KJ, Loy DA, Webster O. Arylsilsesquioxane gels and related materials. New hybrids of organic and inorganic networks. Journal of the American Chemical Society. 1992;**114**(17):6700-6710. DOI: 10.1021/ja00043a014

[49] Jackson CL, Bauer BJ, Nakatami AI, Barnes J. Synthesis of hybrid organic-inorganic materials from interpenetrating polymer network chemistry. Chemistry of Materials. 1996;**8**(3):727-733. DOI: 10.1021/cm950417h

[50] Attar AS, Ghamsari MS, Hajiesmaeilbaigi F, Mirdamadi S. Modifier ligands effects on the synthesized TiO_2 nanocrystals. Journal of Materials Science. 2008;**43**(5):1723-1729. DOI: 10.1007/s10853-007-2244-z

[51] You JH, Hsu KY. Influence of chelating agent and reaction time on the swelling process for preparation of porous TiO_2 particles. Journal of the European Ceramic Society. 2010;**30**(6):1307-1315. DOI: 10.1016/j.jeurceramsoc.2009.10.011

[52] Huang T, Huang W, Zhou C, Situ Y, Huang H. Superhydrophilicity of TiO_2/SiO_2 thin films: Synergistic effect of SiO_2 and chase-separation-induced porous structure. Surface & Coatings Technology. 2012;**213**:126-132

[53] Chang JA, Vithal M, Baek IC, Seok SI. Morphological and phase evolution of TiO_2 nanocrystals prepared from peroxotitanate complex aqueous solution: Influence of acetic acid. Journal of Solid State Chemistry. 2009;**182**(4):749-756. DOI: 10.1016/j.jssc.2008.12.024

[54] Faycal Atitar M, Ismail Adel A, Al-Sayari SA, Bahnemann D, Afanasev D, Emeline AV. Mesoporous TiO_2 nanocrystals as efficient photocatalysts: Impact of calcination temperature and phase transformation on photocatalytic performance. Chemical Engineering Journal. 2015;**264**:417-424. DOI: 10.1016/j.cej.2014.11.075

[55] Mutuma BK, Shao GN, Kim WD, Kim HT. Sol-gel synthesis of mesoporous anatase-brookite and anatase-brookite-rutile TiO_2 nanoparticles and their photocatalytic properties. Journal of Colloid and Interface Science. 2015;**442**:1-7. DOI: 10.1016/j.jcis.2014.11.060

[56] Siwińska-Stefańska K, Zdarta J, Paukszta D, Jesionowski T. The influence of addition of a catalyst and chelating agent on the properties of titanium dioxide synthesized via the sol-gel method. Journal of Sol-Gel Science and Technology. 2015;**75**(2):264-278. DOI 10.1007/s10971-015-3696-2

[57] Ingo GM, Riccucci C, Bultrini G, Dirè S, Chiozzini G. Thermal and microchemical characterisation of sol-gel SiO_2, TiO_2 and $xSiO_2$-(1−x)TiO_2 ceramic materials. Journal of Thermal Analysis and Calorimetry. 2001;**66**(1):37-46. DOI: 10.1023/A:1012471112566

[58] Siwińska-Stefańska K, Paukszta D, Piasecki A, Jesionowki T. Synthesis and physicochemical characteristics of titanium dioxide doped with selected metals. Physicochemical Problems of Mineral Processing. 2014;**50**(1):265-276. DOI: 10.5277/ppmp140122

[59] Fan M, Hu S, Ren B, Wang J, Jing X. Synthesis of nanocomposite TiO_2/ZrO_2 prepared by different templates and photocatalytic properties for the photodegradation of Rhodamine B. Powder Technology. 2013;**235**:27-32. DOI: 10.1016/j.powtec.2012.09.042

[60] Shao GN, Imran SM, Jeon SJ, Engole M, Abbas N, Salman Haider M, Kang SJ, Kim HT. Sol-gel synthesis of photoactive zirconia-titania from metal salts and investigation of their photocatalytic properties in the photodegradation of methylene blue. Powder Technology. 2014;**258**:9-109. DOI: 10.1016/j.powtec.2014.03.024

[61] Kraleva E, Ehrich H. Synthesis, characterization and activity of Co and Ni catalysts supported on AlMe (Me = Zn, Zr, Ti) mixed oxides. Journal of Sol-Gel Science and Technology. 2012;**64**(3):619-629. DOI: 10.1007/s10971-012-2893-5

[62] Kraleva E, Saladino ML, Matassa R, Caponetti E, Enzo S, Spojakina A. Phase formation in mixed TiO_2-ZrO_2 oxides prepared by sol-gel method. Journal of Structural Chemistry. 2011;**52**(2):330-339. DOI: 10.1134/S0022476611020132

[63] Byrappa K, Yoshimura M. Handbook of hydrothermal technology. 1st ed. New York: William Andrew & Sons; 2001. 893 p.

[64] Holzinger M, Maier J, Sitte W. Fast CO_2-selective potentiometric sensor with open reference electrode. Solid State Ionics. 1996;**86-88**:1055-1062. DOI: 10.1016/0167-2738(96)00250-0

[65] Chae SY, Park MK, Lee SK, Kim TY, Kim SK, Lee WI. Preparation of size-controlled TiO_2 nanoparticles and derivation of optically transparent photocatalytic films. Chemistry of Materials. 2003;**15**(17):3326-3331. DOI: 10.1021/cm030171d

[66] Zhang YX, Li GH, Jin YX, Zhang Y, Zhang J, Zhang LD. Hydrothermal synthesis and photoluminescence of TiO_2 nanowires. Chemical Physics Letters. 2002;**365**(3-4):300-304. DOI: 10.1016/S0009-2614(02)01499-9

[67] Caillot T, Salama Z, Chantur N, Cadete Santos Aires FT, Bennici S, Auroux A. Hydrothermal synthesis and characterization of zirconia based catalysts. Journal of Solid State Chemistry. 2013;**203**:79-85. DOI: 10.1016/j.jssc.2013.04.005

[68] Hirano M, Nakahara C, Ota K, Tanaike O, Inagaki M. Photoactivity and phase stability of ZrO_2-doped anatase-type TiO_2 directly formed as nanometer-sized particles by hydrolysis under hydrothermal conditions. Journal of Solid State Chemistry. 2003;**70**(1): 39-47. DOI: 10.1016/s0022-4596(02)00013-0

[69] Kim CS, Moon BK, Park JH, Choi BC, Seo HJ. Solvothermal synthesis of nanocrystalline TiO_2 in toluene with surfactant. Journal of Crystal Growth. 2003;**257**(3-4):309-315. DOI: 10.1016/S0022-0248(03)01468-4

[70] Wen B, Liu C, Liu Y. Solvothermal synthesis of ultralong single-crystalline TiO_2 nanowires. New Journal of Chemistry. 2005;**29**:969-971. DOI: 10.1039/B502604K

[71] Zhu L, Liu K, Li H, Sun Y, Qiu M. Solvothermal synthesis of mesoporous TiO_2 microspheres and their excellent photocatalytic performance under simulated sunlight irradiation. Solid State Sciences. 2013;**20**:8-14. DOI: 10.1016/j.solidstatesciences.2013.02.026

[72] Yang HG, Liu G, Qiao SZ, Sun CH, Jin YG, Smith C, Zou J, Cheng HM, Lu GQ. Solvothermal synthesis and photoreactivity of anatase TiO_2 nanosheets with dominant {001} facets. Journal of the American Chemical Society. 2009;**131**(11):4078-4083. DOI: 10.1021/ja808790p

[73] Oshima K, Nakashima K, Ueno S, Wada S. Synthesis of titanium dioxide nanoparticles by solvothermal method with polymer gel. Transactions of the Materials Research Society of Japan. 2014;**39**(4):451-454. DOI: 10.14723/tmrsj.39.451

[74] Supphasrirongjaroen P, Praserthdam P, Mekasuwandumrong O, Panpranot J, Impact of Si and Zr addition on the surface defect and photocatalytic activity of the nanocrystalline TiO_2 synthesized by the solvothermal method. Ceramics International. 2010;**36**(4):1439-1446. DOI: 10.1016/j.ceramint.2010.02.001

[75] HerrmannJM.Fundamentalsandmisconceptionsinphotocatalysis.JournalofPhotochemistry and Photobiology. 2010;**216**(2-3):85-93. DOI: 10.1016/j.jphotochem.2010.05.015

[76] Lacombe S, Keller N. Photocatalysis: Fundamentals and applications in JEP 2011. Environmental Science and Pollution Research. 2012;**19**(9):3651-365. DOI: 10.1007/s11356-012-1040-8

[77] Langford C. Photocatalysis—A special issue on a unique hybrid area of catalysis. Catalysts. 2012;**2**(3):327-329. DOI: 10.3390/catal2030327

[78] Fujishima A, Rao TN, Try DA. Titanium dioxide photocatalysis. Journal of Photochemistry and Photobiology C. 2000;**1**(1):1-21. DOI: 10.1016/S1389-5567(00)00002-2

[79] Ray M, Ajay K. Heterogeneous photocatalysis in environmental remediation. Developments in Chemical Engineering and Mineral Processing. 2000;**8**:505-550. DOI: 10.1002/apj.5500080507

[80] Ibhadon AO, Fitzpatrick P. Heterogeneous photocatalysis: Recent advances and applications. Catalysts. 2013;**3**(1)189-218. DOI: 10.3390/catal3010189

[81] Friedmanna D, Mendive C, Bahnemann D. TiO_2 for water treatment: Parameters affecting the kinetics and mechanisms of photocatalysis. Applied Catalysis B. 2010;**99**(3-4):398-406, DOI: 10.1016/j.apcatb.2010.05.014

[82] Teoh WY, Amal R, Scott J. Progress in heterogenours photocatalysis: From classical radical chemistry to engineering nanomaterials and solar reactors. Journal of Physical Chemistry Letters. 2012;**3**(5):629-639. DOI: 10.1021/jz3000646

[83] Carp O, Huisman CL, Reller A. Photoinduced reactivity of titanium dioxide. Progress in Solid State Chemistry. 2004;**32**(1-2):33-177. DOI: 10.1016/j.progsolidstchem.2004.08.001

[84] Kosowska B, Mozia S, Morawski AW, Grzmil B, Janus M, Kałucki K. The preparation of TiO_2-nitrogen doped by calcination of $TiO_2 \cdot xH_2O$ under ammonia atmosphere for visiblbe light photocatalysis. Solar Energy Materials and Solar Cells. 2005;**88**(3):269-280. DOI: 10.1016/j.solmat.2004.11.001

[85] Janus M, Inagaki M, Tryba B, Toyoda M, Morawski AW. Carbon-modified TiO_2 photocatalyst by ethanol carbonisation. Applied Catalysis B. 2006:**63**(3-4):272-276. DOI: 10.1016/j.apcatb.2005.10.005

[86] Inagaki M, Nonaka R, Tryba B, Morawski AW. Dependence of photocatalytic activity of anatase powders on their crystallinity. Chemosphere. 2006;**64**(3)437-445. DOI: 10.1016/j.chemosphere.2005.11.052

[87] Li YF, Liu ZP. Particle size, shape and activity for photocatalysis on titania anatase nanoparticles in aqueous surroundings. Journal of the American Chemical Society. 2011;**133**(39):15743-15752. DOI: 10.1021/ja206153v

[88] Tanaka K, Capule MFV, Hisanaga T. Effect of crystallinity of TiO_2 on its photocatalytic action. Chemical Physics Letters. 1991;**187**(1-2):73-76. DOI: 10.1016/0009-2614(91)90486-S

[89] Kumar KNP, Keizer K, Burrgraaf AJ. Textural evolution and phase transformation in titania membranes: Part 1.—Unsupported membranes. Journal of Materials Chemistry. 1993;**3**(11):1141-1149. DOI: 10.1039/JM9930301141

[90] Linsebigler AL, Lu G, Yates J. Photocatalysis on TiO_2 surfaces: Principles, mechanisms, and selected results. Chemical Reviews. 1995;**95**(3):735-758. DOI: 10.1021/cr00035a013

[91] Hoffmann MR, Martin ST, Choi W, Bahnemann DW. Environmental applications of semiconductor photocatalysis. Chemical Reviews. 1995;**95**(1):69-96. DOI: 10.1021/cr00033a004

[92] Hermann JM. Heterogeneous photocatalysis: Fundamentals and applications to the removal of various types of aqueous pollutants. Catalysis Today. 1999;**53**(1):115-129. DOI: 10.1016/S0920-5861(99)00107-8

[93] Lam RCW, Leung MKH, Leung DYC, Vrijmoed LLP, Yam WC, Ng SP. Visible-light-assisted photocatalytic degradation of gaseous formaldehyde by parallel-plate reactor coated with Cr ion-implanted TiO_2 thin film. Solar Energy Materials and Solar Cells. 2007;**91**(1):54-61. DOI: 10.1016/j.solmat.2006.07.004

[94] Venkatachalam N, Palanichamy M, Arabindoo B, Murugesan V. Enhanced photocatalytic degradation of 4-chlorophenol by Zr^{4+} doped nano TiO_2. Journal of Molecular Catalysis A. 2007;**266**(1-2):158-165. DOI: 10.1016/j.molcata.2006.10.051

[95] Di Paola A, Garcia-Lopez E, Ikeda S, Marci G, Ohtani B, Palmisano L. Photocatalytic degradation of organic compounds in aqueous systems by transition metal doped polycrystalline TiO_2. Catalysis Today. 2002;**75**(1-4)87-93. DOI: 10.1016/S0920-5861(02)00048-2

[96] Hagfeldt A, Grätzel M. Light-induced redox reactions in nanocrystalline systems. Chemical Reviews. 1995;**95**:49-68. DOI: 10.1021/cr00033a003

[97] Chatterjee D, Mahata A. Demineralization of organic pollutants on the dye modified TiO_2 semiconductor particulate system using visible light. Applied Catalysis B. 2001;**33**(2):119-125. DOI: 10.1016/S0926-3373(01)00170-9

[98] Chatterjee D, Dasgupta S. Visible light induced photocatalytic degradation of organic pollutants. Journal of Photochemistry and Photobiology C. 2005;**6**(2-3):186-205. DOI: 10.1016/j.jphotochemrev.2005.09.001

[99] Fang X, Zhang Z, Chen Q, Ji H. Dependence of nitrogen doping on TiO_2 precursor annealed under NH_3 flow. Journal of Solid State Chemistry. 2007;**180**(4):1325-1332. DOI: 10.1016/j.jssc.2007.02.010

[100] Oropeza FE, Harmer J, Egdell RG, Palgrave RG. A critical evaluation of the mode of incorporation of nitrogen in doped anatase photocatalysts. Physical Chemistry Chemical Physics. 2010;**12**:960-969. DOI: 10.1039/B914733K

[101] Irie H, Watanabe Y, Hashimoto K. Carbon-doped anatase TiO_2 powders as a visible-light sensitive photocatalyst. Chemical Letters. 2003;**32**(8):772-773. DOI: 10.1246/cl.2003.772

[102] Sakthivel S, Kisch H. Daylight photocatalysis by carbon-modified titanium dioxide. Angewandte Chemie International Edition. 2003;**42**(40):4908-4911. DOI: 10.1002/anie.200351577

[103] Kociołek-Balawejder E, Szymczyk M. Titanium dioxide as pigment and photocatalyst. Przemysł Chemiczny. 2007;**86**(12):1179-1188

[104] Hirano M, Ota K, Inagaki M, Iwata H. Hydrothermal synthesis of TiO_2/SiO_2 composite nanoparticles and their photocatalytic performances. Journal of the Ceramic Society of Japan. 2004;**112**(1303):143-148.

[105] Ren J, Li Z, Liu S, Xing Y, Xie K. Silica-titania mixed oxides: Si-O-Ti connectivity, coordination of titanium, and surface acidic properties. Catalysis Letters. 2008;**124**(3-4):185-194. DOI: 10.1007/s10562-008-9500-y

[106] Ambrus Z, Balázs N, Alapi T, Wittmann G, Sipos P, Dombi A, Mogyorósi K. Synthesis, structure and photocatalytic properties of Fe(III)-doped TiO_2 prepared from $TiCl_3$. Applied Catalysis B: Environmental 2008;**81**(1-2):27-37. DOI: 10.1016/j.apcatb.2007.11.041

[107] Sathasivam S, Bhachu DS, Lu Y, Chadwick N, Althabaiti SA, Alyoubi AO, Basahel SN, Carmalt CJ, Parkin IP. Tungsten doped TiO_2 with enhanced photocatalytic and

optoelectrical properties via aerosol assisted chemical vapor deposition. Scientific Reports. 2015;**5**:1-10 DOI: 10.1038/srep10952

[108] Hwang YJ, Yang S, Lee H. Surface analysis of N-doped TiO_2 nanorods and their enhanced photocatalytic oxidation activity. Applied Catalysis B. 2017;**204**:209-215. DOI: 10.1016/j.apcatb.2016.11.038

[109] Bakar SA, Byzynski G, Ribeiro C. Synergistic effect on the photocatalytic activity of N-doped TiO_2 nanorods synthesised by novel route with exposed (110) facet. Journal of Alloys and Compounds. 2016;**666**:38-49. DOI: 10.1016/j.jallcom.2016.01.112

[110] Szatmáry L, Bakardjieva S, Šubrt J, Bezdička P, Jirkovský J, Bastl Z, Brezová V, Korenko M. Sulphur doped nanoparticles of TiO_2. Catalysis Today. 2011:**161**(1):23-28. DOI: 10.1016/j.cattod.2010.11.082

[111] Rockafellow EM, Stewart LK, Jenks WS. Is sulfur-doped TiO_2 an effective visible light photocatalyst for remediation?. Applied Catalysis B. 2009;**91**(1-2):554-562. DOI: 10.1016/j.apcatb.2009.06.027

[112] Cheng P, Wang Y, Xu L, Sun P, Su Z, Jin F, Liu F, Sun Y, Lu G. High specific surface area urchin-like hierarchical ZnO-TiO_2 architectures: Hydrothermal synthesis and photocatalytic properties. Materials Letters. 2016;**175**:52-55. DOI: 10.1016/j.matlet.2016.03.120

[113] Zhou W, Liu K, Fu H, Pan K, Zhang L, Wang L, Sun C. Multi-modal mesoporous TiO_2-ZrO_2 composites with high photocatalytic activity and hydriophilicity. Nanotechnology. 2008;**19**(3):1-7. DOI: 10.1088/0957-4484/19/03/035610

[114] Yan W, He F, Gai S, Gao P, Chen Y, Yang P. A novel 3D structured reduced graphene oxide/TiO_2 composite: Synthesis and photocatalytic performance. Journal of Materials Chemistry A. 2014;**2**:3605-3612. DOI: 10.1039/C3TA14718E

[115] Thomas RT, Rasheed PA, Sandhyarani N. Synthesis of nanotitania decorated few-layer graphene for enhanced visible light driven photocatalysis. Journal of Colloid and Interface Science. 2014;**428**:214-221. DOI: 10.1016/j.jcis.2014.04.054

[116] Gupta SM, Tripathi M. A review of TiO_2 nanoparticles. Chinese Science Bulletin. 2011;**56**(16):1639-1657. DOI: 10.1007/s11434-011-4476-1

[117] Ismail AA, Bahnemann DW. Mesoporous titania photocatalysts: Preparation, characterization and reaction mechanisms. Journal of Materials Chemistry. 2011;**21**(32):11686-11707. DOI: 10.1039/C1JM10407A

[118] Choi WY, Termin A, Hoffmann MR. The role of metal ion dopants in quantum-sized TiO_2: Correlation between photoreactivity and charge carrier recombination dynamics. The Journal of Physical Chemistry. 1994;**98**(51):13669-13679. DOI: 10.1021/j100102a038

[119] Litter MI. Heterogeneous photocatalysis: Transition metal ions in photocatalytic systems. Applied Catalysis B. 1999;**23**(2-3):89-114. DOI: 10.1016/S0926-3373(99)00069-7

[120] Tayade RJ, Kulkarni RG, Jasra RV. Transition metal ion impregnated mesoporous TiO_2 for photocatalytic degradation of organic contaminants in water. Industrial & Engineering Chemistry Research. 2006;**45**(15):5231-5238. DOI: 10.1021/ie051362o

[121] Khan, M, Cao, W. Preparation of Y-doped TiO_2 by hydrothermal method and investigation of its visible light photocatalytic activity by the degradation of methylene blue. Journal of Molecular Catalysis A. 2013;**376**:71-77. DOI: 10.1016/j.molcata.2013.04.009

[122] Golubović A, Tomić N, Finčur N, Abramović B, Veljković I, Zdravković J, Grujić-Brojčin M, Babić B, Stojadinović B, Šćepanović M. Synthesis of pure and La-doped anatase nanopowders by sol-gel and hydrothermal methods and their efficiency in photocatalytic degradation of alprazolam. Ceramics International. 2014;**40**(8):13409-13418. DOI: 10.1016/j.ceramint.2014.05.060

[123] Kim DH, Choi DK, Kim SJ, Lee KS. The effect of phase type on photocatalytic activity in transition metal doped TiO_2 nanoparticles. Catalysis Communications. 2008;**9**(5):654-657. DOI: 10.1016/j.catcom.2007.07.017

[124] Jing LQ, Sun XJ, Xin BF, Wang BQ, Cai WM, Fu HG. The preparation and characterization of La doped TiO_2 nanoparticles and their photocatalytic activity. Journal of Solid State Chemistry. 2004;**177**(10):3375-3382. DOI: 10.1016/j.jssc.2004.05.064

[125] Mozia S, Heciak A, Morawski AW. Preparation of Fe-modified photocatalysts and their application for generation of useful hydrocarbons during photocatalytic decomposition of acetic acid. Journal of Photochemistry and Photobiology A. 2010;**216**(2-3):275-282. DOI: 10.1016/j.jphotochem.2010.09.016

[126] Zhou M, Yu J, Cheng B. Effects of Fe-doping on the photocatalytic activity of mesoporous TiO_2 powders prepared by an ultrasonic method. Journal of Hazardous Materials. 2006;**137**(3):1838-1847. DOI: 10.1016/j.jhazmat.2006.05.028

[127] Hung WC, Chen YC, Chu H, Tseng TK. Synthesis and characterization of TiO_2 and Fe/TiO_2 nanoparticles and their performance for photocatalytic degradation of 1,2-dichloroethane. Applied Surface Science. 2008;**255**(5):2205-2213. DOI: 10.1016/j.apsusc.2008.07.079

[128] Khan MA, Woo SI, Yang OB. Hydrothermally stabilized Fe(III) doped titania active under visible light for water splitting reaction. International Journal of Hydrogen Energy. 2008:**33**(20):5345-5351. DOI: 10.1016/j.ijhydene.2008.07.119

[129] Litter MI, Navío JA. Photocatalytic properties of iron-doped titania semiconductors. Journal of Photochemistry and Photobiology A. 1996;**98**(3):171-181. DOI: 10.1016/1010-6030(96)04343-2

[130] Hajjaji A, Atyaoui A, Trabelsi K, Amlouk M, Bousselmi L, Bessais B, Khakani MAE, Gaidi M, Cr-doped TiO_2 thin films prepared by means of a magnetron co-sputtering process: Photocatalytic application. American Journal of Analytical Chemistry. 2014;**5**(8):473-482. DOI: 10.4236/ajac.2014.58056

[131] Kafizas A, Parkin IP. Combinatorial atmospheric pressure chemical vapor deposition (cAPCVD): A route to functional property optimization. Journal of the American Chemical Society. 2011;**133**(50):20458-20467. DOI: 10.1021/ja208633g

[132] Treschev SY, Chou PW, Tseng TH, Wang JB, Perevedentseva EV, Cheng CL. Photoactivities of the visible light-activated mixedphase carbon-containing titanium dioxide: The effect of carbon incorporation. Applied Catalysis B. 2008;**79**(1):8-16. DOI: 10.1016/j.apcatb.2007.09.046

[133] Lettmann C, Hildebrand K, Kisch H, Macyk W, Maier W. Visible light photodegradation of 4-chlorophenol with a coke-containing titanium dioxide photocatalyst. Applied Catalysis B. 2001;**32**(4):215-222. DOI: 10.1016/S0926-3373(01)00141-2

[134] Tojo S, Tachikawa T, Fujitsuka M, Majima T. Iodine-doped TiO_2 photocatalysts: Correlation between band structure and mechanism. The Journal of Physical Chemistry C. 2008;**112**(38):14948-14954. DOI: 10.1021/jp804985f

[135] Choi H, Antoniou MG, Pelaez M, Delacruz AA, Shoemaker OA, Dionysiou DD. Mesoporous nitrogen-doped TiO_2 for the photocatalytic destruction of the cyanobacterial toxin microcystin-LR under visible light irradiation. Environmental Science & Technology. 2007;**41**(21):7530-7535. DOI: 10.1021/es0709122

[136] Fang J, Wang F, Qian K, Bao H, Jiang Z, Huang W. Bifunctional N-doped mesoporous TiO_2 photocatalysts. The Journal of Physical Chemistry C. 2008;**112**(46):18150-18156. DOI: 10.1021/jp805926b

[137] Cong Y, Zhang J, Chen F, Anpo M. Preparation, photocatalytic activity, and mechanism of nano-TiO_2 Co-doped with nitrogen and iron(III). The Journal of Physical Chemistry C. 2007;**111**(28):6976-6982. DOI: 10.1021/jp0727493

[138] Liu H, Su Y, Hu H, Cao W, Chen Z. An ionic liquid route to prepare mesoporous ZrO_2-TiO_2 nanocomposites and study on their photocatalytic activities. Advanced Powder Technology. 2013;**24**(3):683-688. DOI: 10.1016/j.apt.2012.12.007

[139] Tian J, Chen L, Dai J, Wang X, Yin Y, Wu P. Preparation and characterization of TiO_2, ZnO, and TiO_2/ZnO nanofilms via sol-gel process. Ceramics International. 2009;**35**(6):2261-2270. DOI: 10.1016/j.ceramint.2008.12.010

[140] Cheng P, Wang Y, Xu L, Sun P, Su Z, Jin F, Liu F, Sun Y, Lu G. High specific surface area urchin-like hierarchikal ZnO-TiO_2 architectures: Hydrothermal synthesis and photocatalytic properties. Material Letters. 2016;**175**:52-55. DOI:10.1016/j.matlet.2016.03.120

[141] Qiu J., Zhang P, Ling M, Li S, Liu P, Zhao H, Zhang S. Photocatalytic synthesis of TiO_2 and reduced graphene oxide nanocomposite for lithium ion battery. ACS Applied Materials & Interfaces. 2012;**4**(7):3636-3642. DOI: 10.1021/am300722d

[142] Ni Y, Wang W, Huang W, Lu C, Xu Z. Graphene strongly wrapped TiO_2 for high-reactive photocatalyst: A new sight for significant application of graphene. Journal of Colloid and Interface Science. 2014;**428**:162-169. DOI: 10.1016/j.jcis.2014.04.022

[143] Tang H, Zhang J, Zhang YJ, Xiong QQ, Tong YY, Li Y, Wang XL, Gu CD, Tu JP. Porous reduced graphene oxide sheet wrapped silicon composite fabricated by steam etching for lithium-ion battery application. Journal of Power Sources. 2015;**286**:431-437. DOI: 10.1016/j.jpowsour.2015.03.185

[144] Wang X, Wang Y, Yang L, Wang K, Lou X, Cai B. Template-free synthesis of homogeneous yolk-shell TiO_2 hierarchical microspheres for high performance lithium ion batteries. Journal of Power Sources. 2014;**262**:72-78. DOI: 10.1016/j.jpowsour.2014.03.081

[145] Wang X, Xi M, Wang X, Fong H, Zhu Z. Flexible composite felt of electrospun TiO_2 and SiO_2 nanofibers infused with TiO_2 nanoparticles for lithium ion battery anode. Electrochimica Acta. 2016;**190**:811-816. DOI: 10.1016/j.electacta.2015.12.123

[146] Tong X, Zeng M, Li J, Li F. UV-assisted synthesis of surface modified mesoporous TiO_2/G microspheres and its electrochemical performances in lithium ion batteries. Applied Surface Science. 2017;**392**:897-903. DOI: 10.1016/j.apsusc.2016.09.094

[147] Choi DW, Choy K-L. Novel nanostructured SiO_2/ZrO_2 based electrodes with enhanced electrochemical performance for lithium-ion batteries. Electrochimica Acta. 2016;**218**:47-53. DOI: 10.1016/j.electacta.2016.08.116

[148] Armstrong MJ, Burke DM, Gabriel T, O'Regan C, O'Dwyer C, Petkovac N, Holmes JD. Carbon nanocage supported synthesis of V_2O_5 nanorods and V_2O_5/TiO_2 nanocomposites for Li-ion batteries. Journal of Materials Chemistry A. 2013;**1**(40):12568-12578. DOI: 10.1039/C3TA12652H

[149] Madian M, Giebeler L, Klose M, Jaumann T, Uhlemann M, Gebert A, Oswald S, Ismail N, Eychmüller A, Eckert J. Self-organized TiO_2/CoO nanotubes as potential anode materials for lithium ion batteries. ACS Sustainable Chemistry and Engineering. 2015;**3**(5):909-919. DOI: 10.1021/acssuschemeng.5b00026

[150] Ma D, Dou P, Yu X, Yang H, Meng H, Sun Y, Zheng J, Xu X. Novel hollow SnO_2 nanosphere@TiO_2 yolk-shell hierarchical nanospheres as anode material for high-performance lithium-ion batteries. Materials Letters. 2015;**157**:228-230. DOI: 10.1016/j.matlet.2015.05.121

[151] Lü X, Yang W, Quan Z, Lin T, Bai L, Wang L, Huang F, Zhao Y. Enhanced electron transport in Nb-doped TiO_2 nanoparticles via pressure-induced phase transitions. Journal of the American Chemical Society. 2014;**136**(1):419-426. DOI: 10.1021/ja410810w

[152] Das SK, Gnanavel M, Patel MUM, Shivakumara C, Bhattacharyya AJ. Anamolously high lithium storage in mesoporous nanoparticulate aggregation of Fe^{3+} doped anatase titania. Journal of the Electrochemical Society. 2011;**158**(12):A1290-A1297. DOI: 10.1149/2.029112jes

[153] Kubiak P, Geserick J, Hüsing N, Wohlfahrt-Mehrens M. Electrochemical performance of mesoporous TiO_2 anatase. Journal of Power Sources. 2008;**175**:510-516. DOI: 10.1016/j.jpowsour.2007.09.044

[154] Mancini M, Kubiak P, Geserick J, Marassi R, Hüsing N, Wohlfahrt-Mehrens M. Mesoporous anatase TiO_2 composite electrodes: Electrochemical characterization and high rate performances. Journal of Power Sources. 2009;**189**(1):585-589. DOI: 10.1016/j.jpowsour.2008.10.050

[155] Kubiak P, Pfanzelt M, Geserick J, Hörmann U, Hüsing N, Kaiser U, Wohlfahrt-Mehrens M. Electrochemical evaluation of rutile TiO_2 nanoparticles as negative electrode for Li-ion batteries. Journal of Power Sources. 2009;**194**:1099-1104. DOI: 10.1016/j.jpowsour.2009.06.021

[156] Jung H-G, Oh SW, Ce J, Jayaprakash N, Sun Y-K. Mesoporous TiO_2 nano networks: Anode for high power lithium battery applications. Electrochemistry Communications. 2009;**11**(4);756-759. DOI: 10.1016/j.elecom.2009.01.030

[157] Zhang F, Zhang Y, Song S, Zhang H. Superior electrode performance of mesoporous hollow TiO_2 microspheres through efficient hierarchical nanostructures. Journal of Power Sources. 2001;**196**(20):8618-8624. DOI: 10.1016/j.jpowsour.2011.06.006

[158] Li Y, Wang Z, Lv X-J. N-doped TiO_2 nanotubes/N-doped graphene nanosheets composites as high performance anode materials in lithium-ion battery. Journal of Materials Chemistry A. 2014;**2**(37):15473-15479. DOI: 10.1039/C4TA02890B

[159] Opra DP, Gnedenkov SV, Sokolov AA, Zheleznov VV, Voit EI, Sushkov YV, Sinebryukhov SL. Enhancing the reversible capacity of nanostructured TiO_2(anatase) by Zr-doping using a sol-gel template method. Scripta Materialia. 2015;**107**:136-139. DOI: 10.1016/j.scriptamat.2015.06.004

[160] Gao L, Li S, Huang D, Shen Y, Wang M. ZnO decorated TiO_2 nanosheet composites for lithium ion battery. Electrochimica Acta. 2015;**182**:529-536. DOI: 10.1016/j.electacta.2015.09.108

[161] Siwińska-Stefańska K, Kurc B. A novel composite TiO_2-SiO_2-ZrO_2 oxide system as a high-performance anode material for lithium-ion batteries. Journal of the Electrochemical Society. 2017;**164**(4):A728-A734. DOI: 10.1149/2.0911704jes

[162] Zhu, Q, Hu, H, Li, G, Zhu, C, Yu, Y. TiO_2 nanotube arrays grafted with MnO_2 nanosheets as high-performance anode for lithium ion battery. Electrochimica Acta. 2015;**156**:252-260. DOI: 10.1016/j.electacta.2015.01.023

[163] Yun YS, Le V-D, Kim H, Chang S-J, Baek SJ, Park S, Kim BH, Kim Y-H, Kang K, Jin H-J. Effects of sulfur doping on graphene-based nanosheets for use as anode materials in lithium-ion batteries. Journal of Power Sources. 2014;**262**:79-85. DOI: 10.1016/j.jpowsour.2014.03.084

[164] Di Lupo F, Tuel A, Mendez V, Francia C, Meligrana G, Bodoardo S, Gerbaldi C. Mesoporous TiO_2 nanocrystals produced by a fast hydrolytic process as high-rate long-lasting Li-ion battery anodes. Acta Materialia. 2014;**69**:60-67. DOI: 10.1016/j.actamat.2014.01.057

[165] Xiu Z, Hao X, Wu Y, Lu Q, Liu S. Graphene-bonded and -encapsulated mesoporous TiO_2 microspheres as a high-performance anode material for lithium ion batteries. Journal of Power Sources. 2015;**287**:334-340. DOI: 10.1016/j.jpowsour.2015.04.086

[166] Li D, Shi D, Liu Z, Liu H, Guo Z. TiO_2 nanoparticles on nitrogen-doped graphene as anode material for lithium ion batteries. Journal of Nanoparticle Research. 2013;**15**(5):1674-1683. DOI: 10.1007/s11051-013-1674-6

[167] Qiu J, Lai C, Wang Y, Li S, Zhang S. Resilient mesoporous TiO_2/graphene nanocomposite for high rate performance lithium-ion batteries. Chemical Engineering Journal. 2014;**256**:247-254. DOI: 10.1016/j.cej.2014.06.116

[168] Zhang JJ, Wei Z, Huang T, Liu Z-L, Yu A-S. Carbon coated TiO_2-SiO_2 nanocomposites with high grain boundary density as anode materials for lithium-ion batteries. Journal of Materials Chemistry A. 2013;**1**(25):7360-7369. DOI: 10.1039/c3ta11137g

[169] Siwińska-Stefańska K, Kurc B. Preparation and application of a titanium dioxide/graphene oxide anode material for lithium-ion batteries. Journal of Power Sources. 2015;**299**:286-292. DOI: 10.1016/j.jpowsour.2015.09.017

[170] Wang J, Ran R, Tade MO, Shao Z. Self-assembled mesoporous TiO_2/carbon nanotube composite with a three-dimensional conducting nanonetwork as a high-rate anode material for lithium-ion battery. Journal of Power Sources. 2014;**254**:18-28. DOI: 10.1016/j.jpowsour.2013.12.090

Section 4

Theoretical Calculations

Chapter 8

DFT-based Theoretical Simulations for Photocatalytic Applications Using TiO_2

Yeliz Gurdal and Marcella Iannuzzi

Additional information is available at the end of the chapter

http://dx.doi.org/10.5772/intechopen.68976

Abstract

TiO_2 has been shown to be a potential candidate for photoinitiated processes, such as dye sensitized solar cells and water splitting in production of H_2. The large band gap of TiO_2 can be reduced by functionalizing the oxide by adsorbing dye molecules and/or water reduction/oxidation catalysts, by metal/nonmetal doping, and by mixing with another oxide. Due to these methods, several different TiO_2-based complexes can be constructed having different geometries, electronic structures, and optical characteristics. It is practically impossible to test the photocatalytic activity of all possible TiO_2-based complexes using only experimental techniques. Instead, density functional theory (DFT)-based theoretical simulations can easily guide experimental studies by screening materials and providing insights into the photoactivity of the complexes. The aim of this chapter is to provide an outlook for current research on DFT-based simulations of TiO_2 complexes for dye sensitized solar cells and water splitting applications and to address challenges of theoretical simulations.

Keywords: density functional theory, ab-initio molecular dynamics, photocatalyst, TiO_2, dye sensitized solar cells, water splitting

1. Introduction

Emergent technologies and the demand for alternative energy sources, which do not produce greenhouse gases as a byproduct lead to a growing awareness in using those renewable sources already provided by nature, such as sunlight. The ideal goal is to emulate the photochemical process with which the plants convert H_2O and CO_2 into O_2 and carbohydrates by absorbing photons in the energy range between 3.3 and 1.5 eV (visible spectrum). Scientists then aim at designing new catalysts that can employ the easily available sunlight and convert it into chemical energy, without depending on the activation energy generated by traditional fuels.

In 1938, for the first time, Goodeve and Kitchener [1] demonstrated the photocatalytic activity of the TiO_2 surface, which can produce oxygen by absorbing UV light thus leading to the photobleaching of dyes. This study has initiated many others toward discovering photocatalytic reactions that can be catalyzed by TiO_2. The success of this material is also justified by its high stability, low cost, no side effects on humans and environment, and ease in large-scale usage. In spite of the promising properties of TiO_2, the photocatalytic activity of the bare surface is not optimal, due to the too large energy gap [2]. This limits the photons' absorption and induces the fast recombination of the photogenerated carriers [3]. While these drawbacks significantly hinder the effective application of the pristine material, several possible solutions have been envisioned by considering surface modifications, such as composite semiconductor coupling, metal/nonmetal doping, and functionalization by means of different types of adsorbates [4]. For instance, Zn-porphyrin adsorbed on the TiO_2 surface reduces the threshold for the photons' absorption, and by allowing the fast electron injection toward the substrate, slows down the charge recombination process [5].

Among many photocatalytic applications of functionalized TiO_2 surface, water splitting for H_2 production and dye sensitized solar cells (DSSCs) are among the most widely studied topics. Although the increasing number of promising studies is going to build photocatalytically efficient and robust several TiO_2-based materials, in this field, the support of the theoretical approach to explore the properties of possible candidate materials is essential. The investigation of proper atomistic models of the systems of interest, possibly including the electronic structure characterization and reproducing the relevant processes, can significantly help the screening of materials. In particular, it is necessary to understand the nature of the adsorbate-substrate interaction, and providing insights into the photoactivity prior to extensive experimental efforts. The aim of this chapter is to review the current progress and challenges in density functional theory (DFT)-based simulations of functionalized TiO_2 surfaces, including rutile, anatase, and TiO_2 nanoparticles, with respect to the applications in photocatalytic water splitting and DSSCs.

2. Overview on density functional theory

DFT is developed by Hohenberg, Kohn, and Sham [6, 7] in 1964 as a minimization problem of the ground state energy as a function of electron density. The approach is to solve any fully interacting problem by mapping it to a noninteracting problem introducing exchange-correlation functional, see Eq. (1).

$$E = T_s[\rho(r)] + J[\rho(r)] + \int v_{ext}(r)\,\rho(r)\,dr + E_{xc}[\rho(r)] \quad (1)$$

where $T_s[\rho(r)]$ is the kinetic energy of the noninteracting system, $J[\rho(r)]$ is the classical Coulomb repulsion energy, $\int v_{ext}(r)\,\rho(r)\,dr$ is the interaction of the external potential acting on the electrons, and $E_{xc}[\rho(r)]$ term is the exchange and correlational energy. All of these terms are called functionals and they depend on the electron density $\rho(r)$, i.e., the number of electrons per unit volume.

Electron density can be expressed in many ways [8]; however, Gaussian and plane wave formalism is shown to be significantly efficient for the description of the orbitals [9]. A localized Gaussian basis set positioned at each atom is used to expand the Kohn-Sham orbitals and an auxiliary plane wave basis set is used to describe the electron density, thus improving the computational performance in the calculation of the Coulomb interactions. This scheme is shown to be a suitable choice for large-scale DFT simulations [10, 11].

Although the achievement in introducing electron density depends on the total energy instead of the electron wave function formalism and providing simple, universal, and self-consistent-field description of the ground-state electronic structure, any practical usage of DFT requires an accurate description to the exchange and correlational effects, $E_{xc}[\rho(r)]$, see Eq. (2).

$$E_{xc}[\rho(r)] = (V_{ee}[\rho(r)] - J[\rho(r)]) + (T[\rho(r)] - T_s[\rho(r)]) \quad (2)$$

where $T[\rho(r)]$ is the kinetic energy of the interacting system and $V_{ee}[\rho(r)]$ is the nonclassical interaction between electrons. Although the exact analytic expression of the exchange-correlation functional is not known, approximations to these terms have been demonstrated to be able capture most of the physical/chemical properties of many systems from the solid state to the liquid state.

Many methods have been proposed to calculate exchange-correlation contribution to the total energy [12]. One of the most commonly used ones is the generalized gradient approximation (GGA), where the exchange and correlation energy depends on both electron density and its gradient [13]. This method includes semiempirical functionals that consist of one or more parameters fitted to experimentally observed quantities. Perdew-Burke-Ernzerhof (PBE) [14] and Becke exchange/Lee-Yang-Parr correlation (BLYP) [15] are the most popular semiempirical functionals. These functionals are successfully applied to many systems from metals to 2D self-assemblies [16]. However, it is shown that they fail to reproduce some of the experimentally observed properties of oxides accurately, due to the incorrect description of electronic localization by standard DFT [17]. The problem is the incomplete cancellation of the Coulomb self-interaction in GGA functionals, which leads to stabilization of electron delocalization [18]. For instance, PBE density functional is shown to be quite good to capture structural properties of both bulk phase and the surface of TiO_2; however, band gap of bulk TiO_2 is predicted as 1.74 eV [19] which significantly underestimates the experimentally measured band gap of 3.2 eV [20]. Therefore, for electronic structure analysis and band alignment of oxides one should go beyond GGA [21].

One of the commonly applied methods to overcome the failure of GGA is to use hybrid density functionals that mix exact exchange from Hartree-Fock exchange [22] and correlation from GGA. Applying hybrid functionals removes some of the self-interaction error and favors localized electronic states by reducing the barrier to the localization [23]. Including orbital analogue of exchange formalism in hybrid functionals often improves the accuracy of the simulations; however, computational cost increases by at least an order of magnitude with respect to the pure GGA formalism. The most popular hybrid functionals are, e.g., HSE06 [24] and PBE0 [25]. The band gap of the bulk TiO_2 is calculated to be 4.21 eV [26] and 3.35 eV [21] by PBE0 and HSE06, respectively. Although larger band gaps are obtained for semiconductors using hybrid functionals, one should carefully choose the exchange-correlation formalism since significantly larger band gaps can also be obtained, e.g., PBE0.

For most of the photocatalytic applications of TiO_2, a photosensitizer or an active catalyst is adsorbed on the surface in which case both chemical and physical interactions play a role on the adsorption geometry or stability of the complex. For instance, it is shown that cis/trans coordination of the ligand with respect to anchoring group adsorbed on anatase TiO_2 (110) surface affects binding mode of the dye which likely arises due to the dispersion interactions [27]. It is well known that the exchange-correlational functionals suffer for a poor description of dispersion interactions or van der Waals interactions. To increase the accuracy of the simulations the missing dispersion interactions can also be incorporated into DFT. One way is to add dispersion energy correction term, Grimme-D3 [28], which calculates pairwise interactions between atomic species and shows usual $1/r^6$ asymptotic behavior, on top of the total energy obtained by DFT.

Using DFT, one can optimize structures, determine the most stable adsorption geometries, calculate corresponding adsorption and interaction energies, and extract electronic properties. Electronic structure can be analyzed by calculating projected density of states, which also provides energy band gaps, schematic representation of molecular orbitals, charge distribution maps, and charge density difference maps. On the other hand, the exploration of the conformational space at finite temperature is obtained by running ab-initio molecular dynamics simulations (AIMD) [29] through the generation of trajectories of several picoseconds. Phase-space trajectories are generated via numerical integration of equations of motion. Due to the advances in the electronic structure calculations, forces can be derived directly from the electrons without any empirical parameters. Within the Born-Oppenheimer approximation [30], electrons are fully decoupled from the nuclear motion at each MD step. Nuclei is subsequently propagated according to the forces obtained from the electronic structure calculation from timestep to timestep.

Although there are significant contributions to the modeling of materials/devices using many simulation methods, such as time-dependent density functional theory (TD-DFT) [31] or quantum mechanics-molecular mechanics (QM/MM) [32], it is beyond the scope of this chapter to give a complete overview of all studies with different simulation techniques. Therefore, the following sections will focus on theoretical simulations of DSSC and water splitting using DFT, sometime in combination with AIMD.

3. Theoretical simulations for photocatalytic applications of TiO_2

The photoinitiated reactions can be performed using TiO_2 as a substrate-like film having 2–3 μm thickness with 4–10 μm surface area [33], supported on another material, or in the form of nanoparticles ranging from 10 to 300 nm [34] depending on the preparation conditions [35]. There are three main stable crystal phases of TiO_2 rutile, anatase, and brookite, that being interested in heterogeneous catalytic processes. We consider the interactions at the TiO_2 surfaces, and that the most stable surfaces, i.e., the exposed surfaces of TiO_2 films or nanoparticles are the rutile (110) and anatase (101) surfaces [36]. Among these surfaces, the most stable one is known to be the rutile and this is also the reason why it is easier to grow high quality single crystal thin films of rutile exposing the (110) surface [37]. Regarding the electronic structures, the band gaps of rutile and anatase are very similar, ≈3.2 eV; in the case of nanoparticles, some effects due to the finite size might appear, for example, the band gap increases slightly and becomes 3.35 eV [38].

For photocatalytic applications of TiO_2, the anatase phase is often preferred to the rutile phase because of its higher activity [39]. This characteristic is attributed to the its larger conduction band (CB) edge energy [40]. On the other hand, the comparison of DSSC performance of rutile and anatase films shows that their voltage and current characteristics are very similar. The overall solar energy conversion efficiency of rutile is only a few percentage lower than the efficiency of the anatase cell [41]. Given the relatively small difference in activity, rutile and anatase phases are the most widely explored phases of TiO_2. Therefore, in the following sections, we are going to consider only the anatase and rutile TiO_2 polymorphs.

3.1. Dye sensitized solar cells

3.1.1. Historical overview

DSSCs are obviously an important and promising application area of TiO_2, converting photon energy to electric energy. The very first examples of DSSCs have been developed in 1954 by Rappaport and coworkers [42]. Gallium arsenide (GaAs)-based p-n junctions, which are formed by joining n-type and p-type semiconductors, were proposed providing a solar energy conversion efficiency of 6%. The main drawback of such a system has always been its high fabrication cost on a large scale, thus preventing its commercialization. In 1991, O'Regan and Grätzel [43] proposed the today's known modern photovoltaic cell based on a n-type highly porous layer of TiO_2 film coated with a monolayer of dye molecules. This system can be produced at a much lower cost and it mimics photosynthesis providing a photon energy conversion efficiency of 7% under sunlight and 12% under diffuse natural light. Dye molecules absorb photons in the visible light region, while TiO_2 nanoparticles are employed to harvest the excited electrons from the dye's conduction band. The large band gap of TiO_2 is overcome by anchoring the dye to the TiO_2 exposed surfaces by means of properly chosen linking groups, dyes to the device are chemically stabilized on oxide nanoparticles by anchoring groups, for instance by carboxylates. Another advantage with respect to the GaAs-based photovoltaic cells is the extremely thin layer of dye molecules on a TiO_2 substrate which facilitates photosensitization and the electron injection into the TiO_2 conduction band with an efficiency of close to unity [44]. The early discovery of these interesting features and of the advances obtained by the efficient combined dye/TiO_2 systems, the research in this area has been strongly supported and it has led to significant progress in the field of photovoltaics.

3.1.2. Working principle of DSSC

A schematic representation of the most common TiO_2-based DSSC is shown in **Figure 1(a)** where the working principle of DSSC is also illustrated. A transparent conductive oxide (TOC) layer, commonly treated with a glass sheet, is the anode that is exposed to photons in the visible spectrum [45]. A mesoporous layer of TiO_2 nanoparticles is deposited over the TCO layer. In order to facilitate the photosensitization, dye molecules having relatively small band gap are

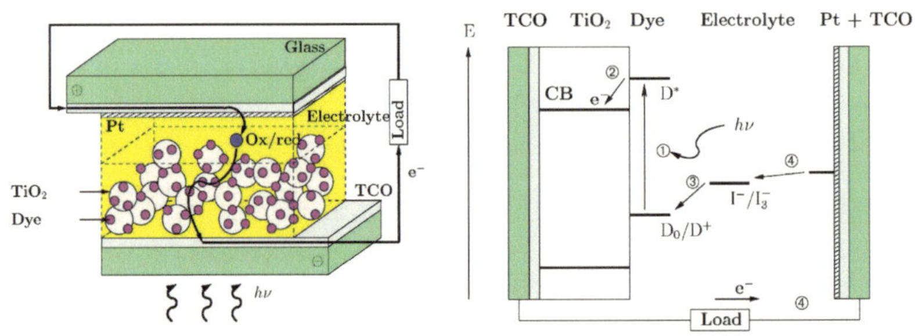

Figure 1. DSSC schematic (left) structure and (right) working principles. TCO stands for transparent conducting oxide. Reprinted with the permission from Ref. [46]. Copyright (2012) American Chemical Society.

chemically bonded to the surface of TiO_2 nanoparticles. The environment is filled with an electrolyte solution, typically consisting of acetonitrile as a solvent and iodine/triiodide (I^-/I_3^-) as a redox couple. The electrolyte is responsible for the regeneration of the dye molecules. The cathode part consists of TOC and an effective catalyst, generally Pt, to collect electrons from the anode.

Exposure to sunlight leads to charge separation in the dye molecules, thus generating excited electrons (e^-) in the conduction band (CB) and electron vacancies, holes (h^+) in the valence band (VB) of the dye. The photogenerated electrons are injected into the TiO_2 CB, thanks to the characteristic ultrafast electron transfer (fs timescale) [47]. Through the oxide, the electrons can be transferred to the TOC at the anode part [48]. This is known as the indirect mechanism. The direct electron injection from the VB of dye into the CB of TiO_2 is also possible which is reported to be faster than an indirect mechanism [49]. Transferring electrons from the anode to the cathode creates an electric current. The photogenerated holes in dye molecules combine with the electrons in $3I^-$, oxidizing $3I^-$ to I_3^-. The electrons at the Pt cathode reduce I_3^- to $3I^-$ so that regeneration of the electrolyte solution is also completed. Thanks to this working principle, solar energy is converted to an electric current in a complete cycle.

In order to be efficient, the described devise has to satisfy some essential conditions. The absorption spectra of the photosensitizers should have a good overlap with the whole solar light spectrum and some of the near-infrared regions. Moreover, the dyes should have high molar extinction coefficient, which is a measure of how strongly a substance absorbs light at a given wavelength per molar concentration. The dye's chemical stability is an another important feature, which affects the lifetime of the device. This can be achieved by chemically linking dyes and semiconductors using anchoring groups. In addition to the chemical stability, properly chosen anchoring groups might also strengthen the degree of coupling between the dyes and oxides thus enhancing the electron injection into the semiconductor CB [50].

It is necessary for a photosensitizer to promote electron injection into TiO_2 layers efficiently. The dye's LUMO has to be aligned correctly if the anchoring group is present, then also its LUMO alignment matters, in particular it has to be energetically close to the dye's LUMO and probably partially overlapping. Besides, the LUMO of the dye should be higher than the CB edge of the oxide to make the electron injection possible. To achieve proper regeneration of the dye, its highest occupied molecular orbital (HOMO) should be below HOMO of the redox mediator, see **Figure 1(b)** for illustration of the complete charge transfer mechanism.

3.1.3. Insights on DSSC components

Dye molecules are relatively well understood with respect to the complex surface structures of TiO_2 phases. The commonly used photosensitizers are either metal-free organic donor-acceptor dyes, such as polythiophenes [51], polypyrroles [52], azobenzene, catechol, and anthraquinone [53] or transition-metal based ligand complexes, such as Ru-polypyridyl molecules [54]. The metal-free organic dyes have large π conjugation leading to π-π^* transitions following the photon absorption. Higher extinction coefficients can be obtained by functionalizing metal-free organic dyes using different donor-acceptor groups, introducing heteroaromatic groups, or changing the number of double bonds in the dye structure thus enabling absorption spectra tuning [33]. Recent advances in designing purely organic dyes adsorbed on nanocrystalline TiO_2 result in achieving solar to power conversion efficiency of 9.5% using indoline dye by Ito et al. [55] and 9.8% using thienothiophene conjugated chromophore by Zhang et al. [56]

Transition metal-based dyes show higher efficiencies with respect to their fully organic counterparts. This is attributed to their capability of absorption in large spectrum range from the visible to the near-infrared regime. The ligand system, i.e., polypyridyl-based or phthalocyanine-based, can be complexed with a chosen transition metal, i.e., Ru, Zn, Mg, and Co. The possibility of long-lived metal-to-ligand (MLCT), d-π^*, or ligand-to-ligand (LLCT), π-π^*, charge transfer once exposed to solar light makes them promising candidates for DSSC applications. However, all these advantages come with a cost, such as the presence of expensive transition metal, i.e., Ru, and rather difficult synthesis and purification steps. Nontoxic and cheap natural dyes that are already present in vegetables and fruits are another possibility to consider. However, unlike synthetic dyes, their recorded efficiency is only around 2% [57].

As already stated, dye molecules are bound to the TiO_2 surface by anchoring groups to enhance chemical stability of the complex. The anchoring groups should be well chosen, since the presence of water in electrolyte solution may weaken the binding strength of the anchors to the TiO_2 surface thus leading to desorption of the dye molecules. Phosphonate groups have been proposed for anchoring Ru-dye/TiO_2 complex and shown to be better candidates than carboxylates [58]. Formic acids are another common example of anchoring groups used in DSSCs [59]. In addition to chemical stability, anchoring groups also affect the degree of coupling between the dye and the semiconductor, which promotes/hinders electron injection into oxide CB. It is shown that strong coupling between Ru-based dyes and TiO_2 nanoparticles anchored by carboxylates leads to an electron injection on femtosecond timescale. On the other hand, SnO_2 and ZnO thin films having lower density of states coupled to a dye/anchoring group have much slower electron injection time with respect to TiO_2 [60].

3.1.4. Simulation studies for DSSC

Up to date, Ru-based dyes such as N3 and N749 are the most widely investigated dyes with recorded efficiency of 11.4% [61]. Labat et al. [62] studied the electronic and structural properties of N3 dye on anatase (101) surface using DFT. As depicted in **Figure 2** while computed orbitals of the HOMO and HOMO-1 are centered on N3 dye and correspond to a $t_{2g} - \pi^*$ interaction between Ru and the ligand, the LUMO is distributed over anatase slab. This picture

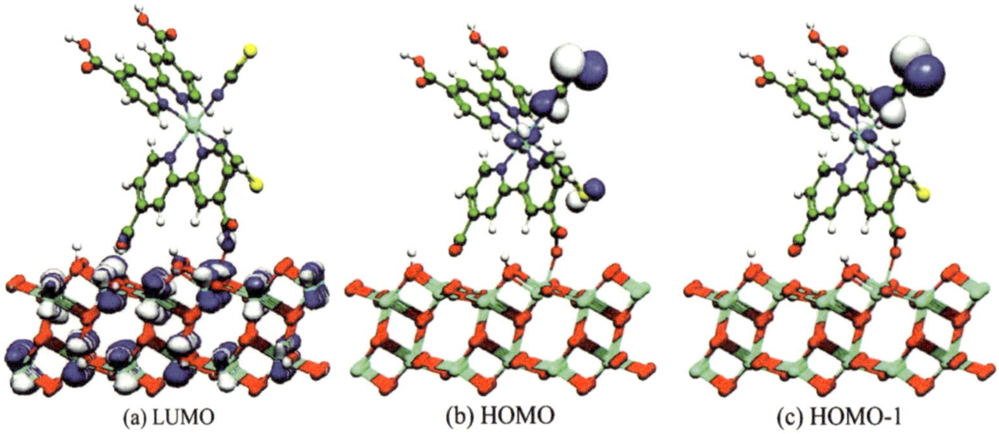

Figure 2. Calculated (a) LUMO, (b) HOMO, and (c) HOMO-1 crystalline orbitals for the bridging adsorption mode of N3/TiO_2\rm_2_2. Isovalues: |0.040| and |0.007| for HOMOs and LUMOs, respectively. Reprinted with the permission from Ref. [62]. Copyright (2010) American Chemical Society.

suggests a favorable electron transfer from the HOMO of the dye to the LUMO of the TiO_2. Schiffmann et al. [27] modeled N3, N712, and N719 dyes adsorbed on anatase (101) surface in the presence of solvent acetonitrile using AIMD. Results indicate that the dyes that are doubly bound to the pristine anatase surface have higher stability. On the other hand, addition of surface protons changes adsorption geometry and mixed bidentate and monodentate binding modes become more favorable. Similarly, the modeling of protonated/deprotonated different N749/$(TiO_2)_{28}$ structures shows that the complexes with deprotonated dye molecules are more favorable with higher adsorption energies [63]. Lundqvist et al. [50] investigated the effects of several spacer unit and anchoring group combinations on the interfacial electronic coupling between a set of Ru^{II}-$(tpy)_2$ dyes and $(TiO_2)_{46}$ nanocrystal. Calculations show that while the strongest coupling is observed using a carboxylic acid anchoring group without a spacer unit, the weakest coupling is observed with the combination of the phosphonic acid anchor group and phenylene spacer group. A good comparison of the calculated structural and electronic characteristics of several Ru-based dyes and their interactions with TiO_2 is available in a recent review [64].

Although Ru-based photosensitizers are very often explored in the literature because of their specific photophysical characteristics and their better performance [65, 66], there are other studies which try to replace expensive Ru metal with cheaper transition metals, such as Fe, Cu, and Co. The experimental study of Ferrere and Gregg [67] investigating Fe(II)bipyridine molecule adsorbed onto nanocrystalline TiO_2 films reveals the substandard performance of Fe-based dyes. The low performance of Fe-based dyes is attributed to the weaker ligand field which prevents MLCT state populations. While the strong ligand field in Ru-based dyes results in having nanoseconds to microseconds lifetime of excited states, excited state lifetime in Fe(II)bipyridine is shown to be only around ps timescale. In spite of the lower performance of the cheaper transition-metal deposited dyes, there are several theoretical studies investigating structural and electronic properties of the most promising ones.

Guillemoles et al. [68] modeled the Os-based polypyridyl dyes with different linker groups (CN^- vs. NCS^-) and in different environments (gas vs. aqueous solution). Their study shows the effects of linker groups and environment on the structural and electronic properties of the dyes. Although in the gas phase the Os-CN^- distance becomes shorter than Os-NCS^- due to the better electron donation character of CN^-, aqueous solution leads to contraction of Os-CN^- (Os-NCS^-) bonds and increases the Os-pyridine distances. Calculated HOMO has contribution from both Os and the linkers, while the LUMO is located on the bipyridyl ligand with π^* character. Shalabi et al. [69] compared the optimized structures of Zn-based porphyrin (ZnPor) and porphyrazin (ZnPz) macrocycles and their energy gaps using DFT. They showed that band gap becomes narrower for ZnPz with respect to ZnPor by more than 0.8 eV, which suggests that light harvesting efficiency can be increased by replacing porphyrin ligand with porphyrazin. Lu et al. [70] compared the electronic structures of several polypyridyl-based dyes with Cu and Ru metal centers. HOMO-LUMO gaps of all considered Cu-based dyes are within the range of 2.41–2.93 eV providing properly aligned energies of the HOMO and LUMO with respect to the electrolyte solution and the CB of TiO_2, respectively. The optical characteristics of Cu and Ru-based dyes are shown to be similar in this study thus providing motivation for the future works on abundant metal-centered dyes.

Fully organic dyes are the alternative of the transition metal deposited ones. One of the widely studied metal-free dyes is catechol. The band gap of catechol, ≈4.2 eV, is larger than the TiO_2 band gap; however, it is shown that the absorption spectra of the catechol/TiO_2 complex have an intense band at 2.95 eV [71]. This observation is explained by a direct electron injection into the CB of TiO_2 from the catechol VB [72]. A DFT study by Persson et al. [73] attributes the reason of direct electron injection to the occupied π level of the catechol at the lower end of the anatase band gap. In terms of adsorption geometry, it is shown that two possible stable catechol monolayer structures present on the rutile TiO_2 (110) surface are monodentate only or mixed monodentate-bidentate configurations [74].

Oprea et al. [75] investigated several coumarin-based dyes adsorbed on different sizes of anatase (101) nanoparticles. Preferential adsorption mode of the coumarin dyes is determined as bidentate bridging which results in localization of the HOMO state on the dyes and the LUMO on the substrate. De Angelis et al. [76] modeled squaraine dye adsorbed on anatase (101) slab in water environment using AIMD. They showed that adsorption mode of squaraine changes from bridged bidentate to a monodentate as simulation time evolves. After 6 ps, however, squaraine is desorbed from the surface due to the strong solvent reorganization which weakens the binding strength of the anchoring group, carboxylic acid.

3.2. Photocatalytic water splitting

Molecular hydrogen has the highest specific energy content as compared to other energy carriers such as electricity and biofuels, conventional fuels such as coal and oil or renewable energy sources as, for example, wind and biomass. Dihydrogen is also environmentally friendly because, the waste product after its reaction with pure dioxygen is only water. These features make H_2 a potential candidate for energy carrier in the near future. One of the dihydrogen generation methods is photochemical water splitting [77]. Several transition metal deposited catalysts such as porphyrin-derived molecules [78], cobaloximes [79], pyrphyrins [80], and molecules with polypyridine ligands [81] have been used for hydrogen evolution in homogeneous environment. Alternatively, to prevent photogenerated charge recombination and stabilize the system and reaction intermediate heterogeneous systems have been widely investigated.

Water splitting on the bare TiO_2 surface is thermodynamically allowed; however, due to the large overpotential for both H_2 and O_2, the reaction is kinetically hindered. Therefore, the pristine TiO_2 surface is inactive for water splitting and needs to be functionalized with metals/nonmetals, large macrocyclic molecules, or mixed with other types of oxides. Several studies deduce the design criteria for an efficient overall water splitting devise based on TiO_2 which can be summarized, as its valence band maximum should be more positive than the oxidation potential of water and its conduction band minimum should be more negative than the reduction potential of a proton [82].

In 1972, Fujishima and Honda [83] carried out the successful and inspiring experiment of water splitting using TiO_2 as a photoanode. In a closed circuit photoelectrochemical cell, photogenerated electron and hole pairs move to the TiO_2 surface where water molecules are oxidized to generate O_2 and resultant protons are combined with electrons at the cathode, Pt, to generate H_2. The correlation has been observed between the band gaps of the photoanode and its water splitting activity. If the band gap becomes closer to the free energy of the water splitting reaction, 1.23 eV, then the efficiency of the photoelectrochemical cell becomes relatively high, around 41%. Efficiency drops to 18% for a band gap of 2 eV. Further increase in the band gap, 3 eV, results in dramatic decrease in the efficiency of the cell, 0.05% [84]. These results engender band gap tuning of TiO_2 by taking advantage of several methods, as discussed below.

An alternative to the photoelectrochemical cells proposed by Fujishima and Honda [83], two other widely explored methods of water splitting using TiO_2 have been recommended. One method is to use transition metal-deposited TiO_2 nanoparticles together with another metal oxide. An example is depicted in **Figure 3(a)** where photon absorption is followed by charge

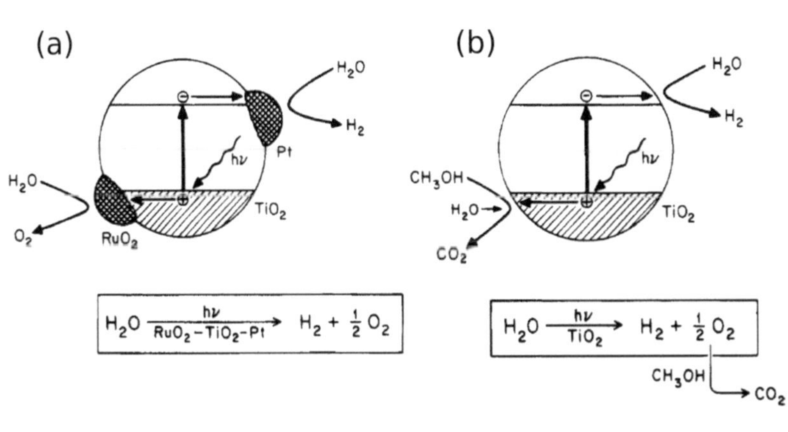

Figure 3. (a) Photosplitting of water on a composite catalyst. (b) Photosplitting of water: sacrificial donor effect. Reprinted with the permission from Ref. [85]. Copyright (1995) American Chemical Society.

separation in TiO_2 and photogenerated electrons and holes are transported into Pt for H_2 generation and RuO_2 for O_2 evolution, respectively. This design results in a substantial decrease in overpotential of the reaction [85]. Another proposed method is to use sacrificial agents to remove one of the photogenerated products. An illustration is shown in **Figure 3(b)** where CH_3OH is used as a sacrificial specie. While photogenerated holes are injected into CH_3OH which afterwards oxidizes water, excited electrons in TiO_2 CB reduces water to H_2 [85]. This system focuses on half reactions, either water reduction via sacrificial electron donor or water oxidation via sacrificial electron acceptor [86].

Prior to reviewing theoretical studies of water splitting on functionalized TiO_2 surfaces, it is useful to first give a closer examination of the interactions between water molecules and the pristine TiO_2 surface. It is shown that water molecules prefer to adsorb on oxygen vacancies of the rutile (110) surface [87]. Water adsorption is followed by a proton transfer to the adjacent two-coordinated surface oxygen thus leading to a formation of hydroxy radical on the surface [88]. Zhao [89] calculated adsorption and decomposition energies of water molecules on different rutile surfaces, such as (110), (100), and (001) using DFT. While dissociative adsorption of water is more favorable on the (110) surface, the activation energy of the water dissociation is largest on the (100) surface. Calculations by Hahn et al. [90] showed that the employed adsorbate coverage affects the favored adsorption mechanism of water molecules. While at low coverage water molecules tend to dissociate on the rutile (110) surface, increasing coverage leads to an associative adsorption. Investigating interactions between water and bare TiO_2 surface is a fundamental research; however, for photocatalytic water splitting TiO_2 surface has to be functionalized, as stated earlier.

Adsorption of large macrocyclic water reduction/oxidation catalysts on the TiO_2 surface is shown to be a promising way of achieving the water splitting reaction. In this regard, Monti et al. [91] proposed a model system of Ru-based catalyst/antenna/TiO_2 complex for water splitting and investigated proton coupled electron transfer dynamics of the catalyst in water solution using AIMD. Results show that a water molecule binds to the Ru center and after 1.9 ps of simulation, time proton transfer from the attached water molecule to the closest one in the solution is achieved. Gurdal et al. [21] modeled adsorption of co-deposited pyrphyrin (CoPyr), a water reduction catalyst [80], on the rutile TiO_2 (110) surface using DFT. The band gap of the complex is decreased by more than 1.5 eV with respect to the bare surface. Additionally, the catalytic properties of CoPyr are preserved upon adsorption since the adsorption process does not lead to strong modifications of the electronic structure at the Co(II) active center.

Alternative complexes, metal/nonmetal deposited TiO_2 structures, are investigated by many groups. Phase stabilities of Fe, Al, Si, and F deposited rutile and anatase phases of TiO_2 are compared using DFT [92]. Results show that while cationic dopants in Ti lattice sites stabilize anatase slightly more than rutile, anionic doping with F substantially reduces the phase stability of the anatase. Modeling Pt, Ru, and Co ions doped anatase TiO_2 (001) surface shows that band gaps of the complexes decrease in the order of deposition of Co (by 1.8 eV), Ru (by 1.7 eV), and Pt (by 0.4 eV) [93]. Nucleation and growth mechanism of Ru clusters on the anatase TiO_2(101) surface reveals that in addition to decrease in the band gap, an electron accumulation at the metal/oxide interface is observed due to the electron flow from Ru to TiO_2 [94].

There are also other studies that focus on designing water splitting systems by mixing TiO_2 with another metal oxide. Graciani et al. [95] modeled water adsorption on the Ce_2O_3 doped rutile TiO_2(110) surface. Results show that the proposed mixed-metal oxide has a promising ability to dissociate water exposing an exothermic dissociation energy of −0.7 eV with a small activation barrier of 0.04 eV. A relatively larger system is investigated by Pastore and De Angelis [96] who modeled TiO_2/Ru-dye/IrO_2 complex using DFT. In the designed complex Ru polypyridyl dye, acting as a linker between the oxides, is attached to both the anatase TiO_2(101) and IrO_2 surfaces via phosphonic acid and malonate groups, respectively. **Figure 4** shows the orientation of the Ru-dye with respect to the oxides. Electronic structure analysis shows that both the HOMO and LUMO are located on IrO_2 showing a metallic-like character. While the occupied molecular orbitals of the dye are located within the IrO_2 VB, its unoccupied orbitals are distributed over the CB of both IrO_2 and TiO_2.

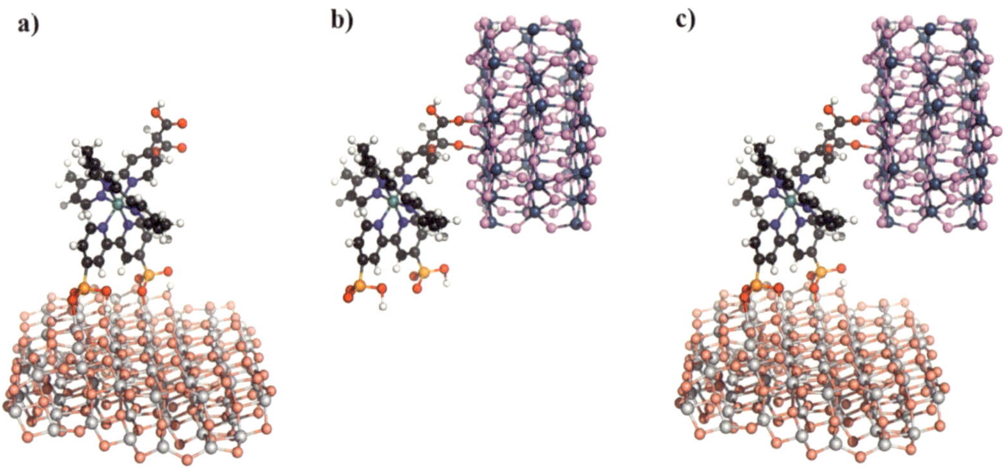

Figure 4. Optimized molecular structure of Ru-dye in its partially deprotonated form, grafted to the (TiO$_2$\rm_2$_2$)$_2$\rm_{82}$_2$ cluster (a), to the (IrO2$_2$\rm_2$_2$)$_2$\rm_{56}$_2$.2H$_2$\rm_2$_2$O nanoparticle (b), and tethered across the TiO$_2$\rm_2$_2$ and IrO$_2$\rm_2$_2$ systems, (c). Reprinted with the permission from Ref.~\cite{pastore2015} Copyright (2015) American Chemical Society

4. Challenges of DFT-based simulations

In order to understand the physical properties of TiO_2 complexes and develop photocatalytically active and efficient devices, one has to use adequate theoretical methods. Although DFT is a powerful tool to analyze and screen TiO_2-based complexes for photocatalytic applications, it also has some limitations. Accurate results can be obtained by increasing system size, i.e., increasing layers of the TiO_2 slab. While the thickness of the slab has to be sufficient to reproduce the correct behavior at the surface and increase the calculated accuracy, increasing system size also increases computational effort. In the case of AIMD runs required computational source becomes even more expensive than DFT calculations. Therefore, one needs to find a compromise between the accuracy and the computational cost. In this regard, Harris and Quong [97] proposed a method named "25% rule" to determine the sufficient slab thickness that should be used in the simulations.

As already stated, the other issue is the uncertain forms of the exchange-correlation functionals which limit the accuracy and cause a dilemma in trusting obtained results. Though, several methods beyond DFT are developed and successfully used to increase the accuracy of the calculations, such as random phase approximation and Green function theory [98, 99].

In spite of several limitations, in both the accuracy of the DFT description and the size of feasible models, this approach seems to be the most appropriate one so far. Constant improvement in computational algorithms together with rapidly increasing computer power, DFT, and AIMD are robust methods used to gain preliminary knowledge on the systems and supervise experimentalists to build promising materials for photocatalytic applications.

5. Conclusions

Search for renewable energy sources leads scientists to benefit from sunlight and convert photon energy to chemical/electric energy using TiO_2-based materials. Although photocatalytic water splitting and DSSC applications are accomplished using TiO_2 surface, large band gap of the oxide limits absorbing photons in the visible spectrum thus hindering device efficiency. Therefore, functionalizing TiO_2 surface by adsorbing photosensitizers and/or water reduction/oxidation catalysts, by metal/nonmetal deposition, or by mixing with other oxides, the optical response of the complex can be shifted from UV to the visible region. This is the crucial requirement in designing promising, robust, and scalable photocatalysts toward water splitting and DSSC applications.

Together with the improvement in the computational power, today, DFT is an important tool to obtain optimized geometries of the complexes, analyze electronic structures, model many spectroscopic techniques, determine intermediate states of the reactions, and so on. In particular, it is powerful for modeling several TiO_2-based materials and testing their physical/chemical/optical characteristics for photocatalytic applications. Although there are some limitations in DFT, several new exchange-correlation density functionals and van der Waals correction schemes have been proposed to increase the flexibility and accuracy of the model. Using DFT, all phases of TiO_2 surfaces, i.e., rutile, anatase, and TiO_2 nanoparticles can be modeled. The outcome of the simulations serves as an initial knowledge on the systems for scientists without experimental effort.

Author details

Yeliz Gurdal* and Marcella Iannuzzi

*Address all correspondence to: yeliz.guerdal@chem.uzh.ch

Department of Chemistry, University of Zurich, Zürich, Switzerland

References

[1] Goodeve CF, Kitchener JA. The mechanism of photosensitisation by solids. Transactions of the Faraday Society. 1938;**34**:902-908. DOI: 10.1039/TF9383400902

[2] Fujiyoshi S, Ishibashi TA, Onishi H. Fourth-order Raman spectroscopy of wide-band gap materials. The Journal of Physical Chemistry B. 2005;**109**:8557-8561. DOI: 10.1021/jp051203i

[3] Iman-Amir M, Julkapli N, Bagheri S, Yousefi AT. TiO_2 hybrid photocatalytic systems: Impact of adsorption and photocatalytic performance. Reviews in Inorganic Chemistry. 2015;**35**:151-178. DOI: 10.1515/revic-2015-0005

[4] Li LL, Diau EWG. Porphyrin-sensitized solar cells. Chemical Society Reviews. 2013; **42**:291-304. DOI: 10.1039/C2CS35257E

[5] Imahori H, Kang S, Hayashi H, Haruta M, Kurata H, Isoda S, et al. Photoinduced charge carrier dynamics of Zn-Porphyrin-TiO_2 electrodes: The key role of charge recombination for solar cell performance. The Journal of Physical Chemistry A. 2011;**115**:3679-3690. DOI: 10.1021/jp103747t

[6] Hohenberg P, Kohn W. Inhomogeneous electron gas. Physical Review. 1964;**136**:B864-B871. DOI: 10.1103/PhysRev.136.B864

[7] Kohn W, Sham LJ. Self-consistent equations including exchange and correlation effects. Physical Review. 1965;**140**:A1133-A1138. DOI: 10.1103/PhysRev.140.A1133

[8] Giannozzi P, Baroni S, Bonini N, Calandra M, Car R, Cavazzoni C, et al. QUANTUM ESPRESSO: A modular and open-source software project for quantum simulations of materials. Journal of Physics: Condensed Matter. 2009;**21**:395502-395520

[9] Lippert G, Hutter J, Parrinello M. A hybrid Gaussian and plane wave density functional scheme. Molecular Physics. 1997;**92**:477-488. DOI: 10.1080/002689797170220

[10] Iannuzzi M, Hutter J. Comparative study of the nature of chemical bonding of corrugated graphene on Ru(0001) and Rh(111) by electronic structure calculations. Surface Science. 2011;**605**:1360-1368. DOI: 10.1016/j.susc.2011.04.031

[11] Mette G, Sutter D, Gurdal Y, Schnidrig S, Probst B, Iannuzzi M, et al. From porphyrins to pyrphyrins: Adsorption study and metalation of a molecular catalyst on Au(111). Nanoscale. 2016;**8**:7958-7968. DOI: 10.1039/C5NR08953K

[12] Becke AD. A new mixing of Hartree-Fock and local density-functional theories. The Journal of Chemical Physics. 1993;**98**:1372-1377. DOI: 10.1063/1.464304

[13] Perdew JP, Yue W. Accurate and simple density functional for the electronic exchange energy: Generalized gradient approximation. Physical Review B. 1986;**33**:8800-8802. DOI: 10.1103/PhysRevB.33.8800

[14] Perdew JP, Burke K, Ernzerhof M. Generalized gradient approximation made simple. Physical Review Letters. 1996;**77**:3865-3868. DOI: 10.1103/PhysRevLett.77.3865

[15] Lee C, Yang W, Parr RG. Development of the Colle-Salvetti correlation-energy formula into a functional of the electron density. Physical Review B. 1988;**37**:785-789. DOI: 10.1103/PhysRevB.37.785

[16] Iannuzzi M, Tran F, Widmer R, Dienel T, Radican K, Ding Y, et al. Site-selective adsorption of phthalocyanine on h-BN/Rh(111) nanomesh. Physical Chemistry Chemical Physics. 2014;**16**:12374-12384. DOI: 10.1039/C4CP01466A

[17] Skorodumova NV, Ahuja R, Simak SI, Abrikosov IA, Johansson B, Lundqvist BI. Electronic, bonding, and optical properties of CeO_2 and Ce_2O_3 from first principles. Physical Review B. 2001;**64**:115108-115116. DOI: 10.1103/PhysRevB.64.115108

[18] Ganduglia-Pirovano MV, Da Silva JLF, Sauer J. Density-functional calculations of the structure of near-surface oxygen vacancies and electron localization on CeO_2(111). Physical Review Letters. 2009;**102**:26101-26104. DOI: 10.1103/PhysRevLett.102.026101

[19] Evarestov RA, Zhukovskii YF. Four-faceted nanowires generated from densely-packed TiO_2 rutile surfaces: Ab initio calculations. Surface Science. 2013;**608**:226-240. DOI: 10.1016/j.susc.2012.10.012

[20] Kavan L, Grätzel M, Gilbert SE, Klemenz C, Scheel HJ. Electrochemical and photoelectrochemical investigation of single-crystal anatase. Journal of the American Chemical Society. 1996;**118**:6716-6723. DOI: 10.1021/ja954172l

[21] Gurdal Y, Luber S, Hutter J, Iannuzzi M. Non-innocent adsorption of co-pyrphyrin on rutile(110). Physical Chemistry Chemical Physics. 2015;**17**:22846-22854. DOI: 10.1039/C5CP02767E

[22] Slater JC. A Simplification of the Hartree-Fock method. Physical Review. 1951;**81**:385-390. DOI: 10.1103/PhysRev.81.385

[23] Da Silva JLF, Ganduglia-Pirovano MV, Sauer J, Bayer V, Kresse G. Hybrid functionals applied to rare-earth oxides: The example of ceria. Physical Review B. 2007;**75**:45121-45130. DOI: 10.1103/PhysRevB.75.045121

[24] Heyd J, Scuseria GE, Ernzerhof M. Hybrid functionals based on a screened Coulomb potential. The Journal of Chemical Physics. 2003;**118**:8207-8215. DOI: 10.1063/1.1564060

[25] Ernzerhof M, Scuseria GE. Assessment of the Perdew–Burke–Ernzerhof exchange-correlation functional. The Journal of Chemical Physics. 1999;**110**:5029-5036. DOI: 10.1063/1.478401

[26] Bahers TL, Rérat M, Sautet P. Semiconductors used in photovoltaic and photocatalytic devices: Assessing fundamental properties from DFT. The Journal of Physical Chemistry C. 2014;**118**:5997-6008. DOI: 10.1021/jp409724c

[27] Schiffmann F, VandeVondele J, Hutter J, Wirz R, Urakawa A, Baiker A. Protonation-dependent binding of ruthenium bipyridyl complexes to the anatase(101) surface. The Journal of Physical Chemistry C. 2010;**114**:8398-8404. DOI: 10.1021/jp100268r

[28] Grimme S, Antony J, Ehrlich S, Krieg H. A consistent and accurate ab initio parametrization of density functional dispersion correction (DFT-D) for the 94 elements H-Pu. The Journal of Chemical Physics. 2010;**132**:154104-154122. DOI: 10.1063/1.3382344

[29] Kirchner B, di Dio PJ, Hutter J. Real-World predictions from ab initio molecular dynamics simulations. In: Kirchner B, Vrabec J, editors. Multiscale Molecular Methods in Applied Chemistry. Berlin, Heidelberg: Springer; 2012. pp. 109-153. DOI: 10.1007/128-2011-195

[30] Barnett RN, Landman U. Born-Oppenheimer molecular-dynamics simulations of finite systems: Structure and dynamics of $(H_2O)_2$. Physical Review B. 1993;**48**:2081-2097. DOI: 10.1103/PhysRevB.48.2081

[31] Burke K, Werschnik J, Gross EKU. Time-dependent density functional theory: Past, present, and future. The Journal of Chemical Physics. 2005;**123**:62206-62214. DOI:10.1063/1.1904586

[32] Lin H, Truhlar DG. QM/MM: What have we learned, where are we, and where do we go from here? Theoretical Chemistry Accounts. 2006;**117**:185. DOI:10.1007/s00214-006-0143-z

[33] Mishra A, Fischer Mâ, Bäuerle P. Metal-free organic dyes for dye-sensitized solar cells: From structure: Property relationships to design rules. Angewandte Chemie International Edition. 2009;**48**:2474-2499. DOI:10.1002/anie.200804709

[34] Liao DL, Liao BQ. Shape, size and photocatalytic activity control of TiO_2 nanoparticles with surfactants. Journal of Photochemistry and Photobiology A: Chemistry. 2007;**187**:363-369. DOI: 10.1016/j.jphotochem.2006.11.003

[35] Barnard AS, Curtiss LA. Prediction of TiO_2 nanoparticle phase and shape transitions controlled by surface chemistry. Nano Letters. 2005;**5**:1261-1266. DOI: 10.1021/nl050355m

[36] Esch TR, Gadaczek I, Bredow T. Surface structures and thermodynamics of low-index of rutile, brookite and anatase – A comparative DFT study. Applied Surface Science. 2014;**288**:275-287. DOI: 10.1016/j.apsusc.2013.10.021

[37] Lazzeri M, Vittadini A, Selloni A. Structure and energetics of stoichiometric TiO_2 anatase surfaces. Physical Review B. 2001;**63**:155409-155417. DOI: 10.1103/PhysRevB.63.155409

[38] Kormann C, Bahnemann DW, Hoffmann MR. Preparation and characterization of quantum-size titanium dioxide. The Journal of Physical Chemistry. 1988;**92**:5196-5201. DOI: 10.1021/j100329a027

[39] Luttrell T, Halpegamage S, Tao J, Kramer A, Sutter E, Batzill M. Why is anatase a better photocatalyst than rutile? - Model studies on epitaxial TiO_2 films. Scientific Reports. 2014;**4**:4043-4050. DOI: 10.1038/srep04043 10.1038/srep04043

[40] Wold A. Photocatalytic properties of titanium dioxide (TiO_2). Chemistry of Materials. 1993;**5**:280-283. DOI: 10.1021/cm00027a008

[41] Park NG, van de Lagemaat J, Frank AJ. Comparison of dye-sensitized rutile and anatase-based TiO_2 solar cells. Journal of Physical Chemistry B. 2000;**104**:8989-8994. DOI: 10.1021/jp994365l

[42] Jenny DA, Loferski JJ, Rappaport P. Photovoltaic effect in GaAs *p-n* junctions and solar energy conversion. Physical Review. 1956;**101**:1208-1209. DOI: 10.1103/PhysRev.101.1208

[43] O'Regan B, Grätzel M. A low-cost, high-efficiency solar cell based on dye-sensitized colloidal TiO_2 films. Nature. 1991;**353**:737-740. DOI: 10.1038/353737a0

[44] Smestad G, Bignozzi C, Argazzi R. Testing of dye sensitized TiO_2 solar cells I: Experimental photocurrent output and conversion efficiencies. Solar Energy Materials and Solar Cells. 1994;**32**:259-272. DOI: 10.1016/0927-0248(94)90263-1

[45] Gong J, Liang J, Sumathy K. Review on dye-sensitized solar cells (DSSCs): Fundamental concepts and novel materials. Renewable and Sustainable Energy Reviews. 2012;**16**:5848-5860. DOI: 10.1016/j.rser.2012.04.044

[46] Labat F, Le Bahers T, Ciofini I, Adamo C. First-principles modeling of dye-sensitized solar cells: Challenges and perspectives. Accounts of Chemical Research. 2012;**45**:1268-1277. DOI: 10.1021/ar200327w

[47] Shklover V, Ovchinnikov YE, Braginsky LS, Zakeeruddin SM, Grätzel M. Structure of organic/inorganic interface in assembled materials comprising molecular components. Crystal structure of the sensitizer Bis[(4,4′-carboxy-2,2′-bipyridine)(thiocyanato)] ruthenium(II). Chemistry of Materials. 1998;**10**:2533-2541. DOI: 10.1021/cm980303g

[48] Duncan WR, Prezhdo OV. Theoretical studies of photoinduced electron transfer in dye-sensitized TiO_2. Annual Review of Physical Chemistry. 2007;**58**:143-184. DOI: 10.1146/annurev.physchem.58.052306.144054

[49] Ghosh HN, Asbury JB, Lian T. Direct observation of ultrafast electron injection from Coumarin 343 to TiO_2 nanoparticles by femtosecond infrared spectroscopy. The Journal of Physical Chemistry B. 1998;**102**:6482-6486. DOI: 10.1021/jp981806c

[50] Lundqvist MJ, Nilsing M, Lunell S. Spacer and anchor effects on the electronic coupling in Ruthenium-bis-terpyridine dye-sensitized TiO_2 nanocrystals studied by DFT. The Journal of Physical Chemistry B. 2006;**110**:20513-20525. DOI: 10.1021/jp064045j

[51] Zhu Y, Xu S, Jiang L, Pan K, Dan Y. Synthesis and characterization of polythiophene/titanium dioxide composites. Reactive and Functional Polymers. 2008;**68**:1492-1498. DOI: 10.1016/j.reactfunctpolym.2008.07.008

[52] Wang D, Wang Y, Li X, Luo Q, An J, Yue J. Sunlight photocatalytic activity of polypyrrole–TiO_2 nanocomposites prepared by 'in situ' method. Catalysis Communications. 2008;**9**:1162-1166. DOI:10.1016/j.catcom.2007.10.027

[53] Jacquemin D, Perpète EA, Scuseria GE, Ciofini I, Adamo C. TD-DFT performance for the visible absorption spectra of organic dyes: conventional versus long-range hybrids. Journal of Chemical Theory and Computation. 2008;**4**:123-135. DOI: 10.1021/ct700187z

[54] Balzani V, Juris A, Venturi M, Campagna S, Serroni S. Luminescent and redox-active polynuclear transition metal complexes. Chemical Reviews. 1996;**96**:759-834. DOI: 10.1021/cr941154y

[55] Ito S, Miura H, Uchida S, Takata M, Sumioka K, Liska P, et al. High-conversion-efficiency organic dye-sensitized solar cells with a novel indoline dye. Chemical Communications. 2008;5194-5196. DOI: 10.1039/B809093A

[56] Zhang G, Bala H, Cheng Y, Shi D, Lv X, Yu Q, et al. High efficiency and stable dye-sensitized solar cells with an organic chromophore featuring a binary π-conjugated spacer. Chemical Communications. 2009;2198-2200. DOI:10.1039/B822325D

[57] Calogero G, Yum JH, Sinopoli A, Marco GD, Grätzel M, Nazeeruddin MK. Anthocyanins and betalains as light-harvesting pigments for dye-sensitized solar cells. Solar Energy. 2012;**86**:1563-1575. DOI: 10.1016/j.solener.2012.02.018

[58] Zakeeruddin SM, Nazeeruddin MK, Pechy P, Rotzinger FP, Humphry-Baker R, Kalyanasundaram K, et al. Molecular engineering of photosensitizers for nanocrystalline solar cells: Synthesis and characterization of Ru dyes based on phosphonated terpyridines. Inorganic Chemistry. 1997;**36**:5937-5946. DOI: 10.1021/ic970008i

[59] Miller KL, Falconer JL, Medlin JW. Effect of water on the adsorbed structure of formic acid on TiO_2 anatase (101). Journal of Catalysis. 2011;**278**:321-328. DOI: 10.1016/j.jcat.2010.12.019

[60] Asbury JB, Hao E, Wang Y, Ghosh HN, Lian T. Ultrafast electron transfer dynamics from molecular adsorbates to semiconductor nanocrystalline thin films. The Journal of Physical Chemistry B. 2001;**105**:4545-4557. DOI: 10.1021/jp003485m

[61] Han L, Islam A, Chen H, Malapaka C, Chiranjeevi B, Zhang S, et al. High-efficiency dye-sensitized solar cell with a novel co-adsorbent. Energy and Environmental Science. 2012;**5**:6057-6060. DOI: 10.1039/C2EE03418B

[62] Labat F, Ciofini I, Hratchian HP, Frisch MJ, Raghavachari K, Adamo C. Insights into working principles of ruthenium polypyridyl dye-sensitized solar cells from first principles modeling. The Journal of Physical Chemistry C. 2011;**115**:4297-4306. DOI: 10.1021/jp108917c

[63] Liu SH, Fu H, Cheng YM, Wu KL, Ho ST, Chi Y, et al. Theoretical study of N749 dyes anchoring on the $(TiO_2)_{28}$ surface in DSSCs and their electronic absorption properties. The Journal of Physical Chemistry C. 2012;**116**:16338-16345. DOI: 10.1021/jp3006074

[64] Fantacci S, De Angelis F. A computational approach to the electronic and optical properties of Ru(II) and Ir(III) polypyridyl complexes: Applications to DSC, OLED and NLO. Coordination Chemistry Reviews. 2011;**255**:2704-2726. DOI: 10.1016/j.ccr.2011.03.008

[65] Hagfeldt A, Grätzel M. Molecular photovoltaics. Accounts of Chemical Research. 2000;**33**:269-277. DOI: 10.1021/ar980112j

[66] Karlsson S, Boixel J, Pellegrin Y, Blart E, Becker HC, Odobel F, et al. Accumulative charge separation inspired by photosynthesis. Journal of the American Chemical Society. 2010;**132**:17977-17979. DOI: 10.1021/ja104809x

[67] Ferrere S, Gregg BA. Photosensitization of TiO_2 by [FeII(2,2′-bipyridine-4,4′-dicarboxylic acid)$_2$(CN)$_2$]: Band selective electron injection from ultra-short-lived excited states. Journal of the American Chemical Society. 1998;**120**:843-844. DOI:10.1021/ja973504e

[68] Guillemoles JF, Barone V, Joubert L, Adamo C. A theoretical investigation of the ground and excited states of selected Ru and Os polypyridyl molecular dyes. The Journal of Physical Chemistry A. 2002;**106**:11354-11360. DOI: 10.1021/jp021517v

[69] Shalabi AS, El Mahdy AM, Assem MM, Taha HO, Soliman KA. Theoretical characterization of highly efficient porphyrazin dye sensitized solar cells. Journal of Nanoparticle Research. 2014;**16**:2579-2595. DOI: 10.1007/s11051-014-2579-8

[70] Lu X, Wei S, Wu CML, Li S, Guo W. Can polypyridyl Cu(I)-based complexes provide promising sensitizers for dye-sensitized solar cells? A theoretical insight into Cu(I) versus Ru(II) sensitizers. The Journal of Physical Chemistry C. 2011;**115**:3753-3761. DOI: 10.1021/jp111325y

[71] Moser J, Punchihewa S, Infelta PP, Grätzel M. Surface complexation of colloidal semiconductors strongly enhances interfacial electron-transfer rates. Langmuir. 1991;**7**:3012-3018. DOI: 10.1021/la00060a018

[72] Abuabara SG, Rego LGC, Batista VS. Influence of thermal fluctuations on interfacial electron transfer in functionalized TiO_2 semiconductors. Journal of the American Chemical Society. 2005;**127**:18234-18242. DOI: 10.1021/ja055185u

[73] Persson P, Bergström R, Lunell S. Quantum chemical study of photoinjection processes in dye-sensitized TiO_2 nanoparticles. The Journal of Physical Chemistry B. 2000;**104**:10348-10351. DOI: 10.1021/jp002550p

[74] Li SC, Wang J, Jacobson P, Gong XQ, Selloni A, Diebold U. Correlation between bonding geometry and band gap states at organic-inorganic interfaces: Catechol on rutile TiO_2(110). Journal of the American Chemical Society. 2009;**131**:980-984. DOI: 10.1021/ja803595u

[75] Oprea CI, Panait P, Cimpoesu F, Ferbinteanu M, Girţu MA. Density functional theory (DFT) study of coumarin-based dyes adsorbed on TiO_2 nanoclusters-applications to dye-sensitized solar cells. Materials. 2013;**6**:2372-2392. DOI: 10.3390/ma6062372

[76] De Angelis F, Fantacci S, Gebauer R. Simulating dye-sensitized TiO_2 heterointerfaces in explicit solvent: Absorption spectra, energy levels, and dye desorption. The Journal of Physical Chemistry Letters. 2011;**2**:813-817. DOI: 10.1021/jz200191u

[77] Junge H, Codolà Z, Kammer A, Rockstroh N, Karnahl M, Luo SP, et al. Copper-based water reduction catalysts for efficient light-driven hydrogen generation. Journal of Molecular Catalysis A: Chemical. 2014;**395**:449-456. DOI: 10.1016/j.molcata.2014.09.005

[78] Lee CH, Dogutan DK, Nocera DG. Hydrogen generation by Hangman metalloporphyrins. Journal of the American Chemical Society. 2011;**133**:8775-8777. DOI:10.1021/ja202136y

[79] Bhattacharjee A, Andreiadis ES, Chavarot-Kerlidou M, Fontecave M, Field MJ, Artero V. A computational study of the mechanism of hydrogen evolution by cobalt(diimine-dioxime) catalysts. Chemistry–A European Journal. 2013;**19**:15166-15174. DOI: 10.1002/chem.201301860

[80] Joliat E, Schnidrig S, Probst B, Bachmann C, Spingler B, Baldridge KK, et al. Cobalt complexes of tetradentate, bipyridine-based macrocycles: Their structures, properties and photocatalytic proton reduction. Dalton Transactions. 2016;**45**:1737-1745. DOI:10.1039/C5DT04426J

[81] Guttentag M, Rodenberg A, Kopelent R, Probst B, Buchwalder C, Brandstätter M, et al. Photocatalytic H_2 production with a rhenium/cobalt system in water under acidic conditions. European Journal of Inorganic Chemistry. 2012;**2012**:59-64. DOI: 10.1002/ejic.201100883

[82] Tran PD, Wong LH, Barber J, Loo JSC. Recent advances in hybrid photocatalysts for solar fuel production. Energy and Environmental Science. 2012;**5**:5902-5918. DOI: 10.1039/C2EE02849B

[83] Fujishima A, Honda K. Electrochemical photolysis of water at a semiconductor electrode. Nature. 1972;**238**:37-38. DOI: 10.1038/238037a0

[84] Baraton MI. Nano-TiO_2 for solar cells and photocatalytic water splitting: Scientific and technological challenges for commercialization. The Open Nanoscience Journal. 2011;**5**:64-77. DOI: 10.2174/1874140101105010064

[85] Linsebigler AL, Lu G, Yates JT. Photocatalysis on TiO_2 surfaces: principles, mechanisms, and selected results. Chemical Reviews. 1995;**95**:735-758. DOI: 10.1021/cr00035a013

[86] Bard AJ, Fox MA. Artificial photosynthesis: Solar splitting of water to hydrogen and oxygen. Accounts of Chemical Research. 1995;**28**:141-145. DOI: 10.1021/ar00051a007

[87] Bikondoa O, Pang CL, Ithnin R, Muryn CA, Onishi H, Thornton G. Direct visualization of defect-mediated dissociation of water on TiO_2(110). Nature Materials. 2006;**5**:189-192. DOI: 10.1038/nmat1592

[88] Wendt S, Matthiesen J, Schaub R, Vestergaard EK, Lægsgaard E, Besenbacher F, et al. Formation and splitting of paired hydroxyl groups on reduced TiO_2(110). Physical Review Letters. 2006;**96**:66107-66110. DOI: 10.1103/PhysRevLett.96.066107

[89] Zhao ZY. Single water molecule adsorption and decomposition on the low-index stoichiometric rutile TiO_2 surfaces. The Journal of Physical Chemistry C. 2014;**118**:4287-4295. DOI: 10.1021/jp500177n

[90] Hahn KR, Tricoli A, Santarossa G, Vargas A, Baiker A. First principles analysis of H_2O adsorption on the (110) surfaces of SnO_2, TiO_2 and their solid solutions. Langmuir. 2012;**28**:1646-1656. DOI:10.1021/la204124p

[91] Monti A, de Ruiter JM, de Groot HJM, Buda F. A dynamic view of proton-coupled electron transfer in photocatalytic water splitting. The Journal of Physical Chemistry C. 2016;**120**:23074-23082. DOI: 10.1021/acs.jpcc.6b08244

[92] Hanaor DAH, Assadi MHN, Li S, Yu A, Sorrell CC. Ab initio study of phase stability in doped TiO2. Computational Mechanics. 2012;**50**:185-194. DOI: 10.1007/s00466-012-0728-4

[93] Zhang Y, Kilin DS. Computational modeling of wet TiO_2 (001) anatase surfaces functionalized by transition metal doping. International Journal of Quantum Chemistry. 2012;**112**:3867-3873. DOI: 10.1002/qua.24289

[94] Zhang ST, Li CM, Yan H, Wei M, Evans DG, Duan X. Density functional theory study on the metal-support interaction between Ru cluster and anatase TiO_2(101) surface. The Journal of Physical Chemistry C. 2014;**118**:3514-3522. DOI:10.1021/jp409627p

[95] Graciani J, Plata JJ, Sanz JF, Liu P, Rodriguez JA. A theoretical insight into the catalytic effect of a mixed-metal oxide at the nanometer level: The case of the highly active metal/ CeO_x/TiO_2(110) catalysts. The Journal of Chemical Physics. 2010;**132**:104703-104710. DOI: 10.1063/1.3337918

[96] Pastore M, De Angelis F. First-principles modeling of a dye-sensitized TiO_2/IrO_2 photoanode for water oxidation. Journal of the American Chemical Society. 2015;**137**:5798-5809. DOI: 10.1021/jacs.5b02128

[97] Harris LA, Quong AA. Molecular chemisorption as the theoretically preferred pathway for water adsorption on ideal rutile TiO_2(110). Physical Review Letters. 2004;**93**:086105-086108. DOI: 10.1103/PhysRevLett.93.086105

[98] Wilhelm J, Seewald P, Del Ben M, Hutter J. Large-scale cubic-scaling random phase approximation correlation energy calculations using a Gaussian basis. Journal of Chemical Theory and Computation. 2016;**12**:5851-5859. DOI:10.1021/acs.jctc.6b00840

[99] Xu H, Zhang RQ, Ng AMC, Djurišić AB, Chan HT, Chan WK, et al. Splitting water on metal oxide surfaces. The Journal of Physical Chemistry C. 2011;**115**:19710-19715. DOI:10.1021/jp2032884

Chapter 9

Quantum Chemistry Applied to Photocatalysis with TiO_2

Sergio Ricardo de Lazaro,

Renan Augusto Pontes Ribeiro and

Luis Henrique da Silveira Lacerda

Additional information is available at the end of the chapter

http://dx.doi.org/10.5772/intechopen.69054

Abstract

Heterogeneous catalysis is a topic very studied in science. Its application in technologies of energy conversion, water purification, chemical synthesis, car catalytic converter and so on is studied. Recently, the TiO2 material in anatase and rutile phases has been used extensively in photocatalytic systems; its band-gap is localized in visible and ultra-violet spectra, proportioning a good material for generation of chemical radicals. Nowadays, the density functional theory (DFT) is shown as a great tool to simulate all types of materials and the possibilities to simulate bulk and surfaces of materials importance in last few decades. Recently, quantum periodic calculations based on DFT methods have been widely used to simulate materials and the main functionals applied are PBE, PBE0 and B3LYP; they are important for doping and adsorption theoretical investigations and are present in various simulation programs, such as, Crystal, Wien, Vasp and others. This methodology has investigate the influence of dangling bonds, cationic and anionic doping, charge transfer, surface energy and more quantum properties. Quantum chemistry tools, in particular, DFT methods, are key points to develop high quality research and technology once theoretical calculations are important to guide and understand the molecular design in photocatalysis.

Keywords: solid state, chemistry of materials, simulation, electronic structure, bulk, surface, optical property

1. Introduction

Over the last few decades, the field of photocatalysis has assumed a remarkable importance in environmental topics, mainly on technologies based on the capability of use of the main supply of energy on Earth: solar irradiation. In this case, the solar light can be used to activate a chemical process of radicals species, destroy undesired compounds or be transformed in chemical energy; for example, hydrogen production from H_2O [1, 2]. The aim of this field was initially motivated by the oil crisis, which promoted the research about alternative energy sources. In addition, the concern about pollution has attracted a recent economical, political and scientific interest because of the toxicological potential of recalcitrant compounds for environmental contamination, as well as for animal and human health. Thus, eco-friendly methods became a fundamental topic, being the advanced oxidation processes (AOPs) the most powerful tool to destroy recalcitrant synthetic species due to high reactive potential of photoinduced radicals, which allows the extinction of a wide range of compounds [1, 3, 4].

Nowadays, heterogeneous photocatalysis proved its efficiency to degrade chemical contaminants through a photoinduced reaction in the presence of a semiconductor photocatalyst. This method was originated in the decade of 1970 when Fujishima and Honda described the water splitting by photoelectrochemical cell containing TiO_2 [5]. Posteriorly, the scientific and technological interest in such methods has exponentially increased, TiO_2 being the widely used semiconductor material for charge carriers generation due to the photostability, chemical and biologically inert nature, availability and low cost. Such material can be found/synthetized in anatase, rutile and brookite phases, anatase being a polymorph, the most prominent candidate for photocatalytic applications. Anatase TiO_2 is an n-type semiconductor with indirect band gap of around 3.2 eV. Several reasons were proposed to explain the higher photocatalytic efficiency of anatase polymorph, such as increased surface area, –OH group concentrations and electronic structure features. In particular, the electronic features have been shown as the key advantages of anatase TiO_2 mainly related to the band edge positions, which allow the reduction or oxidation of protons from water creating –OH radicals as intermediate in the photo-oxidation reactions to break organic compounds because of high oxidation potential [1, 6, 7].

However, due to large band gap (3.2 eV), anatase TiO_2 absorbs only UV light above 400 nm, which amounts for 4–5% of the solar photons. To enable the effective use of solar radiation in photocatalytic process with TiO_2, much effort has been directed to the narrowing of band gap from UV to the visible spectra range. Such efforts include the crystal shape engineering and the doping or co-doping process. Recently, a lot of theoretical and experimental studies have been developed to investigate metals (Fe, Cr, V, Mo, Re, Ru, Mn, Co, Rh) and non-metals (N, S, B, C, F) doping effects on electronic structure of anatase TiO_2 photocatalysts [7–9]. In the metal doping process, the theoretical and experimental reports suggest that a redshift of the band gap occurs due to the insertion of a new electronic band closer to the conduction band minimum (CBM) improving the photocatalytic properties. Despite narrowing of the band gap, foreign cations frequently act as recombination center, suggesting that significant improvements in the photocatalytic efficiency are possible only at low concentration of dopants [10–12]. Alternatively, the coupling of TiO_2 with another semiconductor (SnO_2, WO_3, CdS and others) is commonly used as an approach to improve the photocatalytic efficiency through the band structure mismatch, which allows a physical separation between the photoinduced charge carriers, reducing the recombination rate. On the other hand, for non-metal doping, there are three different main opinions about the band-gap narrowing that can be related to the shift of valence band maximum (VBM), creation of impurity levels or oxygen vacancies [7–9].

Therefore, the main challenge in titania-assisting photocatalytic process remains on the tailoring of their electronic structure in order to improve and stabilize the photoinduced reactions. From a lot of experimental and theoretical researches, it is possible to investigate the understating of the main features of these complex reactions, such as the control and engineering of band gap, band edges, charge transport and recombination rate. In this aspect, theoretical analyses are helpful due to the precision of quantum mechanics-based methods to predict structural-electronic properties and its changes from chemical compounds. For example, recently it was published in manuscripts announcing the reliability of density functional theory (DFT) methods to research the band structure distribution, band-gap evaluation and engineering, electron-hole transport and dielectric constant, which are key properties of semiconductor photocatalytics. Besides the theoretical endeavor to investigate and predict bulk, surface chemistry, electronic structure and other properties have played a fundamental role in the understanding of morphological transformations and crystal growth, as well as for adsorption and reaction phenomena [13–15].

In this chapter, we propose a theoretical point of view about the photocatalytic properties of TiO_2, focusing on the evaluation of electronic structure parameters for both bulk and surface-oriented materials. The following sections are dedicated to explain (i) the mechanism of titania-assisted photocatalytic properties, (ii) benchmark of DFT methods on the investigation of photoinduced properties and (iii) theoretical results and discussion for different TiO_2 models.

1.1. Basic mechanism of TiO_2 photocatalysis

Photo-driven processes, such as used in photocatalytic and photovoltaic devices, are based on the conversion of light energy into other forms of energy such as electricity (solar cells) or chemical compounds (photocatalysis, water splitting, CO_2 reduction and others). These technologies, mainly photocatalytic processes, require the semiconductor electronic structure for the light absorption and conduction of the photoinduced charge carriers. In a molecular point of view, the general mechanism behind such devices can be divided in three steps: (i) light adsorption; (ii) electron-hole dissociation and (iii) charge carriers dynamic [1, 2, 13].

In the first step, commonly described as light adsorption, the light interacts with the electronic structure of the semiconductor. In this case, the light wavelength must be equal to or higher than the band gap, the energy difference between VB and CB, to promote an electron (e^-) from the VB to CB, inducing an electronic vacancy in the VB, denominated as hole ($h^\bullet$). The electron-hole pair interacts through a Coulomb attraction and plays a fundamental role in subsequent steps, controlling the photocatalytic efficiency of semiconductors. Especially for sunlight absorption, an optimum band gap is required due to the light wavelength commonly found in solar radiation. For the solar-driven photocatalysis, the optimum band gap belongs to the range between 1.6 and 2.5 eV. In addition, for photoinduced reactions, another compromise needs to be achieved: besides the band-gap value, the energy of the electron/hole should be high enough to perform the given reactions. These chemical potentials depend on the position of energy levels in the semiconductor, which is one of the key advantages of TiO_2 among other semiconductors because both the reduction of protons ($E_{NHE}(H^+/H_2) = 0.0$ ev) and the oxidation of water ($E_{NHE}(O_2/H_2O) = 1.2$ ev) can be activated simultaneously. Moreover, the superficial –OH groups can act as donor species to generate $OH^\bullet$ radicals that have a very high oxidation potential, which enables the subsequent reactions used in chemical decontamination [1, 6, 9, 13].

Another important feature associated with the optical excitation corresponds to the band-gap nature, which plays an important role in the absorption coefficient of the semiconductor. For semiconductors the band gap can be direct or indirect, depending upon the localization of the VBM and CBM along the Brillouin zone. In photocatalysis, the band-gap nature is important in the recombination of photo-generated electrons and holes due to differences in the electron decay from CBM. For semiconductors with direct band gap, such as rutile TiO_2, the recombination of the charge carriers emits a photon once the CBM and VBM are located in the same k vector. However, for indirect band-gap semiconductors such as anatase, the recombination is assisted by a phonon due to the difference between CBM and VBM, making the direct recombination difficult between excited electrons and holes, which results in an increased electron-hole lifetime. As a key result, the diffusion rate and the reaction time of the excited electron hole in indirect semiconductor also increase, making them promising candidates with superior photocatalytic performance than direct semiconductors [16, 17].

In the next step, the electron and hole have to be dissociated to obtain free charge carriers, which will be used in the electron transport of the device. This dissociation depends on the electron-hole binding energy (E_b) that is inversely proportional to the semiconductor dielectric constant. If we assume that the electron-hole dissociation will be made by the thermal energy, E_b should be lower than k_BT (k_B = Boltzmann constant and T = absolute temperature) around 25 meV at room temperature, and the semiconductor must have a dielectric constant around 10. For TiO_2, both rutile and anatase polymorphs have a superior dielectric constant, which enables a lower binding energy between the electron holes, making the dissociation easier [13, 16].

The final step corresponds to the diffusion of free charge carriers. In this step, the electron and the hole are transported to their active sites where they will be used before the recombination [1, 6, 13, 16]. The diffusion (D) is strictly related to the mobility (μ—Eq. (1)) of the charge carrier which in turn is linked to the effective mass (m^*) and the collision time (τ) of the charge carrier (Eq. (2)). Therefore, D is increased if the effective mass of the photogenerated carriers becomes lighter resulting in enhanced photocatalytic efficiency. Furthermore, the ratio between the effective mass of electrons (m_e^*) and holes (m_h^*) is a powerful tool to predict the electron/hole pair stability with respect to the recombination process. In this case, a larger effective mass difference induces distinct mobility, which reduces the electron-hole pair recombination, increasing the photocatalytic efficiency [13, 14, 17, 18].

$$D = \frac{k_B T}{e} \tag{1}$$

http://dx.doi.org/10.5772/intechopen.69054

$$\mu = e\frac{\tau}{m} \tag{2}$$

However, the simulations involving scattering process are very expensive and the evaluation of mobility of charge carriers requires an alternative approach. The effective mass can be associated with the band curvature at the top of the VB or at the bottom of CB. For a single isotropic and parabolic band, the effective mass can be obtained through the expression:

$$\frac{1}{m} = \frac{1}{\hbar}\frac{\partial^2 E}{\partial k^2} \tag{3}$$

Therefore, m_e^* can be obtained by fitting the bottom of the conduction band, whereas m_h^* corresponds to the fitting along the top valence band. In order to acquire the validity of the parabolic approximation within the CBM and VBM regions, the parabolic fitting must be done within an energy difference around 26 meV near to the CBM and VBM, corresponding to the thermal dissociation energy of carriers at room temperature [18, 19].

2. Theoretical methods and density functional theory

The theoretical methods based on quantum mechanical simulations are an important tool to evaluate material properties, mainly at the molecular level. Historically, the development of new materials to technological applications appears as a difficult task which requests a long time of studies. Meanwhile, the theoretical-computational method plays the traditional role of study of materials already discovered. However, the technological advancement is extremely dependent on development of new materials, once such materials are responsible for the improvement of available technologies. Front of such need, the theoretical methods helped chemists and physicist on the development materials at higher speed; it was possible once the theoretical investigation provides the materials characteristic and the limitations to its applications [20–22]. A large number of computational methods is available for investigation of material properties; however, in the last three decades, the use of DFT has changed the world because it offers an excellent relation between results precision and calculation time. The importance of this theory on material investigation is evidenced by the number of manuscripts based on its application since is very superior than the number of manuscripts based on Hartree-Fock (HF) and semi-empirical methods simulations as for inorganic chemistry as for organic compounds [20, 21, 23].

2.1. Density functional theory

The material properties can be evaluated through several computational approaches, such as molecular dynamics, *ab initio* methods and semi-empirical methods. The calculations based on molecular dynamics evaluated the system properties based on the behavior of ball-and-springs model under application of a force external field to atoms representation. In turn, the *ab initio* and semi-empirical methods employ different approaches to solve the Schrödinger Equation and for the obtainment of system wave function (Ψ) [24]. Particularly, the DFT system interpretation is not based on wave-function (Ψ), once it assumes the system total energy as a single functional of electronic density (ρ). Poorly, this theory can be simplified in two basic postulates [24, 25]:

i. The density functional (ρ) determines exactly and completely all the ground state properties for a system. Thus, ϱ is only dependent on three variables that determine the position (x, y and z);

$$\rho_{(x,\ y,\ z)} = E_0 \tag{4}$$

ii. Any function for electronic density will have energy greater than or equal to the ground state energy for a real system.

$$E_{(v)}\left[\rho_0\right] \geq E_{(0)}\left[\rho_0\right] \tag{5}$$

Nevertheless, the analytical function for electronic density is not yet known and the electronic density is obtained by HF equations for achievement of ϱ by a self-consistent field (SCF) method. Hence, the HF method is very similar to DFT so that the difference lies in the equation's formalism. Although this similarity was observed, the DFT shows highest precision and low time (computational cost) regarding HF simulations due to number of variables in each methodology. The HF and semi-empirical methods employ a high number of variables for a system investigation; the number of variables is in the $4n$ order, where n refers to the number of electrons in the system. In turn, DFT is dependent on three variables [20, 22, 24, 25].

Actually, the quantum calculation based on DFT applied the theorem developed by Kohn and Shan (KS) in 1965; the KS equation describes all the functional theories (Eq. (6)) and their representation for molecular orbitals (Eq. (7)). In such equations, ∇^2 is the kinetic energy for non-interacting electrons; $u(r)$ refers to classic Coulomb potential for a density of n electrons; d refers to the system space and φ corresponds to molecular orbital [26]. The KS equation was applied on two different systems; the first considers that there are no interactions between electrons, whereas the other assumes that such interactions are observed. The obtained results for both systems indicate a significant difference of energy between them. In order to correct this difference, the exchange-correlation term (E_{XC}) was inserted in DFT formulism; the E_{XC}consists of the sum of kinetic and potential energy difference between interacting and non-interacting systems. In terms of system interpretation, this energy refers to 1% of system total energy, and its physical meaning is the interaction between electrons in the investigate compound. Thus, DFT describes 100% of system total energy.

$$E_{KS} = -\frac{1}{2}\sum_i \int d^3 r \phi_i^*(r)\nabla^2\phi_i(r) + \int d^3 r \left[\frac{1}{2}u(r) + V_{ext}(r)\right]\rho(r) + E_{XC}\left[\rho\right] \tag{6}$$

$$\left[-\frac{1}{2}\nabla^2 + V_{eff}(r)\right]\phi_i(r) = E_i^{KS}(r)\phi_i(r) \tag{7}$$

The main characteristic of $E_{XC}$is the possibility of description by several forms dependent on what exchange-correlation functional was employed. Thereby, the choice of a functional to describe E_{XC} has a giant effect on material properties evaluation, as offering better results and as offering a reduction in computational cost. The first functionals are called local functionals and are based on an electron cloud model to represent the system electronic density; among its class of functionals, stands out the local density approximation (LDA)/local spin density approximation (LSDA) and generalized gradient approximation (GGA) [20, 22, 25, 27, 28]. Further ahead, the use of hybrid functionals gains force due to high proximity to experimental results.

2.1.1. *Exchange-correlation functionals*

The E_{XC} term can be described in several forms according to the employed functional. Recently, the functionals are classified as local, non-local and hybrid functionals. Among the local approximations, stands out the LDA that is a general approach to represent the electron gas model and offers a simple description of exchange correlation at local character. The employment of LDA functional results in exact values for system kinetic energy shows better results when applied to a system where ρ slowly changes similar to a uniform electron gas. It is observed because of the LDA evaluates a real system, in which electronic density is $E_{[\rho]}(r)$, by a homogeneous electrons gas system with the same density. Although such representation is valid, the density ρ becomes independent of position since the electrons are evenly distributed in the system. Hence, the investigation of real system by LDA functional is not accurate and over-estimates the correlation energy at 100%. The non-local functionals are developed in order to correct this failure [24, 25]. Furthermore, the LSDA is also a local description of the E_{XC} term and is very similar to LDA; both functionals are based on electron gas model but the LSDA includes the electron spins to system properties determination [20, 24].

The non-local functionals were developed aiming to correct the LDA failures in representation of real system. These approaches are based on a charge gradient and are also commonly known as corrected gradient functional. Among the non-local approximations, stands out the GGA which uses the electronic density gradient; this gradient consists of the first derivative of ϱ as a function of its position. The GGA functional obtains the E_{XC} terms by the sum of correlation energy (E_C) and exchange energy (E_X), both negative values. The GGA functionals offer better results than LDA and LSDA approximations since the relation between ϱ and position is considered. Compared to LDA and LSDA approximations, the GGA shows a deviation of 1% for system E_{XC} once the electronic density distribution is treated in real form and not as an electron gas model [25, 29, 30]. However, the results presented by GGA approximations can be improved from Correlations (Becke Functionals) and Exchange Functionals (LYP and P86) [24].

Ultimately, the other class of functionals is known as hybrid functionals. Such functional employs Hartree-Fock method parameters to determine the E_{XC} energy. The first hybrid functional was the half-half functional which was developed by Becke, and it was developed by a linear description of electronic density. However, this functional shows several limitations. Thus, in 1993, Becke proposes the B3PW91 model that uses three empirical parameters

(a = 0.20, b = 0.72 and c = 0.81) to adequate the theoretical results to experimental results combined to Perdew and Wang correction gradient (PW91). This new functional has excellent results compared to half-half functional due to its more precise description of E_{XC} energy and parameterization [21, 27, 28, 31, 32].

In 1994, Frisch and co-authors adapted the LYP corrections rather than PW91 creating the B3LYP that uses the same empirical parameters of B3PW91. Such functional employs the Slater exchange plus Vosko, Wilk, Nusair (SVWN) to improve the correction proposed by LYP. In the last few years, B3LYP has become one of the most used for computational calculations because of the excellent results obtained. Another factor responsible for the wide use of hybrid functional is its versatility, and once such functional was used to investigate semiconductors, proteins, organic compounds and others [20, 21, 24, 26–28, 31–35].

The hybrid functionals are differentiated according to the percentage of parameters of the Hartree-Fock method employed in the determination of E_{XC}. For example, B3LYP uses 20% of HF parametrization in its formalism. In 1996, Perdew and co-workers increased the HF percentage to 25% in PBE functional and created the PBE0 functionals. The increase in HF aims to minimize the over-estimation of factors arising from the electrons interaction; the PBE0 uses the GGA to evaluate the E_{XC} energy and it does not present adjustable parameters in its formulation (empirical parameters are defined as a = 0.25, b = 0.75 and c = 1). Other differences regarding B3LYP is less sophisticated formulation to PBE0 [27, 36, 37]. Moreover, PBE0 has great efficiency in material properties determination.

Another functional that is largely used is the HSE which was developed by Heyd, Scuseria and Ernzerhof. This functional investigates the E_{XC} energy by slitting it in two parts: a short-range part and a long-range part. Particularly, the HSE hybrid functional employs the parameter ω in function of Bohr radius. Thus, the short-range part is treated in the same form as that in the PBEh global hybrid, that is, it uses 25% short-range exact exchange and 75% short-range PBE exchange; whereas, the long-range part is investigated by PBE. If ω goes to 0, the short-range part dominates and HSE reduces to PBEh; if ω goes to infinity, the short-range part disappears and HSE reduces to PBE. The use of HSE functional reduces the computational cost to obtain the exact exchange-correlation energy [38].

3. Materials modeling based on DFT

The employment of DFT on material modeling offers an excellent relation time × efficiency. The efficiency is denoted by the results precision to prevision of material properties. In case of photocatalysis applications of solid-state materials, a theoretical methodology based on DFT provides important information about band gap, band-gap nature, photoinduced behavior, density and mobility for charge carriers; thus, it is possible to determine the applicability of each material in photoinduced devices and processes.

In order to evaluate these properties, it is necessary to investigate the material electronic structure. In solid-state materials, the electronic structure sorts materials as insulators, conductors and semiconductors. The formation of electronic structure of solid-state materials is

http://dx.doi.org/10.5772/intechopen.69054

originated from interactions between all constituent atoms since the atomic energy levels are perturbed by the neighbor atoms. Thereby, in these materials, the overlap of a large number of valence atomic orbitals results in the formation of molecular orbitals with very close energy, forming a *quasi*-continuous band; that is, an energy band is a continuum of closely spaced electron states [39–41]. Each energy band is occupied by two electrons (α and β) as predicted by Pauli exclusion principle. In solid-state materials, the electrons are observed in the lowest energy states; the last occupied energy level refers to the top or VBM, whereas the first unoccupied energy level is the bottom or CBM. There is no electronic state in the region between the VBM and CBM and it is called the band-gap which has direct influences on conduction properties of a solid [42]. According to band-gap energy, a material is classified as conductor (band gap > 1.0 eV), semiconductor (band gap from 1.0 to 4.0 eV) and insulator (band gap 4.0 eV or higher) [40, 43].

3.1. Density of states (DOS) and band structure

The DOS analysis (**Figure 1**) consists of a graphical representation of packing level of energy states in a quantum system, that is, the number of states in each region of energy. A high amount of energetic states results in a high density of states on projected DOS. In general, the DOS analysis is employed to investigate the energy levels nearest to band gap. As observed in **Figure 1**, the band of lowest energy refers to valence band and the higher energy band is the conduction band. Besides, the DOS analysis can also provide the contribution of each atom to compose the valence and conduction bands; by contribution of this analysis, it is also possible

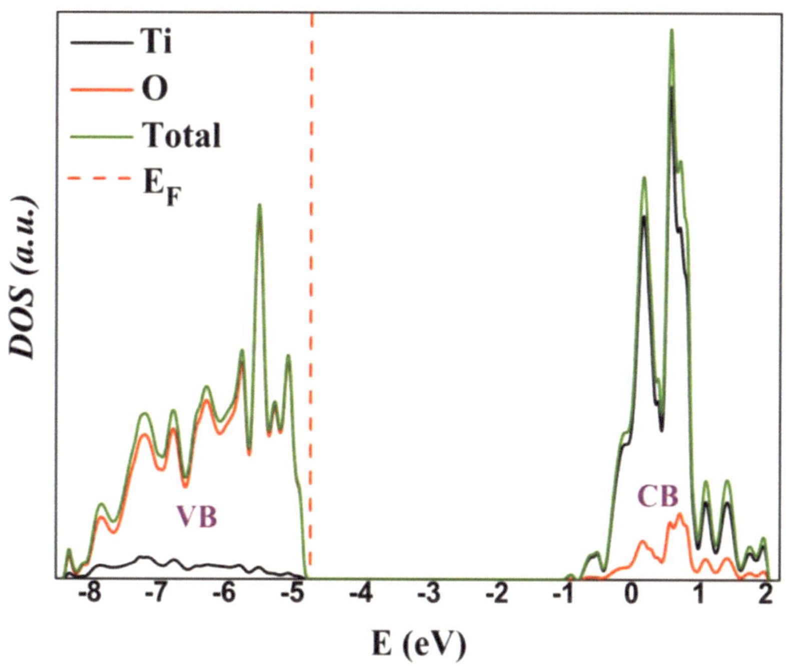

Figure 1. Projected DOS on Ti and O atoms of anatase TiO_2. The region before Fermi energy (E_F) is the valence band and the region after −1 eV is the conduction band.

to clarify the chemical bond composition and predict the electronic configuration of atoms [40–43]. Furthermore, applying the one-third of Simpson rule [44] on projected density of states for pure and doped models, it is possible to evaluate the number of available states in VB and CB. Such methods consist of numeric integration of the area under the DOS curve. Then, the number of states obtained from numeric integration is divided by unit cell volume. The values for CB and VB represent the number of available states for the formation of holes and electrons in electronic structure, respectively. According to the number of available states, a semiconductor material is classified as p-type (greater number of holes) or n-type (greater number of electrons) [44].

However, the energy bands for solid materials are not regular in all crystalline structures, and the same band can show a different energy level at each high symmetry point of Brillouin zone. Such symmetry points vary according to the spatial group of the crystalline structure and are labeled to their coordinates in space. The band structure analysis provides the energy of bands and the position of VBM and CBM. Thus, the band structure evaluation offers the band-gap nature to a material, that is, if the electron excitation occurs directly or indirectly [40]. **Figure 2** presents the band structure for anatase TiO_2; it is observed as an indirect band-gap and that the bands are not regularly distributed at high symmetry points.

The band structure study can also provide the effective mass for a solid-state material. In general, the transfer rate of electron-hole pair is inversely proportional to their effective mass. Thereby, a great effective mass denotes a low transfer rate of carriers, whereas, a small effective mass indicates that the charge carriers are extremely stable. The stability for these carriers also promotes the migration of electron and holes, as well as inhibits their recombination. The

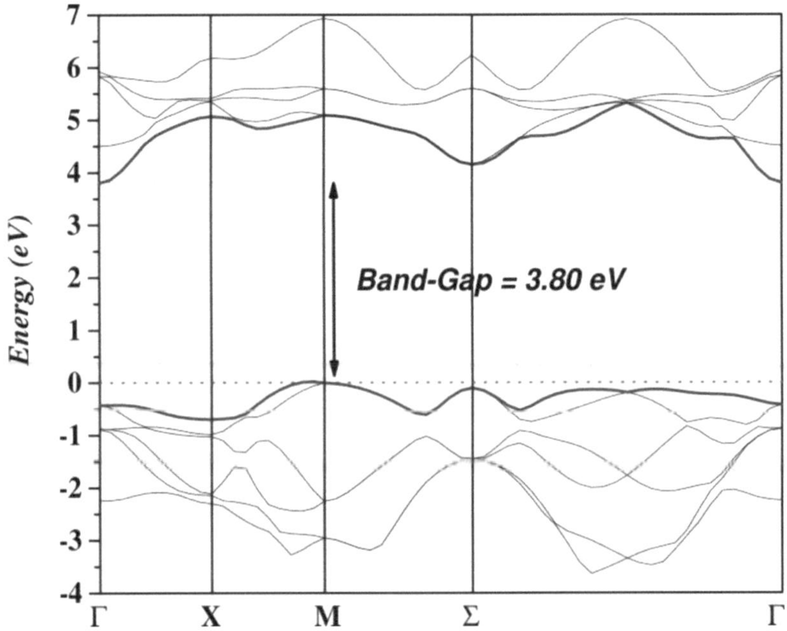

Figure 2. The band structure analysis for anatase TiO_2 shows an indirect band-gap.

effective mass of electrons and holes allows to assess indirectly the rate of charge carriers as presented in Eq. (8); in this formula, m^* is the effective mass of the charge carrier, k is the wave vector, $\hbar$ is the reduced Planck constant and v is transfer rate of photo-generated electrons and holes. Besides, the effective mass for electrons (m_{e^*}) and holes (m_{h^*}) should be investigated regarding the CBM and VBM; such points are very important to determine the E_g for a material. The values for effective mass were calculated by a fitting parabolic function around these points (Eq. (9)); in such an equation, m^* is the effective mass of the charge carrier, k is the wave vector, $\hbar$ is the reduced Planck constant and E refers to the energy of an electron at wave vector k in that band. In order to guarantee the validation of the parabolic approximation within the CBM and VBM, the parabolic fitting is performed considering a difference of 1 meV around the CBM and VBM regions [45–48].

$$v = \hbar k / m^* \tag{8}$$

$$m^* = \hbar^2 \left(\frac{d^2 E}{dk^2} \right)^{-1} \tag{9}$$

3.2. Photoinduced properties

The photoinduced properties of a solid compound determine how it interacts with electromagnetic radiation. In case of solid-state materials, a material interacts only with radiation which has energy equal to or higher than the material band gap [40, 41]. Thus, through evaluation of Eg values, it is possible to determine the characteristic type of radiation that interacts with each investigated material. Allied to the deep electronic structure investigation, it is possible to determine the bulk and surfaces of photoinduced behavior of materials and its applicability on photocatalytic process. Such a condition is essential to form the electron-hole pair (e^--$h^\bullet$), which is the precursor to start the photocatalysis process. Another point is focused on energetic stability of e^--$h^\bullet$ pair, and this factor can be estimated by reduced mass; however, the half-life time is not provided.

A practical example and much discussed in literature is comparing the electronic structures of rutile and anatase TiO_2 to clarify these concepts. Analyzing the band structures of both materials, we found band gaps of 3.50 and 3.80 eV (**Figure 3**), respectively. The minor band gap for anatase structure shows a photoinduction of e^--$h^\bullet$ pair from wavelength close to 326 nm, whereas, for rutile polymorph the e^--$h^\bullet$ pair is photogenerated around to 354 nm. Such difference is possible because of the major disorder associated with the TiO_6 cluster of anatase phase in relation to rutile structure; the displacement of Ti atom from central position of the unit cell in anatase causes modification in electronic structure through overlap between oxygen 2p orbitals and titanium 3d orbitals. This overlap orients to TiO_6 cluster to a TiO_5 cluster configuration, while, in rutile phase this effect is smaller. However, the anatase TiO_2 has an indirect band gap and rutile TiO_2 shows a direct band gap indicating that for anatase the e^--$h^\bullet$ pair recombination process is less possible because it requires a photon-phonon coupling and in rutile structure such phenomenon is direct, that is, localized in the same symmetric region. Thus, for anatase structure as photoinduction energy required to make the e^-–$h^\bullet$ pair as possibility of its recombination are more favorable than rutile phase. Now, let us evaluate the

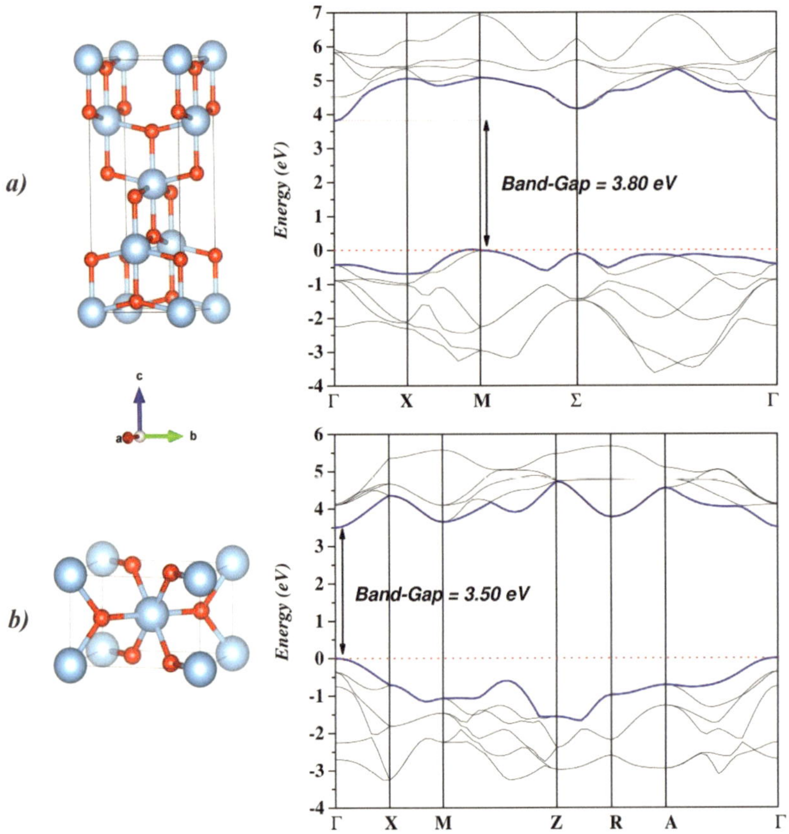

Figure 3. The crystalline structure and band structure for titania materials. (a) anatase crystalline phase and (b) rutile crystalline phase.

energetic stability of the e^{-}-$h^{\bullet}$ pair photoinduced through effective mass. The effective mass for electron and hole charge carriers calculated for anatase phase was 0.46 and 0.14, respectively; then, the electron/hole ratio is 3.29, indicating that e^{-}-$h^{\bullet}$ pair is favorable in anatase phase as carriers. Nevertheless, for rutile structure, this effective mass for electron and hole carriers is 0.65 and 0.32, respectively; likewise, the electron/hole ratio is 2.03 showing that the e^{-}-$h^{\bullet}$ pair is less favorable than anatase structure. Therefore, every electronic structure of anatase TiO_2 material from its crystalline structure is more inclined to photocatalytic process in comparison to rutile TiO_2 polymorph because anatase structure has small and indirect band gap and the e^{-}-$h^{\bullet}$ pair is more favorable.

3.3. Surface properties

Surface chemistry is an interesting and challenging topic in photocatalytic applications. In this case, the molecular level can be described by the combination between the surface structure and the interface region, where the main reactions involving photoinduced charge carriers occur. Therefore, the understanding of surface chemical reactions plays a fundamental role in the description of chemical bond between surface and adsorbed molecules, which are the fundamental basis to clarify the material photocatalytic activity [14, 15].

Recently, a lot of theoretical studies show the benchmark of DFT methods to reproduce the main effects behind the catalytic behavior of transition metal surfaces [49, 50]. For example, Cheng and co-workers systematically investigate the water splitting along rutile (110) surfaces combining DFT and molecular dynamics [51]. In addition, Migani and Blancafort studying the photocatalytic oxidation of methanol on TiO_2 (110) surfaces trough DFT/HSE06 calculations highlighted the role of excitons to produce formaldehyde [52]. Several related reports have appeared recently in the literature, as described by Nolan and co-authors that summarize the main advances in ab initio-based design of photocalytic surfaces [53].

Author details

Sergio Ricardo de Lazaro*, Renan Augusto Pontes Ribeiro and
Luis Henrique da Silveira Lacerda

*Address all correspondence to: srlazaro@uepg.br

Department of Chemistry, State University of Ponta Grossa, Ponta Grossa, Brazil

References

[1] Gaya UI, Abdullah AH. Heterogeneous photocatalytic degradation of organic contaminants over titanium dioxide: A review of fundamentals, progress and problems. Journal of Photochemistry and Photobiology C: Photochemistry Reviews. 2008;**9**:1–12. DOI: 10.1016/j.jphotochemrev.2007.12.003

[2] Ravelli D, Dondi D, Fagnoni M, Albini A. Photocatalysis. A multi-faceted concept for green chemistry. Chemical Society Reviews. 2009;**38**:1999–2011. DOI: 10.1039/b714786b

[3] Fox MA, Dulay MT. Heterogeneous photocatalysis. Chemical Reviews. 1993;**93**:341–357. DOI: 10.1021/cr00017a016

[4] Imamura K, Kominami H. Synthetic applications of titanium(IV) oxide-based photocatalysts. In: Colmenares JC, Xu Y-J, editors. Heterogeneous Photocatalysis: From Fundamentals to Green Applications. Berlin, Heidelberg: Springer; 2016. pp. 283–320. DOI: 10.1007/978-3-662-48719-8_9

[5] Hashimoto K, Irie H, Fujishima A. TiO_2 photocatalysis: A historical overview and future prospects. Japanese Journal of Applied Physics. 2005;**44**:8269. DOI: 10.1143/JJAP.44.8269

[6] Carp O, Huisman CL, Reller A. Photoinduced reactivity of titanium dioxide. Progress in Solid State Chemistry. 2004;**32**:33–177. DOI: 10.1016/j.progsolidstchem.2004.08.001

[7] Hernández-Alonso MD, Fresno F, Suárez S, Coronado JM. Development of alternative photocatalysts to TiO_2: Challenges and opportunities. Energy & Environmental Science. 2009;**2**:1231. DOI: 10.1039/b907933e

[8] Di Valentin C, Pacchioni G. Trends in non-metal doping of anatase TiO_2: B, C, N and F. Catalysis Today. 2013;**206**:12–18. DOI: 10.1016/j.cattod.2011.11.030

[9] Han F, Kambala VSR, Srinivasan M, Rajarathnam D, Naidu R. Tailored titanium dioxide photocatalysts for the degradation of organic dyes in wastewater treatment: A review. Applied Catalysis A: General. 2009;**359**:25–40. DOI: 10.1016/j.apcata.2009.02.043

[10] Bouras P, Stathatos E, Lianos P. Pure versus metal-ion-doped nanocrystalline titania for photocatalysis. Applied Catalysis B: Environmental. 2007;**73**:51–59. DOI: 10.1016/j.apcatb.2006.06.007

[11] Choi W, Termin A, Hoffmann MR. The role of metal ion dopants in quantum-sized TiO_2: Correlation between photoreactivity and charge carrier recombination dynamics. Journal of Chemical Physics. 1994;**98**:13669–13679. DOI: 10.1021/j100102a038

[12] Wang Y, Zhang R, Li J, Li L, Lin S. First-principles study on transition metal-doped anatase TiO_2. Nanoscale Research Letters. 2014;**9**:46. DOI: 12. 10.1186/1556-276X-9-46

[13] Bahers TL, Rérat M, Sautet P, Lyon UD, Claude U, Lyon B, et al. Semiconductors used in photovoltaic and photocatalytic devices: Assessing fundamental properties from DFT. Journal of Physical Chemistry C. 2014;**118**:5997–6008. DOI: 13. 10.1021/jp409724c

[14] Carter EA, Liaw P. New concepts and modeling strategies to design and evaluate photo-electro-catalysts based on transition metal oxides. Chemical Society Reviews. 2013;**42**: 2401–2422. DOI: 10.1039/c2cs35267b

[15] Norskov JK, Bligaard T, Rossmeisl J, Christensen CH. Towards the computational design of solid catalysts. Nature Chemistry. 2009;**1**:37–46. DOI: 10.1038/nchem.121

[16] Luttrell T, Halpegamage S, Tao J, Kramer A, Sutter E, Batzill M. Why is anatase a better photocatalyst than rutile?—Model studies on epitaxial TiO_2 films. Scientific Reports. 2014;**4**:4043. DOI: 16. 10.1038/srep04043

[17] Zhang J, Zhou P, Liu J, Yu J. New understanding of the difference of photocatalytic activity among anatase, rutile and brookite TiO_2. Physical Chemistry Chemical Physics. 2014;**16**:20382–20386. DOI: 10.1039/C4CP02201G

[18] Miglio A, Waroquiers D, Rignanese G-M, Gonze X. How does chemistry influence electron effective mass in oxides? A high-throughput computational analysis. Chemistry of Materials. 2014;**26**. DOI: 10.1021/cm404079a

[19] Jing T, Dai Y, Ma X, Wei W, Huang B. Electronic structure and photocatalytic water-splitting properties of $Ag_2ZnSn(S_{1-x}Se_x)_4$. The Journal of Physical Chemistry C. 2015;**119**:27900–27908. DOI: 10.1021/acs.jpcc.5b09522

[20] Burke K. Perspective on density functional theory. The Journal of Chemical Physics. 2012;**136**(15):150901–150909. DOI: 10.1063/1.4704546

[21] Zhang IY, Su NQ, Brémond ÉAG, Adamo C, Xu X. Doubly hybrid density functional xDH-PBE0 from a parameter-free global hybrid model PBE0. The Journal of Chemical Physics. 2012;**136**(17):174103–174110. DOI: 10.1063/1.3703893

[22] Hautier G, Jain A, Ong S. From the computer to the laboratory: Materials discovery and design using first-principles calculations. Journal of Materials Science. 2012;**47**(21):7317–7340. DOI: 10.1007/s10853-012-6424-0

[23] Atkins P, de Paula J, Friedman R. Quanta, Matter and Change: A Molecular Approach to Physical Chemistry. New York: W. H. Freeman and Company; 2009. p. 986

[24] Lewars EG. Computational Chemistry. Introduction to the Theory and Applications of Molecular and Quantum Mechanics. 2ª Edição ed. Londres: Springer; 2011. p. 665

[25] Morgon NH, Custodio R. Teoria do Funcional de Densidade. Química Nova. 1995;**18**(1):45–55

[26] Janesko BG, Henderson TM, Scuseria GE. Screened hybrid density functionals for solid-state chemistry and physics. Physical Chemistry Chemical Physics. 2009;**11**(3):443–454. DOI: 10.1039/B812838C

[27] Becke AD. Perspective: Fifty years of density-functional theory in chemical physics. The Journal of Chemical Physics. 2014;**140:**18A301. DOI: 10.1063/1.4869598

[28] Zhang IY, Wu J, Xu X. Extending the reliability and applicability of B3LYP. Chemical Communications. 2010;**46**:3057–3070. DOI: 10.1039/C000677G

[29] Perdew JP, Yue W. Accurate and simple density functional for the electronic exchange energy: Generalized gradient approximation. Physical Review B. 1986;**33**(12):8800–8802. DOI: 29. 10.1103/PhysRevB.33.8800

[30] Gunnarsson O, Jones RO. Total-energy differences: Sources of error in local-density approximations. Physical Review B. 1985;**31**(12):7588–7602. DOI: 30. 10.1103/PhysRevB.31.7588

[31] Becke AD. Density-functional thermochemistry. III. The role of exact exchange. Journal of Chemical Physics. 1993;**98**(7):5648–5652. DOI: 10.1063/1.464913

[32] Becke AD. Density-functional exchange-energy approximation with correct asymptotic behavior. Physical Review A. 1988;**38**(6):3098–3100. DOI: 32. 10.1103/PhysRevA.38.3098

[33] Kullgren J, Castleton CWM, Müller C, Ramo DM, Hermansson K. B3LYP calculations of cerium oxides. The Journal of Chemical Physics. 2010;**132**(5):054110. DOI: 10.1063/1.3253795

[34] Zhang Y, Xu X, Goddard WA. Doubly hybrid density functional for accurate descriptions of nonbond interactions, thermochemistry, and thermochemical kinetics. Proceedings of the National Academy of Sciences. 2009;**106**(13):4963–4968. DOI: 10.1073/pnas.0901093106

[35] Stephens PJ, Devlin FJ, Chabalowski CF, Frisch MJ. Ab initio calculation of vibrational absorption and circular dichroism spectra using density functional force fields. The Journal of Chemical Physics. 1994;**98**(45):11623–11627. DOI: 10.1021/j100096a001

[36] Perdew JP, Ernzerhof M, Burke K. Rationale for mixing exact exchange with density functional approximations. The Journal of Chemical Physics. 1996;**105**(22):9982–9985. DOI: 36. 10.1063/1.472933

[37] Adamo C, Barone V. Toward reliable density functional methods without adjustable parameters: The PBE0 model. The Journal of Chemical Physics. 1999;**110**(13):6158–6170. DOI: 10.1063/1.478522

[38] Henderson TM, Paier J, Scuseria GE. Accurate treatment of solids with the HSE screened hybrid. Physica Status Solidi (b). 2011;**248**(4):767–774. DOI: 10.1002/pssb.201046303

[39] West AR. Basic Solid State Chemistry. 2nd ed. Chichester: John Wiley & Sons; 2006. p. 480

[40] Kwok HL. Electronic Materials. Boston: PWS Publishing Company; 1997

[41] Smart LE. Solid State Chemistry: An Introduction. 3rd ed. Boca Raton: Taylor & Francis; 2005. p. 407

[42] Sutton AP. Eletronic Structure of Materials. Oxford: Oxford University Press; 1996. p. 260

[43] Rezende SM. A física de materiais e Dispositivos eletrônicos. 2nd ed. São Paulo: Editora Livraria da Física; 2004. p. 531

[44] Kuno M. Introductory Nanoscience: Physical and Chemical Concept. Oxford: Garland Science, Taylor & Francis Group; 2011. p. 420

[45] Yu J, Zhou P, Li Q. New insight into the enhanced visible-light photocatalytic activities of B-, C- and B/C-doped anatase TiO_2 by first-principles. Physical Chemistry Chemical Physics. 2013;**15**(29):12040–12047. DOI: 10.1039/C3CP44651D

[46] Zhou P, Yu J, Wang Y. The new understanding on photocatalytic mechanism of visible-light response NS codoped anatase TiO_2 by first-principles. Applied Catalysis B: Environmental. 2013;**142–143**:45–53. DOI: 10.1016/j.apcatb.2013.04.063

[47] Ma X, Dai Y, Guo M, Huang B. The role of effective mass of carrier in the photocatalytic behavior of silver halide-based Ag@AgX (X = Cl, Br, I): A theoretical study. Chemical Physics. 2012;**13**(9):2304–2309. DOI: 10.1002/cphc.201200159

[48] Soares GB, Ribeiro RAP, de Lazaro SR, Ribeiro C. Photoelectrochemical and theoretical investigation of the photocatalytic activity of TiO_2: N. RSC Advances. 2016;**6**(92):89687–89698. DOI: 10.1039/C6RA15825K

[49] Keith JA, Anton J, Kaghazchi P, Jacob T. Modeling catalytic reactions on surfaces with density functional theory. In: Modeling and Simulation of Heterogeneous Catalytic Reactions. Wiley-VCH Verlag GmbH & Co. KGaA; Weinheim, Germany, 2011. pp. 1–38. DOI: 10.1002/9783527639878.ch1

[50] Nørskov JK, Abild-Pedersen F, Studt F, Bligaard T. Density functional theory in surface chemistry and catalysis. Proceedings of the National Academy of Sciences. 2011;**108** 937–943. DOI: 10.1073/pnas.1006652108

[51] Cheng J, Sulpizi M, VandeVondele J, Sprik M. Hole localization and thermochemistry of oxidative dehydrogenation of aqueous rutile TiO_2(110). ChemCatChem. 2012;**4**:636–640. DOI: 10.1002/cctc.201100498

[52] Migani A, Blancafort L. Excitonic interfacial proton-coupled electron transfer mechanism in the photocatalytic oxidation of methanol to formaldehyde on TiO_2 (110). Journal of the American Chemical Society. 2016;**138**:16165–16173. DOI: 10.1021/jacs.6b11067

[53] Nolan M, Iwaszuk A, Lucid AK, Carey JJ, Fronzi M. Design of novel visible light active photocatalyst materials: Surface modified TiO_2. Advanced Materials. 2016;**28**:5425–5446. DOI: 10.1002/adma.201504894

Chapter 10

Theoretical Studies of Titanium Dioxide for Dye-Sensitized Solar Cell and Photocatalytic Reaction

Fu-Quan Bai, Wei Li and Hong-Xing Zhang

Additional information is available at the end of the chapter

http://dx.doi.org/10.5772/intechopen.68745

Abstract

This chapter aims to provide researchers in the field of photovoltaics with the valuable information and knowledge needed to understand the physics and modeling of titanium dioxide for dye-sensitized solar cell and photocatalytic reaction. The electronic band structure of titanium dioxide, the treatment of the excited state of titanium dioxide, the molecular dynamics and ultrafast quantum dynamics simulations, and several promising photocatalytic schemes and important considerations for theoretical study are addressed and reviewed. The advanced computational strategies and methods and optimized models to achieve exact simulation are described and discussed, including first principle calculations, nonadiabatic molecular and quantum dynamics, wave function propagation methods, and surface construction of titanium dioxide. These advanced theoretical investigations have become highly active areas of photovoltaics research and powerful tools for the supplement and prediction of related experimental efforts.

Keywords: electronic band structure, first principle calculation, excited state, molecular modeling, photophysical process, photochemistry and photocatalysis

1. Introduction

With the increasing concerns about energy issues and the associated environmental pollution, renewable energy technologies are devoted to being the most promising strategy for sustainable energy supply [1, 2]. Among all of these technologies, including hydro, solar, wind, geothermal heat, and biomass, photovoltaic (PV) technology that converts solar energy into electricity have attracted considerable attention. At present, the solid-state junction devices based on silicon material have dominated PV solar energy converters. However, the light-induced degradation of silicon materials limits the device stability and hinders its effective. Third-generation photovoltaics are able to produce high-efficiency photon to electricity conversion devices at a cheaper production cost. This is a consensus that the price is decreased, and the cell efficiency is maintained for the third-generation solar cell. There are unique features for this kind of cells; the cell should be sensitized by a photoactive material among dye, quantum dots (QDs), and perovskite system [3]. Work on sensitized photovoltaics started during the 1970s with the use of organic dyes as the sensitizer [4]. The similar structure of these solar cells is that in the case of the original design: The cell has three primary parts, glass sheet with transparent conducting oxide coating (ITO or FTO) as anode and the counter electrode (CE) on top and bottom, respectively; and TiO_2 film deposits on the conductive side of the glass sheet which is then immersed in a mixture of a photosensitizer and a solvent for charge transport. The working principle looks also dynamically analogical. But in quantum dot-sensitized solar cells (QDSSCs), that replaced organic dyes in dye-sensitized solar cells (DSSCs) with inorganic sensitizers, the total performance is promoted by the utilization of nano-sized crystals with a short band gap and a high extinction coefficient. Along with this line of thought, since 2009, the perovskite materials as sensitizers have been used initially and works very well with the solid-state hole transfer material. Until now, its efficiency has touched over 21% [5, 6]. For DSSCs, the photoanode components are the dye sensitizer, a mesoporous semiconducting oxide layer and a transparent conducting oxide (TCO). In DSSCs, the transport of charges (electrons) to the external circuit begins when electrons exit the semiconducting network layer and ends when the redox mediator in the charge transport medium returns them to the sensitizers [7]. The photoelectric anode consists of a TCO substrate with a semiconductor oxide layer (usually TiO_2) and a dye sensitizer. In fact, there are two TiO_2 layers. The first layer is a slender and blocking layer to inhibit electron recombination with ionized dyes and/or mediators. The second layer is more than 20–30 nm thickness of mesoporous TiO_2. The larger surface area of the mesoporous TiO_2 area allows a greater amount of dye to be adsorbed on its surface. An electrolyte usually with an iodide/triiodide couple is needed for DSSC. The electrolyte can be in liquid/gel or solid form.

2. The electronic band structure of titanium dioxide

A typical DSSC consists of the wide band gap TiO_2 semiconductor, the dye is employed to absorb light radiation, and the photo-excited electrons are subsequently transferred to the TiO_2 [8]. The fast interfacial electron transfer between the dye and TiO_2 could, in principle, minimize the loss of the utilized photon energy as heat through the electron-phonon relaxation pathway. Through this process, the absorption range can be efficiently extended to the visible region, which dominates the solar spectrum. In the many semiconductor metal oxides, titanium dioxide (TiO_2) is expected to play an import role in PV devices due to its high chemical and optical stability, nontoxicity, and corrosion resistance, low-cost [9]. TiO_2 is a semiconductor material with a band gap of ~3.2 eV, corresponding to a wavelength of ~390 nm. The anatase structure is preferred over other polymorphs for solar cell applications because of its potentially higher conduction band edge energy and lower recombination rate of electron–hole pairs [10]. This is not the case for excellent UV photocatalysts such as rutile or anatase polymorphs of TiO_2 due to the rather high band gap that these materials exhibit (**Figure 1**). This has prompted chemists and material scientists to search for modifications of

http://dx.doi.org/10.5772/intechopen.68745

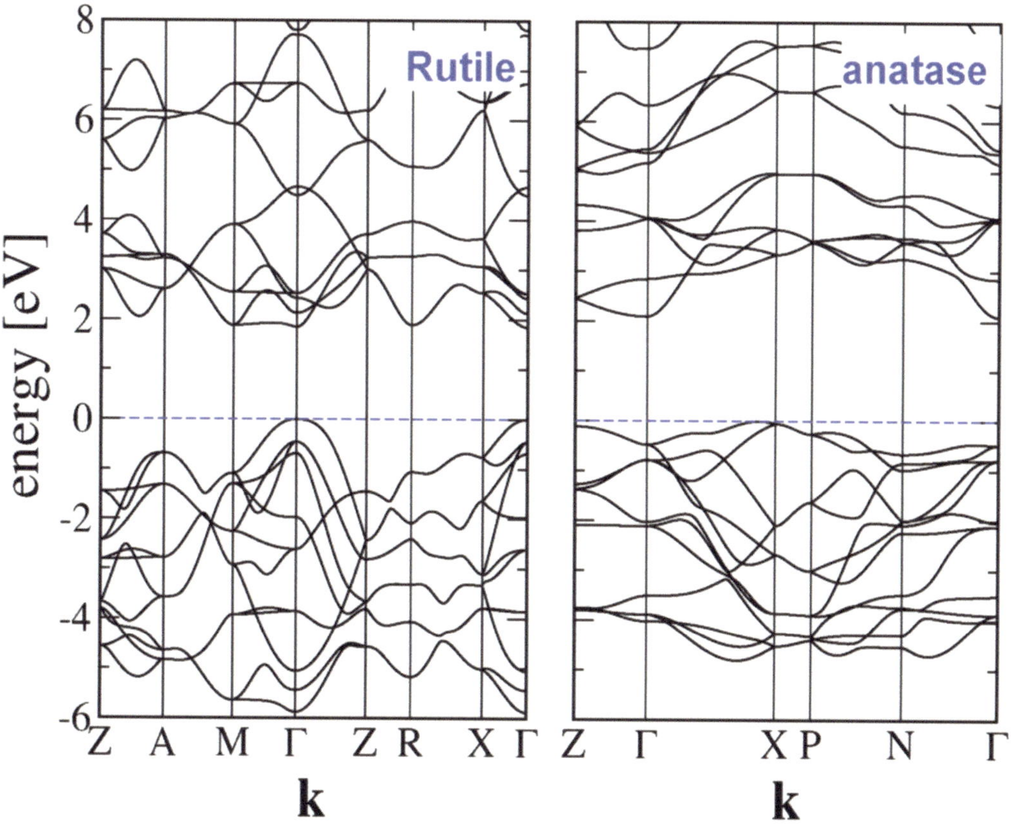

Figure 1. Band structure of rutile and anatase calculated by PBEsol with the respectively relaxed crystal structures. (Copyright 2016 Royal Society of Chemistry).

TiO_2 which could exhibit photocatalytic activity under visible light. Doping seemed to be one of the obvious options.

3. First principle calculations and modeling simulation

In this section, we will provide a comprehensive insight on the different levels of theoretical methods used in the description of the electronic structures, and particularly attention was paid to methods involved in the excited state in the metal oxide. At present, the wave function-based ab initio method and density functional theory (DFT) are the two main types of theoretical method. However, both methods have drawbacks, which are Born-Oppenheimer (BO) approximation break down when treating the excited state potential energy surface (PES). Of course, the BO approximation is good for description of the ground state properties, such as the geometry optimization, electronic structure calculation, since the nuclei are fixed for the ground state. Stochastic methods need to be developed to properly account for the excited state, but this kind of method is usually time-consuming since it requires the evaluation of the second-derivative of

Hessian matrix. In this respect, choosing a particular computational method must take a compromise between accuracy and its feasibility.

3.1. The DFT theory

Quantum mechanics calculations have widely used to address the physical and chemical properties of metal oxide. The level of the theoretical method includes Hartree-Fock (HF) self-consistent field (SCF), second-order Møller-Plesset (MP2) perturbation theory, coupled cluster (CC) theory, as well as the most popular DFT method. DFT has been the standard "answer" for the electronic structure calculations of a series of material systems, from molecules to solids, and from clusters to periodic systems. The one-electron density matrices can be expressed as follows:

$$\rho_0(\vec{X}) = N\int \Psi_0^*(\vec{X_1}, \vec{X_2}, \ldots, \vec{X_N})\Psi(\vec{X_1}, \vec{X_2}, \ldots, \vec{X_N})d_{\vec{X_2}} \ldots d_{\vec{X_N}} \tag{1}$$

where the subindex zero indicates density originates from the ground state wave function. For one N-electron system, DFT approximates the ground state energy of the system as a function of the electron density, that is to say, the ground state energy can be determined as long as the density function is given. However, the exact expression of density functional is still unknown. There are substantial works devoted to exploring the accurate density functionals. Generally, the density functional is given by:

$$E[\rho] = T_s[\rho] + V_{ert}[r] + V_{Coulomb}[\rho] + V_{XC}[\rho] \tag{2}$$

$T_s[\rho]$ is the kinetic energy term of the noninteracting electrons, the $V_{ert}[r]$ is the external potential, and $V_{Coulomb}$ represents the Coulomb repulsion. The last term $V_{XC}[\rho]$ accounts for the exchange-correlation effect, which is difficult to get the exact expression. The success of DFT depends on whether we can find out the appropriate approximations to the exchange-correlation term. Generalized gradient approximation (GGA) and local density approximation (LDA) are mostly used for exchange-correlation functionals in condensed matter systems, and meta-GGA and hybrid functionals are often used in isolated molecule. DFT in the local density approximation (LDA) and various GGA consistently predicts that anatase is more stable than rutile [11], and no significant improvement can be achieved when using more sophisticated DFT approaches, including hybrid functionals (PBE0 and HSE06) [12] and DFT plus the Hubbard U correction (DFT + U) [13, 14], and various dispersion-corrected DFT methods. A similar situation also occurs for the DFT + U method, which predicts rutile to be more stable only if an unphysically large U value is used.

3.2. The treatment of excited state of titanium dioxide

The relative stability of different polymorphic phases of a material is often addressed in terms of the ground state total energy [15, 16], but generally speaking, other degrees of freedom, including, in particular, those of nuclei vibration, may also play a significant role. The contributions of vibration, including the zero-point energy (ZPE) and finite-temperature vibration enthalpy and entropy, can be obtained from the phonon spectrum calculated in the harmonic approximation. We have found the resort that by considering high-order correlations in the adiabatic connection fluctuation–dissipation theory with the random phase approximation (ACFDT-RPA), rutile is correctly predicted to be more stable than anatase, which can be physically attributed to different characters in the electronic band structure of rutile and anatase, including, in particular, that rutile has a smaller band gap than anatase [17]. In the framework of DFT, the most efficient way to approach the excitation problem is its time-dependent formalism (TD-DFT) based on the Runge-Gross theorems, which may be seen as a consequence of the time-dependent Schrödinger equation (TD-SE) to describe the time-dependent phenomena [18]. TD-DFT can be viewed an alternative formulation of time-dependent quantum mechanics but, in contrast to the normal approach that relies on wave functions and on the many-body Schrödinger equation, its basic variable is the one-body electron density. The key exceptions so far that should be mentioned here are the development of extending CIS, TDHF, and TD-DFT to one-dimensional system like a one-dimensional chain of hydrogen molecules by Hirata et al. [19, 20], and the recent implementation of TD-DFT to periodic systems in the CRYSTAL code by Bernasconi et al. [21]. Walker et al. also developed the Turbo TDDFT code to make it is possible to carry out TD-DFT calculations using plane wave basis sets for the large condensed system [22]. However, in spite of using periodic boundary conditions, which is a necessary requirement to date, this formalism is only implemented for finite systems. The introduction of a portion of the Fock exchange is not the only strategy to improve the description of the band gap in the oxide and related materials, where LDA and GGA are even qualitatively ineffective. One common method is to add two parameters to the LDA or GGA exchange-correlation function, as in the Hubbard model, an exchange (J) term between electrons with the same orbital angular momentum, to mimic the effective field Coulomb (U) term [23].

Among the various possible approaches going beyond the DFT band structure calculations to predict the band gap of oxides, a growing interest is nowadays dedicated to the so-called GW quasiparticle approach (Green's function G and the screened Coulomb interaction W). The quasiparticle approach permits simulation of the ejection or the absorption of an electron from an N-electron state to an $N \pm 1$ electron state. And then, the direct comparison of the calculated results with photoemission or inverse-photoemission experiments can be performed based on this in principle. The first formulation of the GW method dates back to the work from 1965. However, the approach was not applied to large-scale, numerical electronic structure calculations before the mid-eighties. The resulting band gaps compare much better with experiment. The GW approximation is an approximation made in order to calculate the self-energy of a many-body system of electrons. The approximation is the expansion of the self-energy in terms of the single particle Green's function G and the screened Coulomb interaction W [24]. The whole formalism is rather complex, and a more detailed description can be found in the review paper of Huang and Carter [25]. Nevertheless, the ever-growing computational facilities have made GW calculations indeed attractive in solid-state computational simulations. The GW approach consists of updating both Green's function and the dielectric potential at every iteration. The amplitudes and the excitation energies, E, can be obtained from the Bethe-Salpeter equation (BSE), which can be written as an eigenvalue equation according to

$$\left[\left(\varepsilon_p^{QP} - \varepsilon_h^{QP}\right) - E\right] c_{ph} + \sum_{p'h'} K_{ph,p'h'} c_{p'h'} = 0 \tag{3}$$

where ε_p^{QP} and ε_h^{QP} are the quasiparticle eigenvalues of the particle and the hole, respectively. The sum in Eq. (3) runs over the kernel $K_{ph,p'h'}$, which can be split into two terms:

$$\boldsymbol{K}_{ph,p'h'} = \boldsymbol{K}^{exc}_{ph,p'h'} + \boldsymbol{K}^{dir}_{ph,p'h'} \tag{4}$$

$\boldsymbol{K}^{exc}_{ph,p'h'}$ is the particle–hole exchange operator, which has the usual form:

$$\boldsymbol{K}^{exc}_{ph,p'h'} = \int dx dx' \Psi_p^*(x) \Psi_h(x) \frac{1}{|x - x'|} \Psi_{p'}(x) \Psi_{h'}^*(x') \tag{5}$$

$\boldsymbol{K}^{dir}_{ph,p'h'}$ is the screened Coulomb operator, where a term depending on the inverse screened dielectric tensor is applied, similarly to the case for GW:

$$\boldsymbol{K}^{dir}_{ph,p'h'} = \int dx dx' \Psi_p^*(x) \Psi_h(x') \eta^{-1}(x, x') \frac{1}{|x - x'|} \Psi_{p'}(x) \Psi_{h'}^*(x') \tag{6}$$

In the same framework, neutral excitations (e.g., optical and energy-loss spectra) are also well described today through the solution of the Bethe-Salpeter equation. A self-contained method is able to perform both the starting self-consistent field calculation in the generalized Kohn-Sham (gKS) framework and the many-body perturbation theory (MBPT) post-treatment, and then that can produce GW quasiparticle energies, optical spectra and the dielectric screening, prior to Eqs. (3)–(6) through a solution of the BSE. Entertainingly, Onida et al. have shown that the BSE is intimately related to the linear response formulation of TDDFT [26].

3.3. The molecular dynamics simulation

The extremely small size of these particles makes them ideal candidates for investigation with molecular dynamics (MD) simulation, which has been a long-established simulation technique to extensively investigate structural and dynamic properties of solids and fluids at the atomic scale. A large number of force fields have been reported in the literature for modeling TiO_2 polymorphs [27–29]. For example, based on the Hubbard-corrected DFT + U and first-principles molecular dynamics (FPMD) simulations, facet-dependent trapping and dynamics of excess electrons at anatase TiO_2 surfaces and aqueous interfaces could be achieved. Whereas no electron trapping is observed on the (101) surface in vacuo, an excess electron at the aqueous (101) interface can trigger water dissociation and become trapped into a stable surface Ti^{3+}-bridging OH complex [30].

4. The titanium dioxide used in DSSC

TiO_2 semiconductor is the widely used electron conductor in Grätzel-type photovoltaic DSSC, because of its low-cost and ease of synthesis. A typical DSSC consists of the wide band gap

TiO_2 semiconductor, the dye is employed to absorb light radiation, and the photo-excited electrons are subsequently transferred to the TiO_2. Two major crystal phases of TiO_2: rutile (a) and anatase (b), are shown in **Figure 2**. The rutile phase is energetically stable since the crystal structure is symmetric. Each Ti atom is sixfold, and coordinated with threefold O atom. The anatase phase has the similar coordination number as rutile but with asymmetrical structure. DFT calculation suggests that the anatase (101) surface is the most active surface since it has unpassivated Ti and O atoms, which has been widely used to simulate the interface electron transfer process [31, 32].

4.1. The rule of titanium dioxide

So far, solid-state devices based on silicon have dominated photovoltaic solar converters. However, light-induced degradation of silicon materials limits device stability, while the relatively high cost (high purity) of photovoltaic power generation hinders its effective competition with fossil energy. Pure silicon has a little use, but doped silicon is the basis for most semiconductors. An attractive alternative to crystalline silicon photovoltaic devices is a unit made of mesoscopic inorganic semiconductors that can be easily prepared and provide a very low-cost manufacturing prospect. In these devices, inorganic semiconductors are primarily used as electron acceptors and provide direct or tortuous paths for electron transport, and in some cases, they can act as scaffolds for absorbing light collectors. Holes reside in the valence band, a level below the conduction band. Doping with an electron acceptor, an atom which may accept an electron, creates a deficiency of electrons, the same as an excess of holes. Among many semiconductor metal oxides, TiO_2 nanomaterials appear to be differentiated candidates because of their high chemical and optical stability, non-toxic, low-cost and corrosion resistance. Anatase TiO_2 has a crystal structure corresponding to the tetragonal system, but the distortion of the TiO_6 octahedron is slightly larger for the anatase phase. The anatase structure with improved electrical properties is superior to other polymorphs used in solar cells because of its potentially higher conduction band edge energy, and it enables

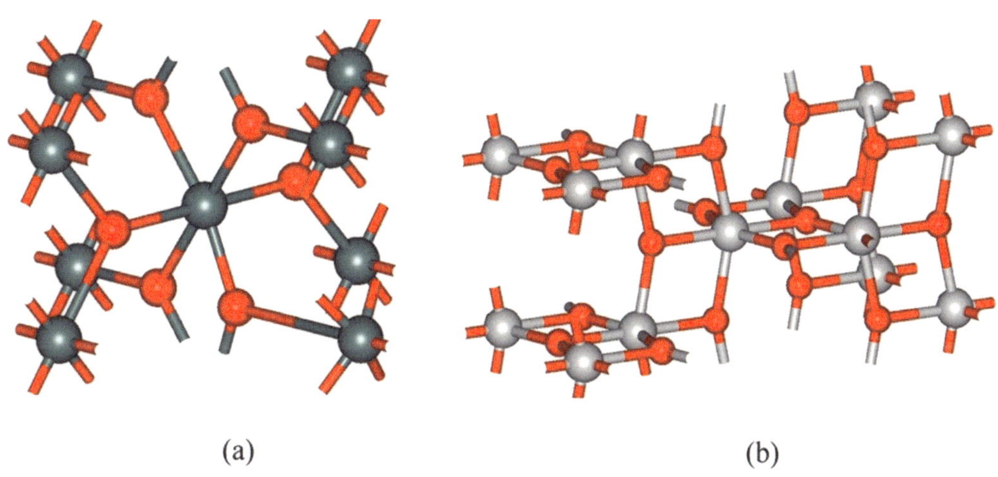

Figure 2. The crystal structure of rutile (a) and anatase (b) TiO_2. (Copyright 2013 American Chemistry Society).

remarkably efficient spatial separation of electron–hole pairs on the submicrosecond time scale. The physical and chemical properties of TiO_2 nanocrystals are affected not only by the intrinsic electronic structure but also by their size, shape, texture, doping and surface properties. The smaller effective mass in anatase is consistent with the high mobility, band like conduction observed in anatase crystals. It is also responsible for the very shallow donor energies in anatase. In addition to the large surface area, TiO_2 nanomaterials should also have high electron mobility, making it possible to efficiently collect electrons injected into TiO_2. Since the defects in the TiO_2 nanomaterials can be used as electron captures and present in the grain boundaries at the junctions between the nano-sized particles, it is expected to use a network structure consisting of TiO_2 monocrystalline nanowires rather than TiO_2 nanoparticles rapid electronic transmission. Besides, the properties of TiO_2 also strongly relay on the modifications of the TiO_2 material host and on the interactions of TiO_2 materials with the environment. The surface modification not only affects the interfacial energy offset but also has significant impact on the charge separation, transport, and recombination processes.

4.2. The charge transfer process and Marcus theory

The Marcus theory is the most fundamental theory for a description of the electron transfer (ET) process in condensed material. Its formula can be given by [33–35]:

$$k(T) = \kappa A\sigma^2 \exp\left\{-\frac{w^r + \frac{\lambda}{4}\left(1 + \frac{\Delta\tilde{G}^o}{\lambda}\right)^2}{k_B T}\right\} \tag{7}$$

$$\Delta\tilde{G}^o = \Delta G^o + w^p - w^r \tag{8}$$

The above equation explicitly takes various contributions to the free energy into account, where $A\sigma^2$represents the collision frequency, and σ is the distance between the reacting species. ΔG^o is the reaction free energy, $\Delta\tilde{G}^o$ is the free energy after work correction. w^r and w^p are the work corresponds to the reactant and product state. λ is the reorganization energy, contains the contributions coming from the reactant and solvent.

$$\lambda = \lambda_r + \lambda_{sol} \tag{9}$$

The reactant and solvent contributions of the reorganization energy are given by:

$$\lambda_r = \sum_i \frac{f_i^{(r)} f_i^{(p)}}{f_i^{(r)} + f_i^{(p)}} (\Delta q_i)^2 \tag{10}$$

$$\lambda_{sol} = (\Delta e)^2 \left(\frac{1}{2R_1} + \frac{1}{2R_2} - \frac{1}{\sigma}\right)\left(\frac{1}{\varepsilon_s} - \frac{1}{\varepsilon_{op}}\right) \tag{11}$$

where $f_i^{(r)}$ and $f_i^{(p)}$ are the force constants of the given normal mode for reactant and product states, respectively; Δq is the changes of the ith nuclear coordinate; and Δe is the amount of transferred charge from reactant to product. ε_s and ε_{op} are dielectric constants of the solvent for static and optical. The original Marcus theory assumes that diabetic PES can correspond to the reactant and product states, and this simplifies the description of the ET.

4.3. The ultrafast quantum dynamics

Nonadiabatic ET involves the participation of the nuclear, the Fermi's Golden rule is usually employed to evaluate the electron transfer rate $K_{I\text{->}F}$ of nonadiabatic ET.

$$k_{I\to F} = \frac{2\pi}{\hbar}|\langle F|H'|I\rangle|^2\delta(E_F - E_I) \tag{12}$$

This formula originates from the first-order time-dependent perturbation theory (TD-PT). Where $I\rangle$ is the initial state, $F\rangle$ is the final state, and H' is the time-dependent perturbation Hamiltonian due to the nuclei vibration. The advantage of FGR is that ET rate can be determined from electronic structure calculation, that is, the coupling between initial and final state depends on the atomic orbital coefficient. This greatly simplifies the calculation of electron transfer for the chromophore-semiconductor system.

A conceptually similar but rougher approach is proposed by May, in which the rate constant derivates to be:

$$k_{i\to\{f\}} = 2\overline{\Gamma} = \frac{2\pi}{\hbar^2}\overline{N}|\overline{V_e}|^2 \tag{13}$$

This equation is based on linear response approaches, in which $\overline{\Gamma}$ is the absorption line broadening. $\overline{V_e}$ is average density of states. Labat and coworkers derivate an even more simple expression for the electron transfer time based on the Newns–Anderson approach [36]:

$$\tau = \frac{658}{\Delta} \tag{14}$$

Δ is the broadening of the dye donor orbital. Generally, this method gives the electron transfer timescale half of the experimental value [37].

The electron transfer rate constant estimated using Marcus and Fermi's golden rule can be calculated at different levels of theory method. But these methods just focus on the description of the single process. However, the electron transfer process might be subject to the influence of environment. Accurate estimation of the electron transfer rate depends on the exact solution of TD-SE, especially the nuclei part at the quantum mechanics level.

$$i\hbar\frac{\partial\tilde{\chi}_i\left(t, R(t)\right)}{\partial t} = \sum_j\left\{\left(T_{nucl} + E_j(R)\right)\delta_{ij} - \hbar^2\frac{d_{ij}^{(2)}}{2M}\right\}\tilde{\chi}_j(t, R(t)) \tag{15}$$

It is challenging to treat nuclei quantum mechanics. This is because a given system can contain a nucleus with different mass and speed. Therefore, it is difficult to developing a universe density functional to treat nuclei vibration. The accurate solution of Eq. (15) only fed with the small system based on the grid method. But as the size of the system increases, the above method becomes impractical. The alternative is to treat the nuclei classically, and then a semi-classical expression can be obtained:

$$i\hbar\frac{\partial\tilde{\chi}_i\left(t, R(t)\right)}{\partial t} = \sum_j\left\{E_j(R)\delta_{ij} - i\hbar^2\frac{p}{M}d_{ij}^{(1)}\right\}\tilde{\chi}_j\left(t, R(t)\right) \tag{16}$$

In the above equation, the classical momenta p is introduced. Surface hopping method is suitable for mimic the wave-packets branching using a classical trajectory. Fewest switches surface hopping (FSSH) is the most popular semi-classical method for nonadiabatic molecular dynamics. The electron transition rate from state i to state j at the small time interval $g(t)$ can be expressed as follows:

$$g_{i\to j}(t) = \max(0, P_{i\to j}(t)) \tag{17}$$

where

$$P_{i\to j}(t) = 2\frac{\int_t^{t+\Delta t}\mathrm{Re}\left(c_i^*(t')c_j(t')\frac{\vec{P}}{M}\vec{d}_{ij}(t')\right)dt'}{c_i^*(t)c_i(t)} \approx 2\frac{\mathrm{Re}\left(c_i^*(t)c_j(t)\frac{\vec{P}}{M}\vec{d}_{ij}(t')\right)\Delta t}{c_i^*(t)c_j(t)} \tag{18}$$

Here, $c_i(t)$ is the time-dependent expansion coefficient. The hopping probability is compared with a random number to determine whether it is hopped to a new state or remain in the current state. Nuclear velocity is rescaled to conserve the total energy. More details of the implementation of FSSH can be found in the literature [38, 39]. It is worthy of noting that classical path approximation (CPA) is employed for treating the large system. CPA assumes that the electron depends on nuclei motion, but a nucleus is independent of the electron, and this is the so-called quantum backreaction problem. CPA is valid in case the heavy nuclei has the large energy than a lighter electron, and this could enable the nuclear evolution is not affected by the electron-nuclear interaction. The use of CPA could lead to the large computational saving since the many stochastic surface hops can be achieved along the precomputed MD trajectory.

The original FSSH formulas take the hop-rejection into consideration, and this is given by the following:

$$g_{i\to j}(t) \to g_{i\to j}(t)b_{i\to j}(t) \tag{19}$$

$$b_{i\to j}(t) = \begin{cases}\exp\left(-\frac{E_j - E_i}{k_B T}\right), E_j > E_i\\ 1, E_j \le E_i\end{cases} \tag{20}$$

The transition probabilities $g_{i\to j}(t)$ in the above equation are rescaled by the Boltzmann factor, taking the detailed balance between transition up and transition down in energy.

http://dx.doi.org/10.5772/intechopen.68745

4.4. Wavefunction propagation methods

The previous section discussed the semi-classical method for nonadiabatic molecular dynamics. That semi-classical method does treat the nuclear classically. However, in some cases, the nuclear quantum effect cannot be neglected, for example, the evaluation of Franck-Condon factor involved in the electron excitation. Explicitly propagation of the nuclear wave function can achieve this goal. Methods like multiconfiguration time-dependent Hartree (MCTDH) method and quantized Hamiltonian dynamics (QHD) are developed for a different level of accuracy. These methods are based on the propagation of Gaussian wavepacket, which has been proposed to study the quantum effects of nuclear dynamics. The MCTDH method is one of the most accurate methods for wave-packet propagation. The nuclei wave function is expressed as follows:

$$|\tilde{\chi}_i(t, R_1, R_2, \ldots R_f)\rangle = \sum_{j_1}^{n_1} \ldots \sum_{jf}^{n_f} A^1_{j_1 \ldots jf}(t) |\chi^1_{j_1}(t, R_1)\rangle \ldots |\chi^1_{j_f}(t, R_f)\rangle \tag{21}$$

MCDTH provides the exact and accurate solution to the quantum description of nuclear behaviors. Therefore, it is extremely time-demanding in computation. Its complexity scales as N^f, where *f* is the number of freedom of the total system, while N^f is the number of the atomic orbital basis function. However, this method is only suitable for the system with limited atoms and has been improved using multilayer MCTDH (ML-MCTDH) technique:

$$|\tilde{\chi}_i(t, R_1, R_2, \ldots R_f)\rangle = \sum_{j_1}^{n_1} \ldots \sum_{j_p}^{n_p} A^1_{j_1 \ldots j_p}(t) |\chi^1_{j_1}(t, R_1^{coll,1})\rangle \ldots |\chi^1_{j_p}(t, R_p^{coll,1})\rangle \tag{22}$$

The concept of collective coordinates ($Rc°ll$) is introduced. In principle, ML-MCTDH theory can be fed with the system contains several thousands of degrees of freedom [33].

In addition to the MCTDH method, quantum dynamics method based on the propagation of electron wave function is also of interest. Time-dependent electron wave function can be expressed as linear combination of atomic orbitals

$$|\varnothing(t)\rangle == \sum\nolimits_{i,\alpha} B_{i,\alpha}(t) |i, \alpha\rangle \tag{23}$$

$|i, \alpha\rangle$ is the corresponding atomic orbital α of atom i. The expansion coefficient can be expressed as follows:

$$B_{i,\alpha}(t) = \sum\nolimits_q Q^q_{i,\alpha} C_q \exp\left(-\frac{i}{\hbar} E_q t\right) \tag{24}$$

Solving the generalized eigenvalue equation:

$$HQ^q = E_q S Q^q \tag{25}$$

S is the overlap matrix in the atomic orbital basis, H is the Hamiltonian matrix which is determined from the extended Hückel theory (EHT) calculations. The extended Hückel calculation is a semi-classical quantum chemistry method. Within the scheme of EHT, the off-diagonal Hamiltonian matrix can be given as follows:

$$H_{ij} = KS_{ij}\frac{H_{ii} + H_{jj}}{2} \tag{26}$$

where K is the Wolfsberg-Helmholtz constant, ~1.75. In EHT, only valence electrons are included; the core electron energies are assumed to be constant.

The population of the electron wave function can be determined from the overlap matrix and atomic orbital basis function,

$$P(t) = \left|\sum_{i,\alpha}^{MOL} \sum_{i,\beta} B^{*}_{i,\alpha}(t) B_{j,\beta}(t) S^{i,j}_{\alpha,\beta}\right| \tag{27}$$

The combination of theoretical methodology enables the realistic description of the ET process in the large system, and we have this method to investigate the ET in dye/TiO_2 interface. We have established the detailed mechanism of electron transfer in DSSC [40].

5. The titanium dioxide used as photocatalyst

Another important application of TiO_2 is a photocatalytic reaction. The titanium dioxide (TiO_2) has been also widely used as a photocatalyst. Numerous studies have been carried out on the popular photocatalysts of TiO_2. The reaction turns out to be very complicated, and its mechanism is still not well understood. It is time to use the theoretical tools to find some important insights out. Photoexcitation of TiO_2 leads to the formation of free electrons and holes in CB and VB, respectively. There are several carrier dynamic pathways that might happen. First, an electron in CB recombines non-radiatively with a hole in VB in the form of heat. Second, electron acceptors are reduced by photogenerated electrons; and third, electron donors are oxidized by photogenerated holes. The photochemical reaction may happen once the electron and hole diffuse to the TiO_2 crystal surface, and then the photogenerated electron and hole can interact with the O_2 or H_2O molecules to achieve the oxidization or reduction process. Some typical oxidization and reduction process are shown in **Figure 3**.

The elementary reaction steps can be expressed as follows:

$$\begin{aligned} &TiO_2 + hv(\lambda < 390nm) \rightarrow e^- + h^+ \\ &h^+ + H_2O \rightarrow OH^\bullet + H^+ \\ &\bar{e} + \bar{O}_2 \rightarrow \bar{O}_2 \end{aligned} \tag{28}$$

5.1. The surface of titanium dioxide

The performance of the TiO_2-based photocatalyst device can be regulated by the surface molecular adsorption, and particularly, the adsorption of O_2 and H_2O on TiO_2 has been widely discussed. Investigation of adsorption of H_2O onto TiO_2 mainly focuses on which sites water molecular adsorption is favored, and also the form of water adsorption, molecular or dissociative. Rutile

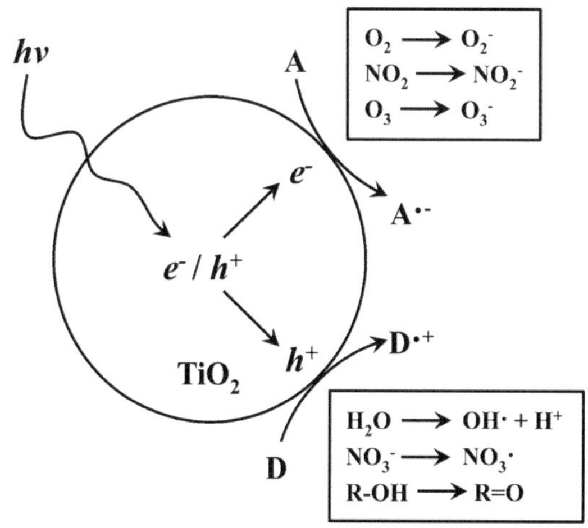

Figure 3. Schematic representation of typical photocatalyst process. (Copyright 2013 American Chemistry Society).

TiO_2 (110) surface, the most stable TiO_2 polymorph, is usually employed to model the surface adsorption. Experimentally, a water molecule was predicted on being molecularly adsorption onto the stoichiometric rutile TiO_2 (110) surface under the low coverage [41–44]. Theoretically, both molecularly and dissociative absorptions have been suggested by the DFT calculation [45–47]. The difference between theory and experiment arising from the exact parameters used in DFT calculation, such as the exchange-correlation function, and from the limitation in the simulation scales, for example, experimental system usually has the diameter up to tens of hundreds of angstrom, which is far beyond the capability DFT can handle at the current stage.

Band-gap engineering is an efficient way to tune the photoactive of the semiconductor. Different heteroatoms like anion nonmetals and cation-metals as well as cation-nonmetals have been used to doping in the semiconductor. Generally, semiconductor doping can introduce gap-state in the band gap, and this leads to the reduction of the optical band gap, improving the photoactive. Electronic structure calculation based on DFT has been used to elucidate the possible effect caused by the dopant. For example, substitute the Ti atom of bulk TiO_2 with metal cations, such as Cr and V, is an n-type doping. This will introduce the electron donor states within the band gap, which allows for the reduction of the optical band gap, thus the better performance for utilizing the photon energy, extending the photocatalytic activity into visible part of the solar spectrum [48–51]. Replace the O with N or C atoms belongs to the p-doping, which introduce the electron acceptor states. Therefore, dopant states have the large extent of overlap between the electronic states of TiO_2, this minimizing the problems arising with the localized states in n-type doping.

5.2. The CO_2 photocatalytic reaction

From the viewpoints of a theoretical worker, one can determine the every property of a given materials using just its coordinates, and this is exactly the idea of first-principles calculation.

The properties include the geometrical structure, the ground state energy, the work function, and the band structure. In order to design the more efficient photocatalyst, it is necessary for theoretical worker knows the basic physics prior to performing first principle calculations.

First, the photocatalyst performance depends on the chosen of the TiO_2 surface. Asymmetric anatase (101) surface has the most uncoordinated Ti and O atoms, and thus, it is suitable for the molecular adsorption of CO_2, thus the better photocatalyst performance (**Figure 4**). Second, proper energy alignment between TiO_2 and the adsorbed molecules is required. This is due to the built-in driving force largely affect the direction of the ET, and a negative driving force should lead to the inefficient ET process, in order to avoid this issue. The TiO_2 CB should be higher than the molecular LUMO orbital, whereas the TiO_2 VB should be lower than the molecular HOMO orbital. Third, the interface details between TiO_2 and adsorbed molecule affect the time scale of interface charge separation. This is because interface charge transfer can be described using the Fermi's Golden rule if the coupling between the electron/hole donor orbital in TiO_2 and acceptor orbital in adsorbed molecule is large, then the ET will be fast. The interface interaction depends on the adsorption details of molecular onto TiO_2 surfaces.

The last, the work function is defined as the minimum energy required transferring an electron from the highest filled level of a solid to a point in the vacuum outside the solid surface. The work function is a property of the surface, which has been widely used to measure how difficult the charge transfer is. Screen the materials with the proper work function is required.

5.3. Other main photocatalytic reactions

The (101)-(001) surface heterojunction constructed on polyhedral TiO_2 nanocrystals has recently been proposed to be favorable for the efficient electron-hole spatial separation due

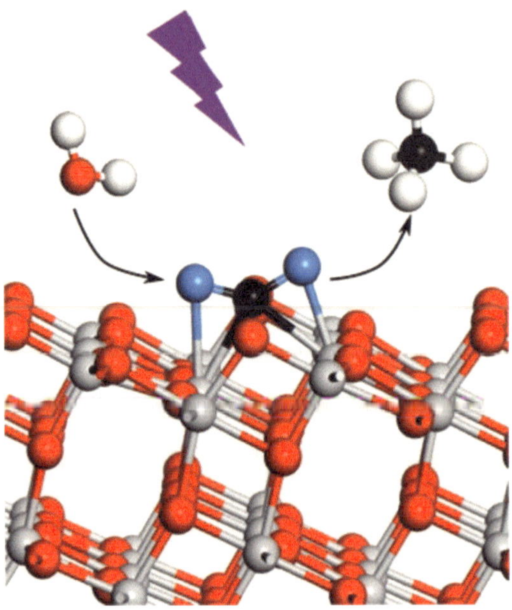

Figure 4. Photocatalytic reduction of CO_2 on the anatase TiO_2 (101) surface (Copyright 2016 American Chemical Society).

http://dx.doi.org/10.5772/intechopen.68745

to the preferential accumulation of electron and hole on (101) and (001) facets, respectively. The formed free electron and hole can promote reactive oxygen species (ROS) production, which potentially can be used for inactivation of bacteria. For the biological system, the theoretical investigation with electronic level is still a challenging for researchers and here only an example for readers a general understanding. A series of truncated octahedral double cone TiO_2 nanocrystals co-planar with (101) and (001) facets are prepared to form (101) to (100) of various ratios for optimizing the electron-(001) surface. All of these polyhedral TiO_2 nanocrystals can produce ROS more prominently than spherical TiO_2 nanocrystals, exhibiting more bacteria against gram-negative (E. coli) and Gram-positive (*Staphylococcus aureus*) bacteria under simulated sunlight high antibacterial activity. Surface heterojunction as a new strategy is expected to develop effective antimicrobial nanomaterials [52].

6. Summary

In this chapter, we aimed to review and summarize of recent important theoretical studies of titanium dioxide for dye-sensitized solar cell and photocatalytic reaction. How to build a suitable TiO_2 model to start a theoretical study for fast interfacial electron transfer between the dye and TiO_2 in DSSC system, and the mechanism for photocatalytic reactions on the TiO_2 surface will be impressive for readers. We need to describe the electronic band structure of excited state or ground state of TiO_2, the surface morphology, active energy for reaction as well as the fast interfacial electron transfer between the adsorbed molecule and TiO_2 accurately for fine theoretical studies. For some situation, we also need to do molecular dynamics for the surface structure of adsorbed TiO_2. May the summary report will help the readership to strengthen the fundamental knowledge or to find more interesting ideas.

Acknowledgements

The authors thank Jilin University and Young Scholar Training Program of Jilin University. The authors gratefully acknowledge support from the National Natural Science Foundation of China (Grant No. 21573088).

Author details

Fu-Quan Bai*, Wei Li and Hong-Xing Zhang

*Address all correspondence to: baifq@jlu.edu.cn

International Joint Research Laboratory of Nano-Micro Architecture Chemistry, Institute of Theoretical Chemistry, Jilin University, Changchun, China

References

[1] Gratzel M. Recent advances in sensitized mesoscopic solar cells. Accounts of Chemical Research. 2009;**42**(11):1788-1798. DOI: 10.1021/ar900141y

[2] Hu S, Majumdar A. Opportunities and challenges for a sustainable energy future. Nature. 2012;**488**(7411):294–303. DOI: 10.1038/nature11475

[3] Kojima A, Teshima K, Shirai Y, Miyasaka T. Organometal halide perovskites as visible-light sensitizers for photovoltaic cells. Journal of the American Chemical Society. 2009;**131**(17):6050-6051. DOI: 10.1021/ja809598r

[4] Desilvestro J, Grätzel M, Kavan L, Moser J, Augustynski J. Highly efficient sensitization of titanium dioxide. Journal of the American Chemical Society. 1985;**107**(10):2988–2990

[5] Lee MM, Teuscher J, Miyasaka T, Murakami TN, Snaith HJ. Efficient hybrid solar cells based on meso-superstructured organometal halide perovskites. Science. 2012;**338**(6107):643–647. DOI: 10.1126/science.1228604

[6] Liu D, Kelly TL. Perovskite solar cells with a planar heterojunction structure prepared using room-temperature solution processing techniques. Nature Photonics. 2014;**8**:133–138. DOI: 10.1038/nphoton.2013.342

[7] Snaith HJ, Schmidt-Mende L. Advances in liquid-electrolyte and solid-state dye-sensitized solar cells. Advanced Materials. 2007;**19**(20):3187–3200. DOI: 10.1002/adma.200602903

[8] Souza C, Tosoni, Illas F. Theoretical approaches to excited-state-related phenomena in oxide surfaces. Chemical Reviews. 2013;**113**:4456–4495. DOI: 10.1021/cr300228z

[9] Pelaez M, Nolan NT, Pillai SC, Seery MK, Falaras P, Kontos AG, Dunlop PSM, Hamilton JWJ, Byrne JA, O'Shea K, Entezari MH, Dionysiou DD. A review on the visible light active titanium dioxide photocatalysts for environmental applications. Applied Catalysis B: Environmental. 2012;**125**:331-349. DOI: 10.1016/j.apcatb.2012.05.036

[10] Na-Phattalung S, Smith MF, Kim K, Du MH, Wei SH, Zhang SB, Limpijumnong S. First-principles study of native defects in anatase TiO_2. Physical Review B: Condensed Matter and Materials Physics. 2006;**73**:125205. DOI: 10.1103/PhysRevB.73.125205

[11] Mehta P, Salvador PA, Kitchin JR. Identifying potential BO2 oxide polymorphs for epitaxial growth candidates. ACS Applied Materials Interfaces. 2014;**6**(5):3630–3639. DOI: 10.1021/am4059149

[12] Curnan MT, Kitchin JR. Investigating the energetic ordering of stable and metastable TiO_2 polymorphs using DFT+U and hybrid functionals. Journal of Physical Chemistry C. 2015;**119**(36):21060–21071. DOI: 10.1021/acs.jpcc.5b05338

[13] Dompablo MEA, Morales-Garcıa A, Taravillo M.: DFT+U calculations of crystal lattice, electronic structure, and phase stability under pressure of TiO_2 polymorphs. Journal of Chemical Physics. 2011;**135**:054503. DOI: 10.1063/1.3617244

[14] Vu NH, Le HV, Cao TM, Pham VV, Le HM, Nguyen-Manh D. Anatase–rutile phase transformation of titanium dioxide bulk material: A DFT + U approach. Journal of Physics: Condensed Matter. 2012;**24**:405501. DOI: 10.1088/0953-8984/24/40/405501

[15] Casadei M, Ren X, Rinke P, Rubio A, Scheffler M. Density-functional theory for f-electron systems: The $\alpha-\gamma$ phase transition in cerium. Physical Review Letters. 2012;**109**(14):146402. DOI: 10.1103/PhysRevLett.109.146402

[16] Peng H, Lany S. Polymorphic energy ordering of MgO, ZnO, GaN, and MnO within the random phase approximation. Physical Review B. 2013;**87**(17):174113. DOI: 10.1103/PhysRevB.87.174113

[17] Cui ZH, Wu F, Jiang H. First-principles study of relative stability of rutile and anatase TiO_2 using the random phase approximation. Physical Chemistry Chemical Physics. 2016;**18**(43):29914–29922. DOI: 10.1039/C6CP04973G

[18] Runge E, Gross EKU. Density-functional theory for time-dependent systems. Physical Review Letters. 1984;**52**(12):997. DOI: 10.1103/PhysRevLett.52.997

[19] Hirata S, Head-Gordon M, Bartlett RJ. Configuration interaction singles, time-dependent Hartree–Fock, and time-dependent density functional theory for the electronic excited states of extended systems. The Journal of Chemical Physics. 1999;**111**(24):10774–10786. DOI: 10.1063/1.480443

[20] Izmaylov AF, Scuseria GE. Why are time-dependent density functional theory excitations in solids equal to band structure energy gaps for semilocal functionals, and how does nonlocal Hartree–Fock-type exchange introduce excitonic effects?. The Journal of chemical physics. 2008;**129**(3):034101. DOI: 10.1063/1.2953701

[21] Bernasconi L, Tomić S, Ferrero M, Rerat M, Orlando R, Dovesi R, Harrison NM. First-principles optical response of semiconductors and oxide materials. Physical Review B. 2011;**83**(19):195325. DOI: 10.1103/PhysRevB.83.195325

[22] Walker B, Saitta AM, Gebauer R, Baroni S. Efficient approach to time-dependent density-functional perturbation theory for optical spectroscopy. Physical Review Letters. 2006;**96**(11):113001. DOI: 10.1103/PhysRevLett.96.113001

[23] Anisimov VI, Aryasetiawan F, Lichtenstein AI. First-principles calculations of the electronic structure and spectra of strongly correlated systems: The LDA+ U method. Journal of Physics: Condensed Matter. 1997;**9**(4):767. DOI: 0953-8984/9/4/002

[24] Hedin L. New method for calculating the One-Particle Green's function with application to the Electron-Gas problem. Physical Review. 1965;**139**:A796. DOI: 10.1103/PhysRev.139.A796

[25] Huang P, Carter EA. Advances in correlated electronic structure methods for solids, surfaces, and nanostructures. Annual Review of Physical Chemistry. 2008;**59**:261–290. DOI: 10.1146/annurev.physchem.59.032607.093528

[26] Onida G, Reining L, Rubio A. Electronic excitations: density-functional versus many-body Green's-function approaches. Reviews of Modern Physics. 2002;**74**(2):601. DOI: 10.1103/RevModPhys.74.601

[27] Matsui M, Akaogi M. Molecular dynamics simulation of the structural and physical properties of the four polymorphs of TiO_2. Molecular Simulation. 1991;**6**(4–6):239–244. DOI: 10.1080/08927029108022432

[28] Swamy V, Gale JD. Transferable variable-charge interatomic potential for atomistic simulation of titanium oxides. Physical Review B. 2000;**62**(9):5406. DOI: 10.1103/PhysRevB.62.5406

[29] Koparde VN, Cummings PT. Molecular dynamics study of water adsorption on TiO_2 nanoparticles. The Journal of Physical Chemistry C. 2007;**111**(19):6920–6926. DOI: 10.1021/jp0666380

[30] Selcuk S, Selloni A. Facet-dependent trapping and dynamics of excess electrons at anatase TiO_2 surfaces and aqueous interfaces. Nature Materials. 2016;**15**(10):1107–1112. DOI: 10.1038/nmat4672

[31] Vittadini A, Selloni A, Rotzinger FP, Grätzel, M. Formic acid adsorption on dry and hydrated TiO_2 anatase (101) surfaces by DFT calculations. The Journal of Physical Chemistry B. 2000;**104**(6):1300–1306. DOI: 10.1021/jp993583b

[32] Grätzel M. Perspectives for dye-sensitized nanocrystalline solar cells. Progress in Photovoltaics: Research and Applications. 2000;**8**(1):171–185. DOI: 10.1002/(SICI)1099-159X

[33] Akimov AV, Neukirch AJ, Prezhdo OV. Theoretical insights into photoinduced charge transfer and catalysis at oxide interfaces. Chemical Reviews. 2013;**113**(6):4496–4565. DOI: 10.1021/cr3004899

[34] Pan D, Klymyshyn N, Hu D. Tip-enhanced near-field Raman spectroscopy probing single dye-sensitized TiO_2 nanoparticles. Applied Physics Letters. 2006;**88**(9):093121. DOI: 10.1063/1.2176865

[35] Marcus RA.: Interaction of theory and experiment: examples from single molecule studies of nanoparticles. Philosophical Transactions of the Royal Society of London A: Mathematical, Physical and Engineering Sciences. 2010;**368**(1914):1109–1124. DOI: 10.1098/rsta.2009.0261

[36] Labat F, Ciofini I, Hratchian HP, Frisch M, Raghavachari K, Adamo C. First principles modeling of eosin-loaded ZnO films: A step toward the understanding of dye-sensitized solar cell performances. Journal of the American Chemical Society. 2009;**131**(40):14290–14298. DOI 10.1021/ja902833s

[37] Persson P, Lundqvist M J, Ernstorfer R, Goddard WA, Willig F. Quantum chemical calculations of the influence of anchor-cum-spacer groups on femtosecond electron transfer times in dye-sensitized semiconductor nanocrystals. Journal of Chemical Theory and Computation. 2006;**2**(2):441–451. DOI: 10.1021/ct050141x

[38] Tully JC. Molecular dynamics with electronic transitions. The Journal of Chemical Physics. 1990;**93**(2):1061–1071. DOI: 10.1063/1.459170

[39] Herman MF. Nonadiabatic semiclassical scattering. I. Analysis of generalized surface hopping procedures. The Journal of Chemical Physics. 1984;**81**(2):754–763. DOI: 10.1063/1.447708

[40] Li W, Rego LGC, Bai FQ. What makes hydroxamate a promising anchoring group in dye-sensitized solar cells? onsights from theoretical investigation. The Journal of Physical Chemistry Letters. 2014;**5**(22):3992–3999. DOI: 10.1021/jz501973d

[41] Kurtz RL, Stock-Bauer R, Msdey TE. Synchrotron radiation studies of H_2O adsorption on TiO_2 (110). Surface Science. 1989;**218**(1):178–200. DOI: 10.1016/0039-6028(89)90626-2

[42] Hugenschmidt MB, Gamble L, Campbell CT. The interaction of H_2O with a TiO_2 (110) surface. Surface Science. 1994;**302**(3):329–340. DOI: 10.1016/0039-6028(94)90837-0

[43] Henderson MA. An HREELS and TPD study of water on TiO_2 (110): The extent of molecular versus dissociative adsorption. Surface Science. 1996;**355**(1–3):151–166. DOI: 10.1016/0039-6028(95)01357-1

[44] Epling WS, Peden CHF, Henderson MA. Evidence for oxygen adatoms on TiO_2 (110) resulting from O_2 dissociation at vacancy sites. Surface Science. 1998;**412**:333–343. DOI: 10.1016/S0039-6028(98)00446-4

[45] Lindan PJD, Harrison NM, Gillan MJ. Mixed dissociative and molecular adsorption of water on the rutile (110) surface. Physical Review Letters. 1998;**80**(4):762. DOI: 10.1103/PhysRevLett.80.762

[46] Harris LA, Quong AA. Molecular chemisorption as the theoretically preferred pathway for water adsorption on ideal rutile TiO_2 (110). Physical Review Letters. 2004;**93**(8):086105. DOI: 10.1103/PhysRevLett.93.086105

[47] Zhang C, Lindan PJD. A density functional theory study of the coadsorption of water and oxygen on TiO_2 (110). The Journal of Chemical Physics. 2004;**121**(8):3811–3815. DOI: 10.1063/1.1775784

[48] Wang F, Di Valentin C, Pacchioni G. Doping of WO_3 for photocatalytic water splitting: Hints from density functional theory. The Journal of Physical Chemistry C. 2012;**116**(16):8901–8909. DOI: 10.1021/jp300867j

[49] Asahi R, Morikawa T, Ohwaki T. Visible-light photocatalysis in nitrogen-doped titanium oxides. Science. 200;**293**(5528):269–271. DOI: 10.1126/science.1061051

[50] Borgarello E, Kiwi J, Gratzel M. Visible light induced water cleavage in colloidal solutions of chromium-doped titanium dioxide particles. Journal of the American Chemical Society. 1982;**104**(11):2996–3002. DOI: 10.1021/ja00375a010

[51] Li M, Zhang J, Guo D. Band gap engineering of compensated (N, H) and (C, 2H) codoped anatase TiO_2: A first-principles calculation. Chemical Physics Letters. 2012;**539**:175–179. DOI: 10.1016/j.cplett.2012.04.057

[52] Liu N, Chang Y, Feng YL, Cheng Y, Sun XJ, Jian H, Feng YQ, Xi Li, Zhang HY. {101}-{001} Surface heterojunction-enhanced antibacterial activity of titanium dioxide nanocrystals under sunlight irradiation. ACS Applied Materials & Interfaces. 2017;**9**(7):5907–5915. DOI: 10.1021/acsami.6b16373